Modeling Biochemical Processes
in
Aquatic Ecosystems

Modeling Biochemical Processes

in

Aquatic Ecosystems

edited by

Raymond P. Canale

Associate Professor of Civil Engineering
University of Michigan
Ann Arbor, Michigan

ANN ARBOR SCIENCE
PUBLISHERS INC
P.O. BOX 1425 • ANN ARBOR, MICH. 48106

PREFACE

The ultimate objective of all mathematical models is to increase knowledge of systems and thereby advance the techniques for solving practical problems. The exercise of developing a mathematical model for a biochemical process in an aquatic ecosystem can have a number of immediate goals. A mathematical model may be written to define a conceptual framework for a chemical or biological transformation. Its purpose may be to help design a laboratory experiment or to facilitate interpretation of existing laboratory data that were obtained either with or without benefit of a mathematical model.

Once out of the laboratory, aquatic ecosystems usually become larger and more complex, and are influenced by the transport regime of the suspending fluid. In such cases mathematical models can be used advantageously to design field monitoring programs that include identification of the parameters to be measured as well as the spatial and temporal structuring for the sampling. Models are also developed directly for the purpose of addressing an aquatic management or engineering problem.

Application of a model requires a synthesis of information contributed by a number of specialists. In explaining the mechanisms and rates of change of biochemical transformations or the reactions of organisms to environmental stimuli, scientists are describing the submodel development. Physical limnologists concentrate on the identification of the forces that promote fluid movement, particle dispersion, and thermal structure. The mathematician and numerical analyst develop the techniques by which mass and energy partial differential equations of continuity are solved for the time and space distribution of the physical, chemical, and biological variables important in a given aquatic ecosystem. The ecosystem modeler, who is often an engineer, is engaged in solving practical problems and may be thought of as a specialist in synthesizing the results of other specialists into a single mathematical framework.

The accompanying figure illustrates the interactions among various steps that must be accomplished before mathematical models can be used to help solve practical problems.

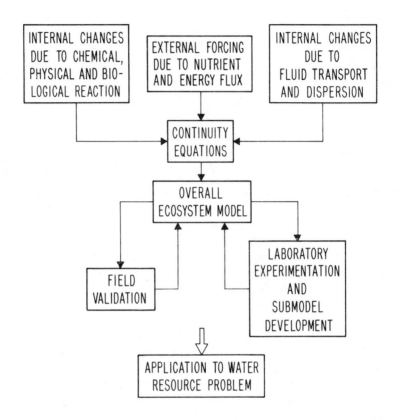

Contributions from each area are necessary in order to make balanced advances in modeling biochemical processes in aquatic ecosystems. This book brings together recent contributions from each of these areas for the purpose of promoting a cooperative exchange that may result in new and significant cases where preservation of our aquatic resources is achieved through utilization of mathematical modeling.

January, 1976   Raymond P. Canale
                                                 Ann Arbor, Michigan

CONTENTS

1

MATHEMATICAL MODEL
OF THE SELECTIVE ENHANCEMENT OF
BLUE-GREEN ALGAE BY NUTRIENT ENRICHMENT

Victor J. Bierman, Jr.[1]

INTRODUCTION

Mathematical modeling techniques can provide a systematic basis for a research approach to the problem of cultural eutrophication and can greatly aid in the comparison of various management options. From an applied standpoint, the general techniques of O'Connor, *et al.* (1973) are quite well-known and have been brought to bear on a variety of different physical systems. In particular, Thomann, *et al.* (1975) and Canale, *et al.* (1973) have investigated phytoplankton-nutrient interactions in the Great Lakes. Chen and Orlob (1972) have also developed such techniques and have used them to investigate the effects of wastewater diversion from Lake Washington. From a research standpoint, work is progressing on a number of systems models and component models in order to gain deeper insight into the chemical-biological processes that occur in natural systems (*e.g.,* Middlebrooks, *et al.* 1973).

The present work is part of the International Joint Commission's Upper Lakes Reference Study involving Saginaw Bay, Lake Huron. The ultimate goal of this work is to develop a mathematical model that can be used both to describe the physical, chemical and biological processes that occur

[1] U.S. Environmental Protection Agency, Grosse Ile Laboratory, Grosse Ile, Michigan.

1

in Saginaw Bay and to predict the effects of reduced waste loadings. Specifically, the modeling effort will focus on phosphorus, nitrogen and silicon loadings to the bay and the resultant production of phytoplankton biomass.

Model development is proceeding along two parallel pathways. The first of these involves the development of research-oriented process models, which include biological and chemical detail but which, for simplicity, do not include any spatial detail. The second pathway involves the development of an engineering-oriented water quality model that mimics, as closely as practicable, the actual physical system, including spatial detail. At any given point in time, the water quality model will contain those chemical and biological processes that have previously been investigated and developed using the spatially-simplified model. There is constant feedback between these two pathways and constant interaction between the entire modeling effort and an ongoing sampling effort on Saginaw Bay.

This chapter deals exclusively with research-oriented process modeling and focuses on several mechanisms of phytoplankton dynamics and competition. Specifically, a multispecies phytoplankton model is used to investigate the relative importance of various processes in providing competitive advantages for the development of blue-green algae under conditions of nutrient enrichment.

The motivation for a multispecies modeling approach is that different classes of algae have very different nutrient requirements; for example, diatoms need silicon, and certain types of blue-greens can fix atmospheric nitrogen. From a water quality standpoint, not all of these classes have the same nuisance characteristics. Diatoms and green algae are grazed by zooplankton, but blue-green algae are not significantly grazed and can form objectionable floating scums.

Preliminary simulations of Saginaw Bay are presented as a first application of the model. These results are then used as a baseline against which the results of various sensitivity analyses are compared. These analyses include effects of changes in phosphorus recycle, phosphorus uptake affinities, cell sinking rates, and zooplankton grazing. A hypothetical batch system is used to investigate the additional effects of changes in hydraulic detention time and phosphorus input dynamics.

MODEL CONCEPTS

The present version of this model has evolved from earlier work that dealt initially with microbial substrate uptake kinetics (Verhoff and Sundaresan 1972; Verhoff, *et al.* 1973), which was expanded to include phytoplankton-growth kinetics modeling (Bierman, *et al.* 1973; Bierman

1974). The compartments in the model are four phytoplankton, two zooplankton, and three nutrients (Figure 1.1). The four functional groups of phytoplankton were chosen primarily for water quality considerations and the fact that phosphorus, nitrogen, and silicon are usually considered

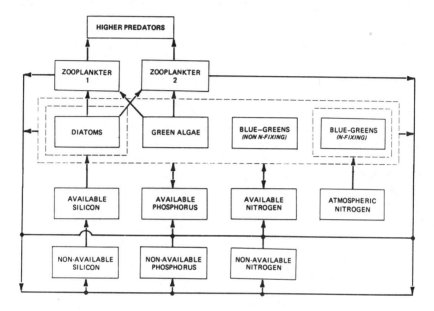

Figure 1.1 Principal compartments of the Saginaw Bay eutrophication model.

to be the major nutrients limiting algal growth. Carbon is not considered important in limiting either the growth rates or the maximum sizes of the algal crops in the simulations. Considerable support for this assumption can be found in the recent *in situ* work by Schindler, *et al.* (1973a, 1973b) in the Canadian Shield and in the recent laboratory work by Goldman, *et al.* (1974).

Each of the nutrients is assumed to exist in two different forms: an available form that the phytoplankton can absorb directly, and an unavailable form that is not directly assimilable. No distinction is made between the dissolved and particulate fractions of the unavailable forms. Two of the phytoplankton groups, diatoms and green algae, are assumed to be grazed by two zooplankton types, differentiated on the basis of their maximum ingestion rates. It is recognized that zooplankton-grazing is a complex phenomenon and involves, among other processes, phytoplankton size-selectivity (McNaught 1975). However, the present version of the model is only intended to investigate gross changes in grazing among the phytoplankton groups.

A unique feature of the model is that cell growth is considered to be a two-step process involving separate nutrient uptake and cell synthesis mechanisms. The motivation for this variable stoichiometry approach is that an increasingly large body of experimental evidence indicates that the mechanisms of nutrient uptake and cell growth are actually quite distinct (Dugdale 1967; Fuhs 1969, 1971; Droop 1973; Azad and Borchardt 1970; Caperon and Meyer 1972a, 1972b; Eppley and Thomas 1969; and Halmann and Stiller 1974). The model includes carrier-mediated transport of phosphorus and nitrogen using a reaction-diffusion mechanism, and possible intermediate storage in excess of a cell's immediate metabolic needs. Specific cell growth rates are assumed to be directly dependent on the intracellullar levels of these nutrients, in contrast to the classical Michaelis-Menten approach that relates these rates directly to extracellullar dissolved phosphorus.

Net specific phosphorus uptake rate is a function of the balance between extracellular and intracellular dissolved phosphorus:

NUTRIENT UPTAKE

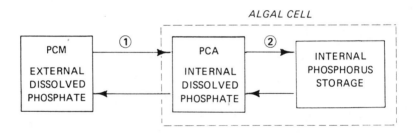

ALGAL CELL

Net specific uptake rate =

$$\text{Maximum uptake rate} \cdot f(T) \, f(I) \left[\frac{1}{1 + (PK1)(PCA)} - \frac{1}{1 + (PK1)(PCM)} \right] \tag{1.1}$$

$$PCA = (PCAMIN)e^{(P/PO - 1)} \tag{1.2}$$

where

PK1 = affinity constant between phosphate and an assumed membrane carrier
P = actual total phosphorus per cell
PO = minimum stoichiometric level of total phosphorus per cell
PCAMIN = minimum possible value of PCA
f(T), f(I) = temperature and light reduction factors, respectively.

An identical approach is used for nitrogen uptake kinetics (Bierman 1974). The quantity PK1 has actual physical significance because it is the equilibrium constant for the reaction between phosphate and an assumed membrane carrier molecule in the cell. Such a molecule has been isolated in the bacterium *Escherichia coli*, and its binding constant with phosphate has been measured (Medveczky and Rosenberg 1970, 1971). Since insufficient knowledge of complex biochemical processes exists at this time, attempts to develop mechanistic theories of these processes usually lead to empirical assumptions at some point. Equation (1.2) relates actual intracellular phosphorus concentration, PCA, to a minimum value, PCAMIN, which is assumed to be a small number. This is not a rate equation but is an empirical device for instantaneously calculating PCA as a function of the cell's total phosphorus storage. Intracellular phosphorus concentration must be determined in some manner so that a feedback effect can be obtained.

A unique value for net specific uptake rate does not exist for a given concentration of phosphorus in solution; instead, there exists a family of values each corresponding to a different level of intracellular phosphorus (Figure 1.2). Negative values for net specific uptake rate correspond to

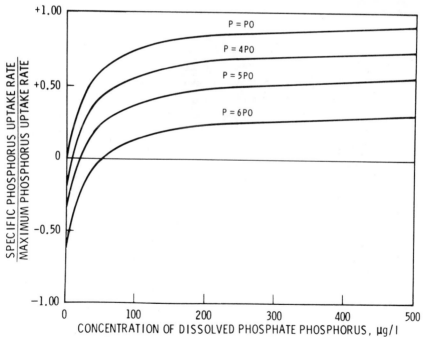

Figure 1.2 Degree of saturation of the phosphorus uptake system for the blue-green algae in the model as a function of both extracellular and intracellular phosphorus.

the rate of phosphorus leakage back to solution that can occur under certain conditions. If the cells are assumed to be starved, that is, if P=PO, then the net specific uptake rate becomes half-maximum at 30 $\mu g \cdot P/l$ in the example in Figure 1.2. Since the traditional Michaelis-Menten approach to nutrient-uptake kinetics does not include a feedback mechanism, Michaelis-Menten kinetics is actually a special case of this kinetics theory in which the cell's nutritional state is assumed to be constant. As PCA approaches PCAMIN in Equation 1.1, the first term in brackets approaches unity and the equation for net specific uptake rate reduces to the familiar hyperbolic form of the Michaelis-Menten equation.

The two-step approach presented here is far from a rigorously correct treatment of phytoplankton kinetics; however, it is much more realistic and consistent with the experimental evidence than traditional single-step approaches. Further, in contrast to completely empirical attempts to develop two-step approaches (*e.g.*, Huff *et al.* 1973), the carrier-mediated mechanism used here is only one application of a more general substrate uptake theory (Verhoff, *et al.* 1973), which provides a systematic methodology for investigating alternative and more sophisticated uptake mechanisms and which can serve as a framework for future research investigations.

MODEL IMPLEMENTATION

A major problem in attempting to implement a complex chemical-biological process model is that such models usually contain coefficients for which direct measurements do not exist. It is possible that more than one set of model coefficients could produce an acceptable fit between the model output and a given data set. In the transition from single-class to multiclass models, this problem becomes particularly acute because it is no longer sufficient to ascertain a range of literature values for a given coefficient. Multiclass models necessitate the definition of class distinctions within this range. Given such circumstances, many of the coefficients in multiclass models must simply be estimated.

The primary operational differences among the phytoplankton types in the model are summarized in Table 1.1. Class difference in phosphorus uptake affinities are the least well established of these differences. Bush and Welch (1972) and Hammer (1964) have produced strong circumstantial evidence indicating that blue-green algae have higher phosphorus uptake affinities than other classes of algae. Shapiro (1973) has reported that the phosphorus uptake mechanisms of blue-green algae reach half-saturation at significantly lower extracellular phosphorus concentrations than do the phosphorus uptake mechanisms of green algae over a wide range of environmental conditions.

Table 1.1 Operational Differences Among Phytoplankton Types

| Characteristic Property | Phytoplankton Type | | | |
	Diatoms	Greens	Blue-Greens (non n-fixing)	Blue-Greens (n-fixing)
Nutrient requirements	phosphorus, nitrogen, silicon	phosphorus, nitrogen	phosphorus, nitrogen	phosphorus
Relative growth rates under optimum conditions at 25°C	high	moderately high	low	low
Phosphorus uptake affinity	low	low	high	high
Sinking rate	high	high	low	low
Grazing pressure	high	high	none	none

The computer program which actually implements the model is written in FORTRAN IV and is structured in such a form that an arbitrary number of phytoplankton and zooplankton types can be simulated, along with an arbitrary set of food web interactions among these groups. Working equations, along with a list of symbols, are referenced in Appendices A and B, respectively. The version of the model in Figure 1.1 consists of 20 simultaneous differential equations. Solutions are obtained using a fourth-order Runge-Kutta method with a time step of 30 minutes for the nutrient kinetics equations and of 6 hours for the growth equations. For a 270-day simulation, approximately 2.5 minutes of CPU time are required on an IBM 370/158 computer.

PRELIMINARY SAGINAW BAY SIMULATIONS

Methods

Detailed background information on Saginaw Bay appears in Freedman (1974). During 1974, an extensive water quality survey of Saginaw Bay was conducted by Cranbrook Institute of Science. The author is indebted to V. Elliott Smith, Cranbrook Institute, for providing the unpublished chemical-biological data used here. Saginaw Bay is highly eutrophic by some standards, but not eutrophic at all by others. For example, corrected chlorophyll a values of up to 70 μg/l have been observed, yet there is no significant oxygen depletion because of the shallow nature of the bay. The primary hydrological influence on Saginaw Bay is Lake Huron and the primary source of nutrient loading is the Saginaw River. The predominant flow pattern in the bay is counterclockwise, with Lake Huron water flowing in along the north shore and a mixture of Lake Huron water and Saginaw River water flowing out of the bay along the south shore (Figure 1.3).

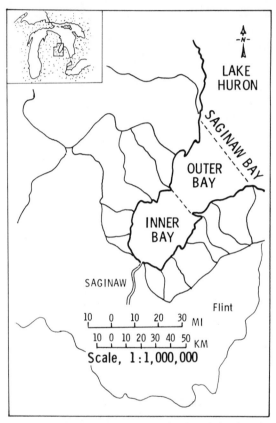

Figure 1.3 Saginaw Bay watershed indicating distinctions between inner and outer portions of the bay.

The first step in the development of a comprehensive water quality model is to apply a spatially simplified version of the model to the system in question. The successful results of this application can then be incorporated in a spatially segmented version. This approach is used for the inner portion of Saginaw Bay (Figure 1.3), which is assumed to be a completely mixed reactor. This portion is subject to a steady hydraulic washout by Lake Huron and to nutrient loading from the Saginaw River, which is considered to be an external point source.

An approximate annual-average hydraulic detention time of two months was used for the inner bay (Richardson 1974). External nutrient loadings and boundary concentrations used in the simulations are listed in Tables 1.2 and 1.3, respectively. Saginaw River loadings were calculated from data collected by the Michigan Department of Natural Resources. Silicon loading was arbitrarily assumed to be in the available form because separate measurements of total and dissolved silicon were not available for the entire year. Boundary concentrations represent seasonal averages from three field stations nearest to the water inflow to the inner portion of the bay.

Table 1.2 Nutrient Loadings to Saginaw Bay from Saginaw River (kg/day)

	Spring	Summer-Fall
Available nutrients		
Phosphorus (WPCM)	3750	795
Nitrogen (WNCM)	47600	5900
Silicon (WSCM)	108000	15550
Unavailable nutrients		
Phosphorus (WTOP)	7500	1590
Nitrogen (WTON)	23800	2950
Silicon (WTOS)	–	–

Table 1.3 Boundary Concentrations Used Between Inner and Outer Portions of Saginaw Bay (μg/l)

	Spring	Summer-Fall
Available nutrients		
Phosphorus (PCMBD)	1.5	6.7
Nitrogen (NCMBD)	450	175
Silicon (SCMBD)	350	510
Unavailable nutrients		
Phosphorus (TOPBD)	3.0	13.4
Nitrogen (TONBD)	225	88
Silicon (TOSBD)	350	510

The principal phytoplankton coefficients used are listed in Table 1.4. Bierman (1974) presents the minimum stoichiometric nutrient requirements, nitrogen parameters, and cell sizes used. A conversion factor of 20 μg·chlorophyll a/mg dry weight biomass is used for all of the phytoplankton types. This value was calculated from fresh weight biomass and chlorophyll a data for Saginaw Bay (Vollenweider, et al. 1974) assuming that dry weight biomass is 20% of fresh weight biomass (Kuenzler and Ketchum 1962).

Table 1.4   Principal Phytoplankton Parameters

Parameter	Diatoms	Greens	Blue-Greens (non N-fixing)	Blue-Greens (N-fixing)
		Phytoplankton Type		
R1PM (day)$^{-1}$	0.024	0.133	0.042	0.059
PK1 (l/mol)	0.50×10^6	0.50×10^6	0.10×10^7	0.90×10^6
PCAMIN (mol/l cell vol)	0.215×10^{-7}	0.215×10^{-7}	0.107×10^{-7}	0.107×10^{-7}
R1NM (day)$^{-1}$	0.015	0.060	0.040	0.040
NK1 (l/mol)	0.10×10^7	0.10×10^7	0.10×10^7	0.10×10^7
NCAMIN (mol/l cell vol)	0.267×10^{-6}	0.267×10^{-6}	0.267×10^{-6}	0.267×10^{-6}
KNCELL (mol N/cell)	0.14×10^{-13}	0.14×10^{-13}	0.14×10^{-13}	0.23×10^{-13}
RADSAT (foot candles)	1000	1000	500	500
Max. growth rate (at 25°C)(day)$^{-1}$	2.1	1.9	0.80	0.80
PO (mol P/cell)	0.20×10^{-14}	0.20×10^{-14}	0.583×10^{-15}	0.134×10^{-14}
NO (mol N/cell)	0.520×10^{-13}	0.520×10^{-13}	0.520×10^{-13}	0.853×10^{-13}
FACT (mg dry wt/cell)	0.150×10^{-6}	0.270×10^{-7}	0.250×10^{-7}	0.410×10^{-7}
ASINK (m/day)	0.40	0.40	0.15	0.15

A traditional Michaelis-Menten kinetics function with a half-saturation constant of 100 μg·S_iO_2/l is used to calculate silicon-dependent growth rate in the diatoms (Lewin and Guillard 1963), and these cells are assumed to be 50% silica by dry weight (Lund 1965). Phytoplankton death is assumed to be second order with respect to cell concentration (DePinto, et al. 1975) and a coefficient of 0.0015 [day-°C-(mg/l)]$^{-1}$ is used. Nutrient conversion from unavailable to available forms is assumed to be a first order process and a coefficient of 0.0005 (day-°C)$^{-1}$ is used.

The two zooplankton ingestion rates are 0.50 and 0.35 mg algae/mg zooplankton day. Assimilation efficiencies are assumed to be 60% of ingestion rates and death rates are assumed to be 0.10 (day)$^{-1}$. The zooplankton ingestion rates for diatoms and green algae both reach half-saturation at algal concentrations of 0.5 mg dry weight/l.

Results

Preliminary simulations are presented using only chlorophyll a and dissolved nutrient data because phytoplankton and zooplankton class data are not yet available. These simulations are attempts to describe an existing data set and are not intended to be predictive in nature. Rather, the purpose of this analysis is to provide a somewhat realistic context for the subsequent sensitivity analyses.

Each of the data points in Figures 1.4-1.7 represents the average of values from 38 field stations in the inner portion of the bay, and the bars on each point indicate plus or minus one-half standard deviation. The solid lines on all plots indicate model output.

The model output describes the chlorophyll a pattern reasonably well during the spring and late summer-fall periods, but there is a significant discrepancy during the month of June (Figure 1.4). Richardson (1975) has shown that two distinct seasonal flow regimes exist in Saginaw Bay with a turbulent transition period occurring between them in June. This suggests

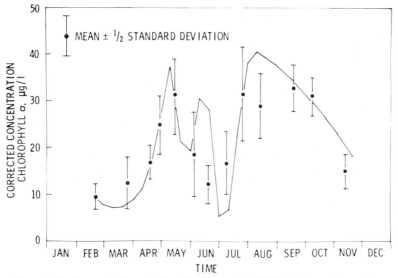

Figure 1.4   Seasonal chlorophyll a distribution for 1974 in Saginaw Bay, inner portion, as compared to model output (solid line).

that the discrepancy is due in large part to the assumption of steady hydraulic flow between the inner and outer bay.

The phosphorus data are too widely scattered for the model output to be meaningful; they show little or no inverse correlation with the spring chlorophyll *a* peak (Figure 1.5). However, there is an apparent correlation

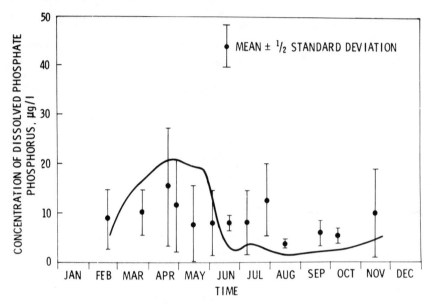

Figure 1.5 Seasonal dissolved orthophosphate phosphorus distribution for 1974 in Saginaw Bay, inner portion, as compared to model output (solid line).

for the late summer-fall peak. The data also suggest that there are important distributed sources of phosphorus other than Lake Huron inflow and the Saginaw River. In contrast, the data for nitrogen (Figure 1.6) and silicon (Figure 1.7) follow more well-defined patterns, and the model describes these patterns reasonably well.

The first attempt to approximate the inner portion of Saginaw Bay with a spatially-simplified model has been partially successful in describing the trends in the data. Work is continuing on a spatially-segmented approach and on developing a finer time scale for the forcing functions such as flows and loadings.

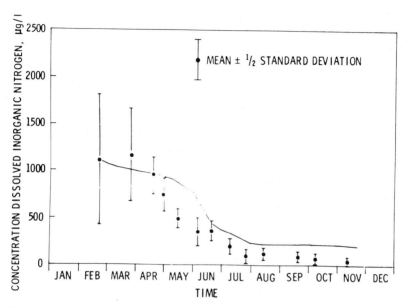

Figure 1.6 Seasonal dissolved inorganic nitrogen (ammonia plus nitrate) distribution for 1974 in Saginaw Bay, inner portion, as compared to model output (solid line).

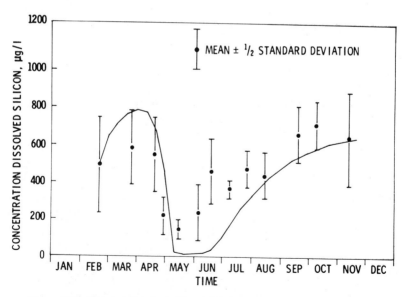

Figure 1.7 Seasonal dissolved silicon distribution for 1974 in Saginaw Bay, inner portion, as compared to model output (solid line).

SENSITIVITY ANALYSES

Approach

The class composition of the Saginaw Bay model output indicates that the early phytoplankton crops are dominated by diatoms and green algae and that the broad summer-fall peak is dominated by blue-green algae (Figure 1.8). This successional pattern has been observed in the inner portion of the bay by Vollenweider, *et al.* (1974), and Chartrand (1973) has reported significant late summer crops of *Alphanizomenon,* a filamentous blue-green algae, in the outer portion of the bay.

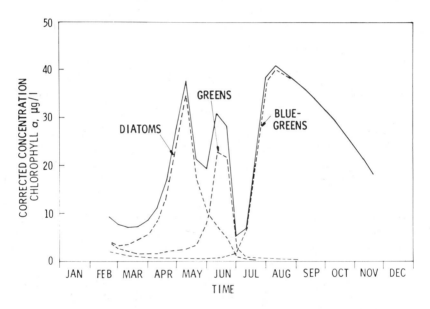

Figure 1.8 Phytoplankton class composition of model output in Figure 1.5. Baseline case for sensitivity analyses.

The development of multiclass phytoplankton models creates serious research problems due to their comprehensive data requirements. The approach adopted here is to perform sensitivity analyses on several of the assumed competitive advantages enjoyed by blue-green algae to establish their relative importance. The coefficients used in the model result in the blue-green algae having higher phosphorus uptake affinities, lower sinking rates, and freedom from grazing, as compared to the diatoms and green algae. These advantages will be removed, individually, from the blue-green, and the resulting simulations will be compared to the assumed baseline case (Figure 1.8). Although the use of recycled phosphorus is not restricted exclusively to blue-green algae, the importance of this process will also be investigated.

Effects due to class differences in nitrogen dynamics will not be investigated. In developing the model to this point, it is assumed that class differences in nitrogen dynamics are not as important as differences in phosphorus dynamics and that the primary mechanism for nitrogen competition is the ability of some blue-greens to fix atmospheric nitrogen (DePinto, et al., 1975).

Results

Phosphorus recycle occurs in the model as a result of phytoplankton death and zooplankton excretion (Figure 1.1). Since the phytoplankton cells are assumed to have a variable phosphorus stoichiometry, the recycle mechanism has two components: recycle of the cell's minimum stoichiometric phosphorus level to the unavailable pool, and recycle of all cellular phosphorus above this minimum level directly to the available phosphorus pool. Such a two-component nutrient release upon cell death has been reported by Foree, et al. (1970). If both of these phosphorus recycle components are turned off, the early crops of diatoms and green algae are little affected, but the subsequent blue-green crop shows a significant decrease, compared to the baseline case (Figure 1.9).

The choice of coefficients used in the model for the phosphorus uptake mechanisms is based on the assumption that competitive differences are primarily the result of differences in phosphorus affinities rather than in maximum phosphorus uptake rates. The actual situation is probably some combination of these factors. Maximum phosphorus uptake rates have simply been chosen in proportion to the minimum stoichiometric requirements of each of the four classes of algae. Assuming the cells are phosphorus-starved, the uptake mechanisms of the diatoms and green algae reach half-saturation at 60 μg·P/l and the uptake mechanisms of the blue-greens reach half-saturation at approximately 30 μg·P/l. The phosphorus affinity of the nonnitrogen-fixing blue-greens is set slightly higher than that for the nitrogen-fixing blue-greens, based on Fitzgerald (1969). This range of values is consistent with the data of Fuhs (1971), Healey (1973) and Rhee (1973).

When the phosphorus affinities of both blue-green algae are reduced so that their phosphorus uptake mechanisms now half-saturate at 60 μg·P/l, this causes a significant reduction in the late summer-fall total crop (Figure 1.10a) and also changes the class successional pattern. The

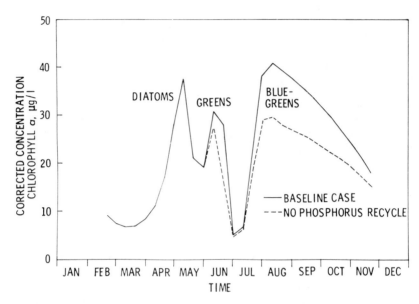

Figure 1.9 Comparison between baseline case and case for which no phosphorus is recycled upon phytoplankton death or zooplankton excretion.

reduction in the blue-green component of the total crop (Figure 1.10b) is more significant than is apparent from Figure 1.10a alone because the diatom-green component develops an extra peak in August (Figure 1.10c).

The sinking rate used for the diatoms and green algae is 0.40 m/day. This value is consistent with the laboratory data of Smayda (1974) and with the range of sinking rates used by Thomann, *et al.* (1975) for simulation of Lake Ontario phytoplankton. Blue-greens are arbitrarily sunk at 0.15 m/day. If both of the blue-greens are also sunk at 0.40 m/day, their contribution to the total crop becomes completely negligible, and diatoms and green algae dominate the simulations throughout the growing season (Figure 1.11). Similarly, if all four algal types are grazed in the same manner as the diatoms and green algae, the blue-green component of the total crop again becomes negligible (Figure 1.12).

Finally, since all of the simulations to this point involved the Saginaw Bay system with a hydraulic detention time of two months, it was decided to investigate some of the above mechanisms in a system with totally different hydraulic characteristics. In order to approximate a system whose hydraulic detention time is extremely long compared to the length of a growing season, a hypothetical batch system was used in which there were assumed to be no hydraulic inputs. Simulations were

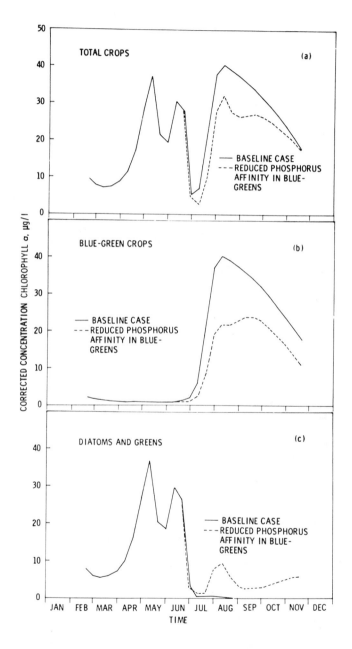

Figure 1.10 (a) Comparison of total crops between baseline case and case for which phosphorus uptake affinity has been reduced in the blue-green algae. (b) Comparison of the blue-green components of the total crops for the cases in (a). (c) Comparison of the non-blue-green components of the total crops for the cases in (a).

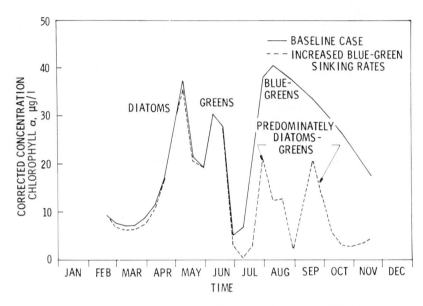

Figure 1.11 Comparison between baseline case and case for which blue-green sinking rates have been increased.

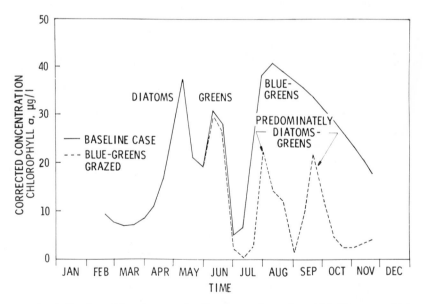

Figure 1.12 Comparison between baseline case and case for which blue-green algae are subject to the same grazing pressure as the diatoms and green algae.

run using nitrogen and silicon values that were high enough to be non-limiting, and the effects of changes in phosphorus input dynamics, phosphorus recycle, and phosphorus affinities were investigated using sets of model coefficients identical to the Saginaw Bay system. Since at this point, grazing and cell sinking effects have already been shown to be extremely important, the purpose of the batch simulations is to investigate only some of the more subtle effects of phosphorus dynamics.

The two sets of simulations in the hypothetical batch system are identical in almost every respect, including the fact that the total amount of available phosphorus supplied to the system in each set is 60 μg·P/l. The only difference between them is that in Figure 1.13a all of the 60 μg·P/l is present on the first day of the simulation and there are no further inputs, whereas in Figure 1.13b, only 10 μg·P/l is present on the first day

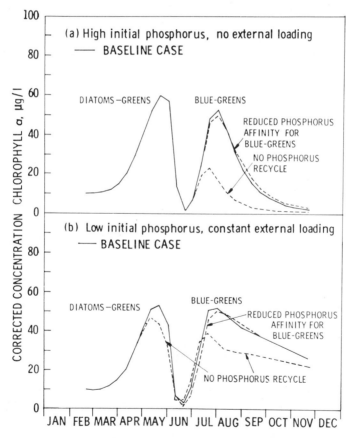

Figure 1.13 Comparative simulations in a hypothetical batch system using the Saginaw Bay model coefficients.

of the simulation and an additional 50 μg·P/l is supplied during the remainder of the simulation as a constant external point load.

Comparison of the baseline curves in Figures 1.13a and 1.13b shows that the above difference in phosphorus input dynamics significantly affects blue-green production, as measured by the areas under the total crop curves for the second half of the growing season. The duration of the large blue-green crop is substantially longer for the case involving the chronic external input. Comparison of the same figures for the case where there is no phosphorus recycle indicates that phosphorus recycle is much more important to the development of blue-green algae in a system where there are no significant phosphorus inputs during the growing season, relative to the phosphorus present at the beginning of the season. Finally, in sharp contrast to the Saginaw Bay system, a reduction in the phosphorus affinities of the blue-green algae has only a slight effect, regardless of the phosphorus input dynamics.

DISCUSSION

A general picture emerges in which blue-green algae, although they have slower maximum growth rates than other groups of algae and tend to occur at the end of the typical successional pattern in eutrophic environments, are nonetheless armed with a variety of competitive advantages that can lead to their dominance under conditions of nutrient enrichment. Freedom from grazing and lower sinking rates were the most important of the competitive advantages considered here, although more subtle effects involving phosphorus dynamics were also significant in certain circumstances. In addition, differences in hydraulic characteristics resulted in changes in the relative importance of some of these factors. The conclusions presented, however, must be qualified by noting that they only apply strictly to the ranges of coefficients used and to the particular structure of the model.

Phosphorus recycle was shown to be an important factor in the development of late-growing blue-green algae, even in the Saginaw Bay system where there are continuous sources of external nutrient supply. Since little quantitative information is available on the mechanisms of phytoplankton decay, zooplankton excretion, and the role of bacteria in nutrient recycle, this is an area that deserves more research attention. Other studies have also identified the importance of nutrient recycle (Thomann, et al. 1975; DePinto, et al. 1975). Gunnison and Alexander (1975) have suggested that blue-green algae as a group are more susceptible than other algal types to bacterial decomposition. A complete treatment of phytoplankton growth and nutrient recycle may even necessitate the inclusion of heterotrophic activity as a separate state variable in systems models.

The simulations comparing the effects of changes in phosphorus affinities between algal types, even though they were based on incomplete information, showed that these differences could be important for both total crops and class successional patterns. The results suggest the importance of conducting short-term phosphorus uptake studies on a wide variety of algal types to provide a basis for systematic comparisons. Such studies should investigate the effects of temperature, light and pH both to understand the actual mechanisms involved and to obtain rate coefficients.

Relatively small differences in sinking rates between blue-green algae and non-blue-green algae in the simulations showed significant effects. Modelers frequently think in terms of order-of-magnitude differences, but when blue-green sinking rates were reduced by less than a factor of three in the simulations, previously large blue-green crops were reduced to almost negligible sizes. While it is unfortunate that sinking rates are so difficult to measure accurately in the natural environment, laboratory studies with a variety of algal types under different sets of conditions could at least provide useful comparative data.

The significant effects of differences in hydraulic detention time demonstrate that the entire question of nutrient enrichment is extremely complicated. The difference in the responses of the blue-green algae to reduced phosphorus uptake affinities between the Saginaw Bay system and the hypothetical batch system suggest that the degree of stress on a phytoplankton crop can determine the relative importance of some of its competitive characteristics. The blue-greens were subject only to sinking and cell death in the batch case but were further subject to hydraulic washout in the Saginaw Bay case, in which their increased affinity for phosphorus was shown to be more important.

The comparative simulations within the hypothetical batch system suggest that it is not sufficient to consider only the total amount of phosphorus supplied to a system but that the time-dependence of this supply must also be considered. The important factor in the enhancement of the blue-green crops was the amount of phosphorus supplied to the system during the time when the blue-greens were best able to exploit their particular dynamic characteristics. However, the importance of phosphorus recycle indicates that this supply should be thought of as the net sum of both external and internal sources.

Carbon has been ignored in the present study as a possible explanation for phytoplankton class succession. It has frequently been contended (*e.g.*, King 1970) that species shifts in phytoplankton from greens to blue-greens as pH is increased is a function of class differences in the ability to absorb carbon in the form of dissolved CO_2 or perhaps HCO_3^-.

However, if a similar assumption is made for the case of phosphorus absorption, that is, that phytoplankton prefer to absorb phosphorus in the forms of the dissolved species with the lowest electrical charges (*i.e.*, $H_2PO_4^-$ or $HPO_4^=$), then class succession as a function of increased pH levels could also be explained on the basis of class differences in phosphorus uptake efficiencies. Goodman and Rothstein (1957) have studied the pH dependence of phosphate uptake in yeast cells, and their data strongly suggest that $H_2PO_4^-$, which is present in very low proportions at high pH values, is selectively absorbed in preference to $HPO_4^=$, which is more abundant at high pH levels. Until comprehensive studies of phosphorus-uptake kinetics in phytoplankton are conducted, definitive conclusions cannot be drawn regarding species shifts and changes in pH. It is possible that such phenomena are caused by a combination of both of the above factors, which can allow blue-green algae to better adapt themselves to the low-nutrient, high-pH conditions frequently existing late in the growing season in eutrophic environments.

ACKNOWLEDGMENTS

The author would like to thank V. Elliott Smith, Cranbrook Institute of Science, for providing unpublished chemical and biological data for Saginaw Bay. W. L. Richardson, EPA, Grosse Ile Laboratory, compiled the Saginaw River loading data. In addition, the author would like to thank T. T. Davies, N. A. Thomas, D. M. Dolan, and J. K. Crawford, all from EPA, Grosse Ile Laboratory, for reviewing the manuscript. The Editor, Raymond P. Canale, provided a thorough critique of the manuscript and offered many valuable suggestions.

REFERENCES

Azad, H. S. and J. A. Borchardt. "Variations in Phosphorus Uptake by Algae," *Environ. Sci. Technol.* 4, 737 (1970).

Bierman, V. J., Jr., F. H. Verhoff, T. L. Poulson, and M. W. Tenney. "Multi-Nutrient Dynamic Models of Algal Growth and Species Competition in Eutrophic Lakes. Modeling the Eutrophication Process," Proceedings of a Workshop, September 5-7, 1973; Utah Water Research Laboratory and Division of Environmental Engineering, Utah State University, Logan, National Eutrophication Research Program, Environmental Protection Agency (1973), p. 89.

Bierman, V. J., Jr. "Dynamic Mathematical Model of Algal Growth in Eutrophic, Freshwater Lakes," Ph.D. thesis, University of Notre Dame, Notre Dame, Indiana (1974).

Bush, R. M. and E. F. Welch. "Plankton Associations and Related Factors in a Hypereutrophic Lake," *Water, Air, Soil Pollution* 1, 257 (1972).

Calane, R. P., S. Nachiappan, D. J. Hineman, and H. E. Allen. "A Dynamic Model for Phytoplankton Production in Grand Traverse Bay,"

Proceedings, Sixteenth Conference on Great Lakes Research, International Association for Great Lakes Research, Huron, Ohio, (April 16-18, 1973), p. 21.

Caperon, J. and J. Meyer. "Nitrogen-Limited Growth of Marine Phytoplankton—I. Changes in Population Characteristics with Steady-State Growth Rate," *Deep Sea Res.* **19**, 601 (1972a).

Caperon, J. and J. Meyer. "Nitrogen-Limited Growth of Marine Phytoplankton—II. Uptake Kinetics and their Role in Nutrient Limited Growth of Phytoplankton," *Deep Sea Res.* **19**, 619 (1972b).

Chartrand, T. A. "A Report on the Taste and Odor in Relation to the Saginaw—Midland Supply at Whitestone Point in Lake Huron," Saginaw Water Treatment Plant, Saginaw, Michigan (1973).

Chen, C. W. and G. T. Orlob. "Ecologic Simulation for Aquatic Environments," Water Resources Engineers, Inc., Walnut Creek, California, prepared for Office of Water Resources Research, U.S. Department of the Interior (1972).

DePinto, J. V., V. J. Bierman, Jr. and F. H. Verhoff. "Seasonal Phytoplankton Succession as a Function of Phosphorus and Nitrogen Levels," paper presented at the 169th Meeting of the American Chemical Society, Philadelphia, Pennsylvania (April 6-11, 1975).

Droop, M. R. "Some Thoughts on Nutrient Limitation in Algae," *J. Phycology* **9**, 264 (1973).

Dugdale, R. C. "Nutrient Limitation in the Sea: Dynamics, Identification, and Significance," *Limnol. Oceanog.* **12**, 685 (1967).

Eppley, R. W. and W. H. Thomas. "Comparison of Half-Saturation Constants for Growth and Nitrate Uptake of Marine Phytoplankton," *J. Phycology* **5**, 375 (1969).

Fitzgerald, G. P. "Some Factors in the Competition or Antagonism Among Bacteria, Algae, and Aquatic Weeds," *J. Phycology* **5**, 351 (1969).

Foree, E. G., W. J. Jewell, and P. L. McCarty. "The Extent of Nitrogen and Phosphorus Regeneration from Decomposing Algae," in *Advances in Water Pollution Research*, Vol. 2, III-27 (Oxford and New York: Pergamon Press, 1970).

Freedman, P. L. "Saginaw Bay: An Evaluation of Existing and Historical Conditions," Environmental Protection Agency, Region V Enforcement Division, Great Lakes Initiative Contract Program Report No. EPA-905/9-74-003 (1974).

Fuhs, G. W. "Phosphorus Content and Rate of Growth in the Diatoms *Cyclotella nana* and *Thalassiosira fluviatilis*," *J. Phycology* **5**, 312 (1969).

Fuhs, G. W., S. D. Demmerle, E. Canelli, and M. Chen. "Characterization of Phosphorus-Limited Planktonic Algae," Nutrients and Eutrophication: The Limiting Nutrient Controversy, Proceedings of a Symposium, American Society of Limnology and Oceanography and Michigan State University, East Lansing, Michigan (February 11-12, 1971), p. 113.

Goldman, J. C., W. J. Oswald, and D. Jenkins. "The Kinetics of Inorganic Carbon Limited Algal Growth," *J. Water Poll. Control Fed.* **46**, 554 (1974).

Goodman, J. and A. Rothstein. "The Active Transport of Phosphate into the Yeast Cell," *J. Gen. Physiol.* **40**, 915 (1957).

Gunnison, D. and M. Alexander. "Resistance and Susceptibility of Algae to Decomposition by Natural Microbial Communities," *Limnol. Oceanog.* **20**, 64 (1975).

Halmann, M. and M. Stiller. "Turnover and Uptake of Dissolved Phosphate in Freshwater. A Study in Lake Kinneret," *Limnol. Oceanog.* **19**, 774 (1974).

Hammer, V. T. "The Succession of 'Bloom' Species of Blue-Green Algae and Some Causal Factors," *Internat. Assoc. Theoretical Appl. Limnol.* **15**, 829 (1964).

Healey, F. P. "Characteristics of Phosphorus Deficiency in *Anabaena*," *J. Phycology* **9**, 383 (1973).

Huff, D. D., J. F. Koonce, N. R. Ivarson, P. R. Weiler, E. H. Dettman, and R. F. Harris. "Simulation of Urban Runoff, Nutrient Loading, and Biotic Response of a Shallow Eutrophic Lake," in *Modeling the Eutrophication Process*, Proceedings of a Workshop, Utah Water Research Laboratory and Division of Environmental Engineering, Utah State University, Logan, National Eutrophication Research Program, Environmental Protection Agency (September 5-7, 1973), p. 33.

King, D. "The Role of Carbon in Eutrophication," *J. Water Poll. Control Fed.* **42**, 2035 (1970).

Kuenzler, E. J. and B. H. Ketchum. "Rate of Phosphorus Uptake by *Phaeodactylum tricornutum*," *Biol. Bull.* **123**, 134 (1962).

Lewin, J. C. and R. L. Guillard. "Diatoms," *Ann. Rev. Microbiol.* **17**, 373 (1963).

Lund, J. W. G. "The Ecology of Freshwater Phytoplankton," *Biol. Rev.* **40**, 231 (1965).

McNaught, D. C. "A Hypothesis to Explain the Succession from Calanoids to Cladocerans During Eutrophication," *Verh. Internat. Verein. Limnol.* **19** (in press).

Medveczky, N. and H. Rosenberg. "The Phosphate-Binding Protein of *Escherichia coli*," *Biochim. Biophys. Acta* **211**, 158 (1970).

Medveczky, N. and H. Rosenberg. "Phosphate Transport in *Escherichia coli*," *Biochim. Biophys. Acta* **241**, 494 (1971).

Middlebrooks, E. J., D. H. Falkenburg, and T. E. Maloney, Eds. *Modeling the Eutrophication Process*, Proceedings of a Workshop, Utah Water Research Laboratory and Division of Environmental Engineering, Utah State University, Logan, National Eutrophication Research Program, Environmental Protection Agency (September 5-7, 1973).

O'Connor, D. J., R. V. Thomann, and D. M. DiToro. "Dynamic Water Quality Forecasting and Management," Environmental Protection Agency Ecological Research Series EPA-660/3-73-009 (1973).

Rhee, G. "A Continuous Culture Study of Phosphate Uptake, Growth Rate and Polyphosphate in *Scenedesmus* Sp.," *J. Phycology* **9**, 495 (1973).

Richardson, W. L. "Modeling Chloride Distribution in Saginaw Bay," presented at the Seventeenth Conference on Great Lakes Research, International Association for Great Lakes Research, Hamilton, Ontario (in press).

Richardson, W. L. "An Evaluation of the Transport Characteristics of Saginaw Bay Using a Mathematical Model of Chloride," presented at the 169th Meeting of the American Chemical Society, Philadelphia, Pennsylvania (April 6-11, 1975).

Schindler, D. W., H. Kling, R. V. Schmidt, J. Prokopovich, V. E. Frost, R. A. Reid, and M. Capel. "Eutrophication of Lake 227 by Addition of Phosphate and Nitrate: the Second, Third, and Fourth Years of Enrichment, 1970, 1971, and 1972," *J. Fisheries Res. Board, Canada* **30**, 1415 (1973a).

Schindler, D. W. and E. J. Fee. "Diurnal Variation of Dissolved Inorganic Carbon and its Use in Estimating Primary Production and CO_2 Invasion in Lake 227," *J. Fisheries Res. Board, Canada* **30**, 1501 (1973b).

Shapiro, J. "Blue-Green Algae: Why They Become Dominant," *Science* **179**, 382 (1973).

Smayda, T. J. "Some Experiments on the Sinking Rates of Two Fresh-water Diatoms," *Limnol. Oceanog.* **19**, 628 (1974).

Thomann, R. V., D. M. DiToro, R. P. Winfield, and D. J. O'Connor. "Mathematical Modeling of Phytoplankton in Lake Ontario. I. Model Development and Verification," Environmental Protection Agency, Ecological Research Series EPA-660/3-75-005 (1975).

Verhoff, F. H. and K. R. Sundaresan. "A Theory of Coupled Transport in Cells," *Biochim. Biophys. Acta* **255**, 425 (1972).

Verhoff, F. H., J. B. Carberry, V. J. Bierman, Jr., and M. W. Tenney. "Mass Transport of Metabolites, Especially Phosphorus in Cells," *Amer. Inst. Chem. Engin., Symp. Series 129*, **69**, 227 (1973).

Vollenweider, R. A., M. Munawar, and P. Stadelmann. "A Comparative Review of Phytoplankton and Primary Production in the Laurentian Great Lakes," *J. Fisheries Res. Board, Canada* **31**, 739 (1974).

APPENDIX A

State Variables

PSA(L), NSA(L)	— Moles phosphorus (nitrogen) per milligram cell dry weight for phytoplankter, L
PCM, NCM, SCM	— Moles phosphorus (nitrogen, silicon) per liter in solution
A(L)	— Milligrams dry weight per liter of phyto-plankter, L
Z(K)	— Milligrams dry weight per liter of zooplankter, K
TOP, TON, TOS	— Moles total unavailable phosphorus (nitrogen, silicon) per liter in solution

For each state variable the model equations are written in the form of a mass balance differential equation:

(Rate of change of state variable) = (Rate of change due to water circulation, Q)

+ (Rate of change due to interactions in system volume, V)

State Variable Equations

For each phytoplankton type, L, the rate of change of intracellular phosphorus is given by:

$$V \frac{d\ A(L) \cdot PSA(L)}{dt} = Q(ABD(L) \cdot PSABD(L) - A(L) \cdot PSA(L))$$

$$+ V \cdot A(L) \cdot R1PM(L) \cdot f(T) \cdot f(I) \cdot (0.322 \times 10^{-4}\ \text{moles/mg})$$

$$x \left[\frac{1}{1 + PK1(L) \cdot PCA(L)} - \frac{1}{1 + PK1(L) \cdot PCM} \right]$$

$$- V \cdot A(L) \cdot PSA(L) \left[RAGRZD(L) + RLYS(L) \cdot T \cdot TCROP \right.$$

$$\left. + \frac{ASINK(L)}{DEPTH} \right]$$

(1.3)

Expanding by the chain rule for derivatives:

$$V \frac{d\ A(L) \cdot PSA(L)}{dt} = V \left[A(L) \frac{d\ PSA(L)}{dt} + PSA(L) \frac{d\ A(L)}{dt} \right]$$

(1.4)

For each phytoplankton type, L, the rate of change of biomass is given by:

$$V \frac{d\ A(L)}{dt} = Q(ABD(L) - A(L))$$

$$+ V \cdot A(L) \left[SPGR(L) - RAGRZD(L) - RLYS(L) \cdot T \cdot TCROP \right.$$

$$\left. - \frac{ASINK(L)}{DEPTH} \right]$$

(1.5)

In Equation (1.5), it is assumed that the contribution to the derivative due to changes in intracellular phosphorus is negligible.

Setting the right-hand sides of Equations (1.3) and (1.4) equal and substituting Equation (1.5) gives the following equation for the state variable PSA(L):

$$V A(L) \frac{d\ PSA(L)}{dt} = Q \cdot ABD(L) \cdot (PSABD(L) - PSA(L))$$

$$+ V \cdot A(L) \cdot R1PM(L) \cdot f(T) \cdot f(I) \cdot (0.322 \times 10^{-4}\ \text{mole/mg})$$

$$x \left[\frac{1}{1 + PK1(L) \cdot PCA(L)} - \frac{1}{1 + PK1(L) \cdot PCM} \right]$$

$$- V \cdot A(L) \cdot PSA(L) \cdot SPGR(L)$$

(1.6)

An identical approach is used for the state variable NSA(L).

The equation for phosphorus concentration, PCM, is given by:

$$V \frac{d\ PCM}{dt} = Q\ (PCMBD - PCM) \tag{1.7}$$

$$- V \cdot f(T) \cdot f(I) \cdot (0.322 \times 10^{-4} \text{ moles/mg})$$

$$x \sum_{L=1}^{Naspec} \left\{ R1PM(L) \cdot A(L) \left[\frac{1}{1 + PK1(L) \cdot PCA(L)} \right. \right.$$

$$\left. \left. - \frac{1}{1 + PK1(L) \cdot PCM} \right] \right\} + (V) \sum_{L=1}^{Naspec} \left\{ [RLYS(L) \cdot T \cdot A(L) \cdot TCROP \right.$$

$$+ RAEXC(L) \cdot A(L)] \ [PSA(L) - PSAMIN(L)] \Big\}$$

$$+ V \cdot RDCMP \cdot T \cdot TOP + WPCM$$

The equation for nitrogen concentration, NCM, is functionally identical to the above equation for PCM.

The equation for silicon concentration, SCM, is given by:

$$V \frac{d\ SCM}{dt} = Q\ (SCMBD - SCM) \tag{1.8}$$

$$- (V) \sum_{Diatoms} \Big\{ A(L) \cdot SPGR(L) \cdot SSA(L) \Big\}$$

$$+ V \cdot RDCMP \cdot T \cdot TOS + WSCM$$

The equation for Z(K), the concentration of zooplankter K, is given by:

$$V \frac{d\ Z(K)}{dt} = Q\ (ZBD(K) - Z(K)) \tag{1.9}$$

$$+ V \cdot Z(K)\ [RZ(K) - ZDETH(K)]$$

The equation for TOP, total unavailable phosphorus concentration, is given by:

$$V \frac{d\,TOP}{dt} = Q\,(TOPBD - TOP) \tag{1.10}$$

$$+ \; V \cdot T \cdot TCROP \sum_{L=1}^{Naspec} \left\{ RLYS(L) \cdot A(L) \cdot PSAMIN(L) \right\}$$

$$+ \; (V) \sum_{K=1}^{Nzspec} \left\{ RZPEX(K) \cdot Z(K) \right\} - V \cdot TOP \left[RDCMP \cdot T + \frac{TOSINK}{DEPTH} \right]$$

$$+ \; WTOP$$

The equations for TON and TOS are functionally identical to the preceding equation.

Process Rate Equations

Specific growth rate of phytoplankter L, SPGR(L), is equal to the minimum value of the following three functions:

Maximum growth rate $\cdot$ f(T) $\cdot$ f(I) $[1 - EXP(-0.693(P/PO-1))]$

Maximum growth rate $\cdot$ f(T) $\cdot$ f(I) $[(N\text{-}NO)/(KNCELL + (N\text{-}NO))]$

Maximum growth rate $\cdot$ f(T) $\cdot$ f(I) $[SCM/(KSCM + SCM)]$

Rate of growth of zooplankter K:

$$RZ(K) = RZMAX(K) \cdot ZASSIM(K) \sum_{L=1}^{Naspec} \left\{ \frac{ZEFF(K,L) \cdot A(L)}{KZSAT(K,L) + \sum\limits_{L=1}^{Naspec} ZEFF(K,L) \cdot A(L)} \right\}$$

Rate at which phytoplankter L is ingested by zooplankton:

$$RAGRZD(L) = \sum_{K=1}^{Nzspec} \left\{ \frac{RZMAX(K) \cdot Z(K) \cdot ZEFF(K,L)}{KZSAT(K,L) + \sum\limits_{L=1}^{Naspec} ZEFF(K,L) \cdot A(L)} \right\}$$

Rate at which phytoplankter L is excreted by zooplankton (used to calculate phosphorus excreted to available phosphorus pool):

$$RAEXC(L) = \sum_{K=1}^{Nzspec} \left\{ \frac{(1\text{-}ZASSIM(K)) \cdot RZMAX(K) \cdot ZEFF(K,L) \cdot Z(K)}{KZSAT(K,L) + \sum\limits_{L=1}^{Naspec} ZEFF(K,L) \cdot A(L)} \right\}$$

Rate at which phosphorus is excreted to the unavailable pool by zooplankter K:

$$RZPEX(K) = (1-ZASSIM(K)) \; RZMAX(K)$$

$$x \sum_{L=1}^{Naspec} \left\{ \frac{ZEFF(K,L) \cdot A(L) \cdot PSAMIN(L)}{KZSAT(K,L) + \sum\limits_{L=1}^{Naspec} ZEFF(K,L) \cdot A(L)} \right\}$$

Miscellaneous Functions

$$f(T) = \Theta^{(T-20)} \quad \text{where:} \quad \Theta = 1.07 \text{ Diatoms}$$
$$\Theta = 1.08 \text{ Greens}$$
$$\Theta = 1.10 \text{ Blue-Greens}$$

$$\text{Temperature (T)} = TMAX \left\{ 0.50 - 0.50 \; SINE \left[\frac{6.28 \; TIME + \phi}{360} \right] \right\}$$

where ϕ is chosen such that SINE = 0 on November 1.

$$f(I) = \frac{1}{ke \cdot DEPTH} \; [e^{-\alpha 1} - e^{-\alpha 0}] \quad \text{O'Connor } et \; al.((1973)$$

$$\text{where} \quad \alpha 1 = \frac{Ia}{Is} e^{-(ke \cdot DEPTH)}$$

$$\alpha 0 = \frac{Ia}{Is}$$

$$ke = 1.9/\text{Secchi Depth} + 0.17 \cdot TCROP$$

$$= 0.633 + 0.17 \cdot TCROP$$

$$Ia = 2000 \text{ foot candles}$$

$$\text{Photoperiod} = 0.50$$

APPENDIX B

List of Symbols

A	phytoplankton concentration (mg dry wt/liter)
ASINK	phytoplankton sinking rate (m/day)
FACT	mg dry weight per phytoplankton cell
f(I), f(T)	light and temperature reduction factors, respectively
KNCELL	intracellular half-saturation constant for nitrogen-dependent growth (mol N/cell)
KSCM	Michaelis constant for silicon-dependent growth (mol Si/l)
KZSAT(K,L)	half-saturation concentration of phytoplankter L for grazing by zooplankter K
Naspec	number of phytoplankton species
Nzspec	number of zooplankton species
P, N	moles phosphorus (nitrogen) per phytoplankton cell
PCA, NCA	intracellular phosphorus (nitrogen) concentration (mol/l cell volume)
PCM, NCM, SCM	nutrient concentrations (phosphorus, nitrogen, silicon) in solution (mol/l)
PSA, NSA	phosphorus (nitrogen) storage in phytoplankton cells (mol/mg dry wt)
PK1, NK1	affinity constant for phosphorus (nitrogen) uptake mechanism (l/mol)
PO, NO	minimum stoichiometric level of phosphorus (nitrogen) per phytoplankton cell (mol/cell)
Q	water circulation rate (vol/day)
R1PM, R1NM	maximum phosphorus (nitrogen) uptake rate $(day)^{-1}$
RADSAT(L)	saturation light intensity for growth of phytoplankter L
RAEXC(L)	rate at which phytoplankter L is excreted by zooplankton $(day)^{-1}$
RAGRZD(L)	rate at which phytoplankter L is grazed (ingested) by zooplankton $(day)^{-1}$
RDCMP	decomposition rate from unavailable to available nutrient pools $(day-^{\circ}C)^{-1}$
RLYS	algal death rate $[day-^{\circ}C\text{-}(mg/l)]^{-1}$
RZ	zooplankton specific growth rate $(day)^{-1}$
RZMAX	zooplankton maximum ingestion rate $(day)^{-1}$
RZPEX(K), RZNEX(K), RZSEX(K)	phosphorus (nitrogen, silicon) excretion by zooplankter K to unavailable nutrient pool (mol/mg zooplankter day)
SPGR	phytoplankton specific growth rate $(day)^{-1}$
SSA	silicon stoichiometry for diatoms (mol/mg dry wt)
T	temperature $(^{\circ}C)$
TCROP	total phytoplankton biomass (mg dry wt/l)

TOP, TON, TOS	concentration of unavailable phosphorus (nitrogen, silicon) (mol/l)
TOSINK	sinking rate of nonliving organic material (m/day)
V	system volume
WPCM, WNCM, WSCM	external point loading rates of available phosphorus (nitrogen, silicon) (mol/day)
WTOP, WTON, WTOS	external point loading rates of unavailable phosphorus (nitrogen, silicon) (mol/day)
Z	zooplankton concentration (mg dry wt/l)
ZASSIM	zooplankton assimilation efficiency
ZEFF(K,L)	ingestion efficiency of zooplankter K for phytoplankter L
ZDETH	zooplankton death rate $(day)^{-1}$

Note: the addition of the suffix "BD" to a variable name refers to the boundary value of that variable.

2

A PLANKTON-BASED FOOD WEB MODEL
FOR LAKE MICHIGAN

Raymond P. Canale, Leon M. DePalma, and Allan H. Vogel[1]

INTRODUCTION

The value of Lake Michigan as a water resource has declined during recent years because of eutrophication and deterioration of the fishery. Mathematical models of the ecosystem may help formulate effective management strategies to reverse this trend. However, these models must be taxon-specific because eutrophication favors specific algal taxa over others, and the invasion of the Great Lakes by marine fish has altered the original food web by selective removal of zooplankton and native species of fish. These two problems are related because unedible blue-green algae eventually sink, die, and decompose, reducing dissolved oxygen in the hypolimnion, further stressing the remaining zooplankton and fish. Overgrazing on the larger zooplankton by these marine invaders can also affect the phytoplankton composition by reducing the grazing pressure on the larger algae. The combined effects of these two problems has hastened the decline of water quality in the lake, and models that fail to include these complex interactions, such as those developed by Hydroscience (1973) and Canale et al. (1974), may result in inaccurate estimates of phytoplankton and fish standing stocks.

---

[1] Department of Civil Engineering, University of Michigan; Sea Grant Program, University of Michigan; and Sea Grant Program, University of Michigan, respectively.

33

PROPOSED FOOD WEB

The proposed model divides the phytoplankton into diatoms and non-diatoms to account for silicon requirements. Because green algae, in general, are grazed upon by zooplankton, and blue-greens are not, this type of separation is also necessary in models of aquatic systems that are affected by a long-term increase in trophic condition. Therefore, questions concerning the effect of eutrophication may require a tripartite division of the phytoplankton. A more comprehensive Great Lakes productivity model might also separate the blue-green algae into nitrogen-fixers and nonfixers since nitrogen limitation exists in some regions (Vanderhoff *et al.* 1974). Flagellates are another taxon that might be added since they appear to require higher nutrient levels than other algae and may exhibit heterotrophic characteristics. Including these taxa would be essential in modeling certain areas of the Great Lakes such as Saginaw Bay, where flagellates become the dominant summer algae. However, it is felt that favorable conditions for nitrogen-fixers and flagellates do not exist in the open waters of Lake Michigan; therefore these taxa are not included in the model discussed in this chapter.

If the model were seeking to answer only questions concerning eutrophication, only two groups of zooplankton would be necessary, predatory and herbivorous. Division of the zooplankton into two groups in a eutrophication model is desirable because the predators reduce the grazing of herbivores upon the phytoplankton. A eutrophication model including three phytoplankton states, the herbivorous zooplankton, and the predatory zooplankton has the virtue of being easy to handle and of simulating most critical plankton interactions under increased eutrophication. However, there are two difficulties with such a model, the first being the presence of omnivorous zooplankton in Lake Michigan. The second problem is that the forage fish of Lake Michigan, particularly the alewife, graze upon the zooplankton differentially. Alewives are size-selective; whereas all the predators are large enough to be eaten, the omnivores and herbivores are not. Thus a more realistic model would have to provide for a division based on forage-fish predation.

Another complication in the food web arises because the alewife apparently prefers cladocerans to copepods. Therefore, the small herbivores must be divided into cladocerans and copepods. The cannibalism of *Cyclops*, a small omnivore, requires further division of the small herbivorous copepods into *Diaptomus* and the nauplii. The above division of zooplankton leads to the model proposed herein with assignment of taxonomic names to the seven groups of zooplankton:

1) Predators = *Leptodora, Polyphemus*, and *Mesocyclops*
2) Large Omnivores = *Limnocalanus, Epischura,* and *Senecella*
3) Small Omnivores = *Cyclops*
4) Large Herbivorous Cladocerans = *Daphnia* and *Diaphanosma*
5) Small Herbivorous Copepods = *Diaptomus* and *Eurytemora*
6) the Nauplii
7) Small Herbivorous Cladocerans = *Bosmina, Holopedium, Ceriodaphnia,* and the chydorids.

The diatoms have already been separated from the other phytoplankton on the basis of nutrient requirements. However, further separation of the diatoms into large and small forms is required by the presence of different zooplankton herbivores. As the model is presently organized, only the omnivores and herbivorous copepods can graze upon the large diatoms, while the small diatoms are used by all the herbivores. The interactions in the final model are shown in Figure 2.1. While the effects of selective feeding by the alewife are included in the zooplankton equations, the alewife feeding rate and standing stock are assumed known and entered into the model as time variable forcing functions.

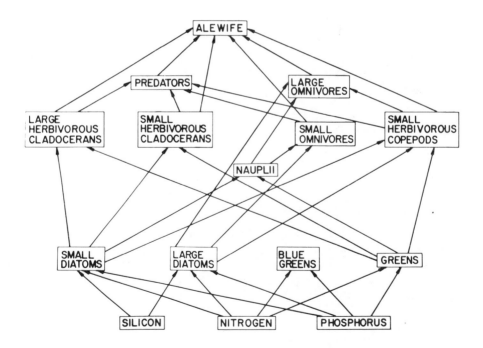

Figure 2.1   Complex Lake Michigan food web model.

SPATIAL AND TEMPORAL HYPOTHESES

The ultimate objective of our research is to describe both temporal and spatial variation of system variables in Lake Michigan. However, three factors limit a detailed study of the spatial behavior of food web variables. (1) Extensive data that define spatial variations of chemical and biological parameters are not available. (2) It is appropriate to examine a model with complex chemical and biological mechanisms but having simple spatial detail before attempting a model with complex spatial resolution. (3) The model described below has 25 water quality variables and two vertical layers. If the lake were divided into more numerous spatial segments, a computationally infeasible problem would arise quickly.

The two layer trophic level model will be verified against the average of data from four stations in the west arm of Grand Traverse Bay. These stations are shown in Figure 2.2. Zooplankton samples were taken by vertical hauls from the bottom. Phytoplankton was sampled discreetly

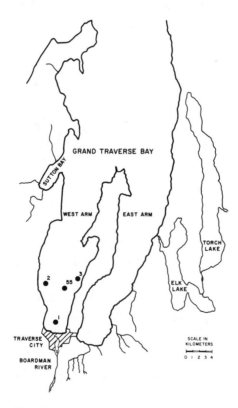

Figure 2.2 Location of Grand Traverse Bay sampling stations.

at 2 meters and 20 meters, and average concentrations in the 20-meter column were calculated. Nutrients were sampled discreetly at a number of depths between the surface and the bottom with averages for the epilimnion and the hypolimnion calculated. Additional details regarding sample collection, handling, storage and analysis are discussed in another report.[1]

MODEL MECHANISMS

The mechanisms that are thought to contribute to predominant changes in the state of the system have been incorporated into the model. Some of the mechanisms considered are given much coverage in the literature and have been formulated mathematically by other researchers. However, because the model is taxon-specific, some of the mechanisms have not been studied in detail, and previous mathematical descriptions may not be available. Where a mechanism is known to be important, and where previous research has contributed a sound structural formulation for the mechanism, the difficulty of estimating coefficients for the formulation still persists. The heterogeneity of the biological states makes coefficient estimation a serious problem. Age and size distribution, as well as the spatial distribution of a species in the lake, often suggest a broad range of scientifically acceptable coefficient values. The inherent problem here, of course, is a problem often confronted in population modeling, *i.e.,* model mechanisms must portray the population, not the individual. All the references that were used to develop these mechanisms are discussed in Canale, *et al.* (1975); however, for the sake of brevity, only the most important ones have been included here.

Zooplankton

Eating Rate

The growth of a zooplankton state, z, is determined by its present abundance, its rate of grazing, and the efficiency with which it converts food carbon into zooplankton carbon. These factors are assumed to operate independently of one another, in that the positive term determining growth of the population is

$$\left[\text{growth}\right]_z = \left[\begin{array}{c}\text{assimilation}\\ \text{efficiency}\end{array}\right]_z * \left[\begin{array}{c}\text{eating}\\ \text{rate}\end{array}\right]_z * \quad c_z$$

$$\frac{\text{mg } z \text{ c}}{\text{mg food c}} \qquad \frac{\text{mg food c}}{\text{mg } z \text{ c· day}} \qquad \frac{\text{mg } z \text{ c}}{1} \qquad (2.1)$$

---

[1] University of Michigan Sea Grant report in preparation.

The eating mechanism differs greatly enough among species to warrant modeling three types of eating behavior: raptorial, selective filtering, and nonselective filtering. Table 2.1 lists the feeding category for each zooplankton state. The eating rate attainable by a particular zooplankton state, regardless of classification, is temperature-dependent. The temperature dependence is denoted as $\Phi_z(T)$ for state z. Figure 2.3 shows the assumed behavior of $\Phi_z(T)$ for the zooplankton as well as for the phytoplankton that will be discussed in a later section. This function is normalized to a value of unity at $20^\circ C$. The maximum snatching rate for raptors, the maximum filtering rate for selective filterers, and the filtering rate for nonselective filterers, all at $20^\circ C$, are defined as $A7_z$.

Table 2.1   The State of the System

State Number	Description	Classification			Units
1	*Leptodora* & *Polyphemus*	} Raptors			
2	*Cyclops*				
3	*Cyclops* nauplii				
4	*Diaptomus* nauplii				
5	*Limnocalanus* & *Epischura* nauplii	Selective filterers	Zooplankton		
6	*Diaptomus*				mg c / ℓ
7	*Limnocalanus* & *Epischura*				
8	*Daphnia*	Nonselective filterers			
9	*Bosmina* & *Holopedium*				
10	Small diatoms				
11	Large diatoms	Phytoplankton			
12	Blue-greens				
13	Greens				
14	Detrital nitrogen				
15	Dissolved organic nitrogen	Nitrogen	Epilimnion nutrients		mg N / ℓ
16	Ammonia				
17	Nitrate				
18	Detrital phosphorus				
19	Dissolved organic phosphorus	Phosphorus			mg P / ℓ
20	Dissolved inorganic phosphorus				
21	Detrital silicon	Silicon			mg Si / ℓ
22	Dissolved silicon				
23	Total nonaccessible nitrogen		Nutrient mass held in hypolimnion and sediments		mg N
24	Total nonaccessible phosphorus				mg P
25	Total nonaccessible silicon				mg Si

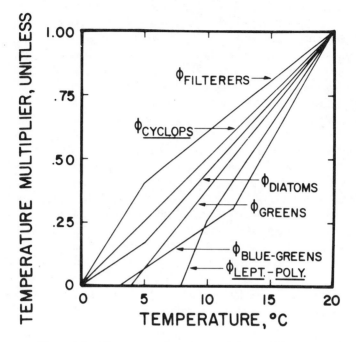

Figure 2.3   Temperature-dependence of the growth rates.

Raptorial species obtain their food by selecting, snatching, and devouring. This operation is time-consuming, and even when food is abundant the raptor is limited by the rate at which he can consume food. Since the eating rate of a raptor increases with food concentration, but reaches a saturation level, a Michaelis-Menton type expression has been utilized in the mathematical formulation of the mechanism. For a raptorial state z, then

$$\begin{bmatrix} \text{eating} \\ \text{rate} \end{bmatrix}_z = A7_z * \Phi_z(T) * \left(\frac{\sum\limits_{\text{prey}} c_i}{\sum\limits_{\text{prey}} c_i + KFOOD_z} \right) \qquad (2.2)$$

$$\frac{\text{mg food c}}{\text{mg z c· day}}$$

where

$\sum\limits_{\text{prey}} c_i$ = sum of concentrations of all states that can serve as food for raptor state z

$KFOOD_z$ = half-saturation food level for state z.

Two important assumptions have been made in the above formulation and will be made for filter-feeding species also. (1) There is no lower food threshold below which raptors cannot find their prey. That is, the model does not account for refuge sites that may be available to the prey to protect it from the predator although Gause (1934) and Frost (1974) have discussed the importance of refuge sites in predator-prey systems. (2) Although preferences must play a role in determining the allocation of the eating rate toward each prey species, they do not affect the rate of intake of food carbon. The animal is assumed to have a basic food requirement that must be satisfied regardless of the species present. As long as those species are preyed upon by the zooplankton in question, they will be utilized to meet the basic food requirement.

Selective filterers obtain food by grazing, but have developed an ability to vary their filtering rate. As the concentration of food decreases, a maximum filtering rate is approached, a rate fixed by the physical characteristics of the species. In an environment where food concentration is high, the selective filterer responds by adjusting its particle-size selectivity and lowering its filtering rate. In this way, the animal is able to make more efficient use of food and its own energy. A proposed formulation of this mechanism becomes

$$
\left[\begin{array}{c}\text{eating}\\\text{rate}\end{array}\right]_z = A7_z * \Phi_z(T) * \left(\frac{A9 * \sum\limits_{\text{prey}} c_i + A10}{\sum\limits_{\text{prey}} c_i + A10}\right) * \sum\limits_{\text{prey}} c_i \qquad (2.3)
$$

$$
\frac{1}{\text{mg z c} \cdot \text{day}} \qquad\qquad \frac{\text{mg food c}}{1}
$$

where

$A9$ = minimum filtering rate multiplier
$A10$ = food level where multiplier is ½ (1 + A9)

It has been assumed that, even at high food concentrations, the organism will be filtering at some minimum level. Thus, no provision has been made for extreme eutrophic situations where food concentrations may be so high that the ability of the organism to filter is hampered.

Nonselective filterers, unlike selective filterers, filter at a uniform rate, and thus operate below maximum efficiency. A formulation of their eating mechanism is

$$
\left[\begin{array}{c}\text{eating}\\\text{rate}\end{array}\right]_z = A7_z * \Phi_z(T) * \sum\limits_{\text{prey}} c_i \qquad (2.4)
$$

$$
\frac{1}{\text{mg z c} \cdot \text{day}} \qquad \frac{\text{mg food c}}{1}
$$

As discussed earlier, it may be more appropriate to model the eating rate as a function of a weighed sum of prey concentration rather than total food concentration. Nonselective filterers cannot lower their filtering rate when the plankton content of filtered water increases. Therefore they operate below maximum possible efficiency. A simple but unconfirmed formulation is

$$\left[\begin{array}{c} \text{assimilation} \\ \text{efficiency} \end{array} \right]_z = A11N^* \left(\frac{A24}{\sum\limits_{\text{prey}} c_i + A24} \right) \qquad (2.5)$$

$$\frac{\text{mg z c}}{\text{mg food c}}$$

where

A11N = maximum efficiency possible
A24 = half-maximum efficiency food level

A formulation similar to this seems justifiable, since nonselective filterers such as *Daphnia* and *Bosmina* require more euthrophic waters to flourish than do selective filterers such as *Diaptomus* and *Limnocalanus*. The maximum efficiency possible has not been reported in the literature, but it may be very close to 100% since it is defined at near-zero food concentrations.

Before considering other mechanisms, it is worthwhile to compare the dependence of growth rate on food concentration for the three classes of zooplankton. In Figure 2.4, plots of the eating rate and assimilation rate for the three groups are presented (assimilation rate = eating rate * assimilation efficiency) as functions of food concentration. The curves have been normalized so that the eating rate is unity at 0.20 mg food c/l. The region between the curves indicates the rate of egestion and rejection of unassimilated matter.

Preferred Diet

The mathematical formulations of eating rate discussed earlier involved only total food carbon available. The assumption was made that food intake is controlled by the amount of edible food in the environment, not by the relative abundance of the various species eaten. However, given that a population has a certain food intake rate—determined by the characteristics of the population, the amount of food present, and the temperature—it will prefer certain species over others. The rate at which state z consumes prey state k is in general given by Equation 2.6.

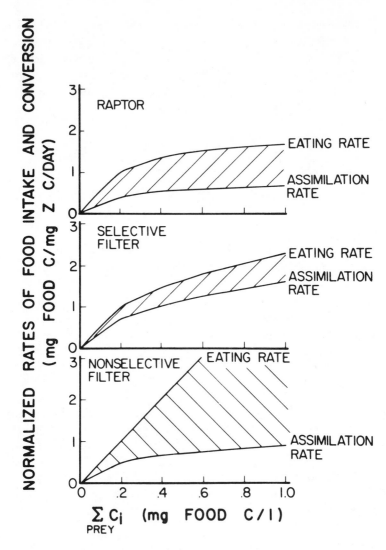

Figure 2.4 Growth mechanisms of the zooplankton groups.

$$\begin{bmatrix} \text{predation of species k} \\ \text{by species z} \end{bmatrix} = \begin{bmatrix} \text{preference of species z} \\ \text{for species k} \end{bmatrix} * \begin{bmatrix} \text{eating} \\ \text{rate} \end{bmatrix}_z \quad * \quad c_z \qquad (2.6)$$

$$\frac{\text{mg k c}}{\text{l}\cdot\text{day}} \qquad\qquad\qquad \frac{\text{mg food c}}{\text{mg z c}\cdot\text{day}} \quad \frac{\text{mg z c}}{\text{l}}$$

The mathematical formulation of the preference factor involves a simple assumption: the preference of species z for food species k is proportional to the product of the electivity of species z for species k, and the concentration of species k. The electivity, α_k^z, is defined as the fraction of the species z diet that would be composed of food species k if all food species were present in equal concentrations. This leads to the formulation

$$\left[\begin{array}{c} \text{preference of species } z \\ \text{for species } k \end{array}\right] = \frac{\alpha_k^z * c_k}{\sum_i (\alpha_i^z * c_i)} \tag{2.7}$$

When all food concentrations are equal, the above expression reduces to α_k^z, since $\sum_i \alpha_i^z = 1$ by definition. The above form differs somewhat from food preference models proposed by Kitchell *et al.* (1973) but uses the approach of O'Neill (1969).

Given equal concentrations of prey species, the raptorial zooplankton prefers larger organisms to smaller ones because they provide more food carbon for about the same energy expenditure (Cummins *et al.* 1969). Similarly, there is a natural preference for slow-moving organisms over rapidly moving ones. The electivities of the filter-feeding species are determined by food size, shape, and palatability (Arnold 1971; Frost 1972). The nauplii, averaging under 0.5 mm in length, are assumed to be herbivorous filter feeders. No studies have been done to establish directly the electivities of zooplankters in Lake Michigan. Thus the electivities used in the model are unverified.

The diatom groupings have been chosen so that the nauplii are restricted to eating only the small form. Some blue-greens possess a mucilage coating or have large filaments that make both filtering and digesting of them difficult for most zooplankton species. The omnivorous copepods, *Limnocalanus* and *Epischura*, adjust their filtering parts so that larger organisms are ingested when available. Smaller organisms are only taken when larger forms are scarce (Main 1962). The nonselective cladocerans, *Daphnia, Bosmina,* and *Holopedium*, graze upon small diatoms and green algae. However, Arnold (1971) contends that the survival of blue-green fed *Daphnia* is low. The web of electivities incorporated into the lower food web model is shown in Figure 2.5.

These results are based on an extensive review of the literature and are discussed in detail by Canale *et al.* (1975). However the results are subject to modification as researchers collect data and complete experiments on predator-prey relations.

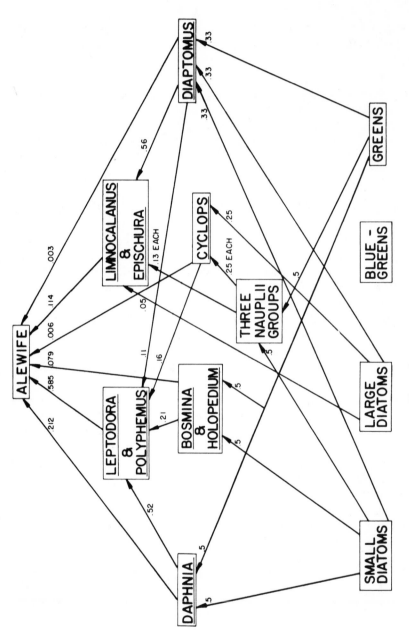

Figure 2.5   Web of electivities.

Respiration

Zooplankton respiration research reported in the literature must be analyzed with care. Several factors such as crowding, nutritional state, thermal shocking, and seasonality affect experimental results. Available information implies different respiration rates for the zooplankton states of the model; however, little information is actually available on taxon-specific rates. At this stage it is reasonable to assume the same temperature-dependence for all species and to differentiate only between the respiration rate of adults and nauplii. A linear temperature-dependence has been commonly used in plankton production models (Canale *et al.* 1974). Although some evidence supports an exponential dependence (Warren 1971), it is thought that a linear approximation for the temperature range of interest is justifiable. Therefore

$$\left[\begin{matrix} \text{respiration} \\ \text{loss} \end{matrix}\right]_z = A12_z {}^* (T/20.) {}^* c_z \tag{2.8}$$

$$\text{day}^{-1} \qquad \frac{\text{mg z c}}{1}$$

where

$A12_z$ = rate at $20°C$ for zooplankton state z (one rate for adults and one rate for nauplii)

Phytoplankton

Considerable research has been conducted by DiToro and his colleagues at Hydroscience on modeling phytoplankton dynamics (DiToro *et al.* 1971; Hydroscience Inc. 1973). Canale and others at the University of Michigan (1974) have utilized similar techniques to construct a dynamic model for the natural association of phytoplankton in Grand Traverse Bay. After accounting for taxon-specific behavior, many of the mechanisms utilized in these models can be applied to the present effort.

Production Rate

The production of phytoplankton carbon for a particular state p is governed predominantly by water temperature, light intensity, and nutrient abundance. Under optimum light conditions and saturating nutrient concentrations, production proceeds at a rate fixed by temperature. Canale and Vogel (1974) have surveyed the wealth of literature on this subject to ascertain the variation of specific growth rate with temperature for diatoms, greens, blue-greens and flagellates. The average maximum water temperature observed in Lake Michigan is about $17°C$, which is well below the optimum temperature for the taxa considered in the model. Thus, instead of introducing the maximum saturated growth rate as a coefficient, the model deals with the saturated growth rate at $20°C$, which is denoted as Al_p. Productivity for phytoplankton state p, then, may be formulated

$$\begin{bmatrix} \text{growth} \end{bmatrix}_p = Al_p * \Phi_p(T) * \begin{bmatrix} \text{light} \\ \text{reduction} \end{bmatrix}_p * \begin{bmatrix} \text{nutrient} \\ \text{reduction} \end{bmatrix}_p * c_p \qquad (2.9)$$

$$\text{day}^{-1} \qquad\qquad\qquad \frac{mg\ c}{l}$$

The temperature multiplier $\Phi_p(T)$ depends on temperature in the manner shown in Figure 2.3. Ryther (1956) has shown that an optimum light intensity exists for phytoplankton photosynthesis. Steele (1965) has formulated a light reduction factor describing the effect of nonoptimum light on the production rate:

$$\begin{bmatrix} \text{light reduction} \\ \text{at a point} \end{bmatrix} = \frac{I}{IS} e^{\left(-\frac{I}{IS} + 1\right)} \qquad (2.10)$$

where

I = light intensity at the point
IS = optimum light intensity

This reduction factor describes light limitation at a point. However, since the model considers a completely mixed euphotic zone, the reduction factor must be averaged over depth, which accounts for light extinction in the water column. If light attenuation is uniform throughout the column, the light at depth d can be expressed by the familiar formula: $I(d) = I(o)*e^{-KE*d}$ where $I(o)$ is the surface intensity and KE the extinction coefficient. DiToro et al. (1971) have integrated the reduction factor over depth with pointwide intensity given by $I(d)$. The result expresses the average reduction factor as a function of $I(o)$ and KE at an instant of time. To account for daily cycling in a tractable manner, surface light intensity is assumed uniform over the daylight hours at a value IA, equal to the average over the daylight hours. With this simplifying assumption, the average daily reduction factor has been calculated by DiToro et al. (1971):

$$\begin{bmatrix} \text{light reduction averaged} \\ \text{over depth and time} \end{bmatrix}_p = \frac{e*PHOTO}{KE*DEPTH} \left(e^{-\frac{IA}{IS_p} e^{-KE*DEPTH}} - e^{-\frac{IA}{IS_p}} \right) \qquad (2.11)$$

where

e   = 2.718
PHOTO = daylight fraction of day
DEPTH = depth of euphotic zone

The difference in nutrient utilization between diatoms and the green and blue-green algae groups must be modeled carefully. Until recently, phytoplankton blooms in Lake Michigan may have been controlled by the availability of inorganic phosphorus. Low phosphorus levels limit the production rate of all three phytoplankton groups in a similar manner. An analysis of the seasonal variation of species in Lake Michigan suggests that diatoms utilize essentially all available inorganic phosphorus by the end of June, and that green and blue-green populations, which are not cold-water adapted, are forced to utilize recycled phosphorus in midsummer. The low amounts of recycled phosphorus have kept the greens and blue-greens from reaching undesirable concentrations in midsummer.

However, Schelske and Stoermer (1972) have found an interesting phenomenon developing in southern Lake Michigan, where phosphorus has reached 1930 Lake Erie levels. Diatoms, which require silicon as an essential nutrient, appear to have an excess of phosphorus and are now limited by silicon. However, since greens and blue-greens do not require silicon, they can utilize the inorganic phosphorus left by diatoms to reach higher midsummer peaks. The above sequence of events may be typical in lakes undergoing eutrophication, making modeling of the underlying causes critical to the development of any control scheme. It is possible to extend the above sequence further by considering the ability of certain blue-green species to fix dissolved nitrogen gas. When phosphorus reaches a critical point, inorganic nitrogen will begin to limit the green algae, whereas certain blue-greens may switch over to nitrogen fixation, thereby surpassing the greens and nonnitrogen-fixing blue-greens in midsummer.

The mechanisms for nutrient limitation described above have been formulated by other workers (Dugdale 1967; Eppley *et al.* 1969; Di Toro *et al.* 1971). To date, most useful models have assumed that the productivity of the population follows Michaelis-Menton kinetics as a nutrient becomes limiting. Under this assumption, the nutrient reduction factors for nitrogen, phosphorus, and silicon are respectively

$$\frac{c_{ammonia} + c_{nitrate}}{c_{ammonia} + c_{nitrate} + KN} , \quad \frac{c_{inorg. P}}{c_{inorg. P} + KP} \quad \frac{c_{diss. Si}}{c_{diss. Si} + KS}$$

where *KN*, *KP*, and *KS* are the half-saturation levels of the three nutrients and the state variables are in units of mg basic element/l. As the nitrogen

factor indicates, the plankton can utilize both ammonia and nitrate.
When ammonia is available in the same concentration as nitrate, ammonia
may be preferred over nitrate, a mechanism to be discussed further in a
subsequent section of this chapter. Only the total usable nitrogen determines the reduction due to nitrogen, and the relative amounts of the
two inorganic nitrogen states have no bearing on the growth reduction.

Some evidence (*e.g.*, Herbes 1974 has an extensive bibliography) indicates that organic phosphorus may be available to phytoplankton. This
research, however, indicates that organics provide a small fraction of the
phosphorus used by plankton. Until experimental work can confirm the
selectivity for inorganics over organics, organics will be assumed unavailable to the plankton. The calculated growth rate of the diatoms in
response to silicon limitation has been modified by correcting for unavailable silicon as suggested by the work of Paasche (1973a, 1973b). The
model assumes that 35 μgSi/l is unavailable. The ability of the blue-green
state to fix nitrogen gas has been ignored since this mechanism only becomes critical under severe nitrogen-limiting environments.

Bierman *et al.* (1973) have recently offered an alternative to the classical
Michaelis-Menton approach to modeling nutrient uptake. The alternative
is based upon the concept of luxury uptake, whereby the algal cell does
not use all absorbed nutrients immediately. An intermediate state, or
internal nutrient pool, is considered to control cell growth, with transport
between water and this state governed by a reversible kinetic mechanism.
This refinement in the description of phytoplankton growth, according to
Bierman *et al.* (1973), becomes a necessity when considering the competition between green and blue-green taxa. The transport from water to
internal nutrient state occurs more rapidly in blue-greens than in greens,
especially at low nutrient levels, affording blue-greens the opportunity to
take up more nutrients than they actually need for immediate growth.
Although this approach is promising, the mathematics of the mechanism
have not yet been verified against extensive data. For this reason a
complex growth mechanism beyond the Michaelis-Menton formulation
seems premature for this study.

Respiration

Endogenous respiration acts along with predation to lower phytoplankton carbon during and after a bloom. Most models, including those of
DiToro *et al.* (1971) and Canale *et al.* (1974), have modeled the effect
of temperature on phytoplankton respiration linearly. Riley (1965) has
suggested, and many have agreed, that the effect is probably exponential.
However, if the proposed exponential dependence is used, with a 20°C

rate as reported in most of the literature, the respiration rate below 5°C appears too high to permit persistence of the phytoplankton through the winter months. It is therefore hypothesized that the plankton manifest an exponential dependence at higher temperatures, but that in the lower range of temperatures characteristic of Lake Michigan a linear approximation may be substituted. The temperature-varying linear kinetics may then be formulated

$$\left[\begin{array}{c} \text{respiration} \\ \text{loss} \end{array} \right]_p = A3*(T/20.)*c_p \qquad (2.12)$$

$$\text{day}^{-1} \qquad \frac{\text{mg c}}{1}$$

where A3 = phytoplankton respiration rate at 20°C.

Sinking

Although sinking is an important contribution to phytoplankton loss during certain seasons, the mechanism is difficult to model. A myriad of parameters, including temperature, internal nutrient concentrations, internal monovalent-divalent concentration ratios, the presence of gas and/or oil vacuoles, and turbulence, affect the net upward or downward velocity of a species. With the present level of sophistication in the area, a reasonable approach is to measure sinking velocities throughout the year and use these data as a forcing function. If the depth-averaged velocity of a species is given by *VS*, then the sinking rate is given by *VS/DEPTH*, where *DEPTH* is the depth of the euphotic zone. If $A6_p$ is the maximum sinking rate of state p, and the function SINK(t) exhibits the seasonal variation, then

$$\left[\begin{array}{c} \text{sinking} \\ \text{loss} \end{array} \right]_p = A6_p *SINK(t)*c_p \qquad (2.13)$$

Unfortunately, comprehensive studies have not been made on the variation of sinking with time in Lake Michigan. Burns and Pashley (1974) have measured sinking velocities in central Lake Ontario at four times of the year. The results indicate that sinking is prevalent in March and September, and that in July and October plankton are generally rising to the surface.

Nutrient Cycles

The mass balance equations written for each of the three nutrient cycles under consideration acquire simple forms when one assumes that carbon, nitrogen, and phosphorus occur in the same ratios in all

plankton.   In general, it is known that these ratios can vary among
phytoplankton species and are even critically dependent upon relative
concentrations of the nutrients in the surrounding water and upon the
life history of the cells. DiToro *et al.* (1971) have summarized the data
of Strickland (1965) on the dry weight percentage of carbon, nitrogen,
and phosphorus in various taxa of phytoplankton. The great variability
indicated by these data certainly must be accepted as evidence that taxon-
specific nutrient ratios must be further investigated. However, in this
model, as in the Massdale model of DiToro *et al.* (1971) and the Lake
Erie model contributed by Hydroscience (1973), zooplankton cells are
assumed to have the same nitrogen-to-carbon and phosphorus-to-carbon
ratios as their phytoplankton counterparts. Values used in these previous
models are consistent with those values adopted in the present study and
are listed in Appendix B.

Diatoms, unlike the other phytoplankton groups considered, require the
element silicon for continued production. Since green algae, blue-green
algae, and zooplankton require only trace amounts of silicon, the silicon
cycle involves only the diatoms, detrital (particulate) silicon forms and
dissolved silicon. The silicon-to-carbon ratio for diatoms follows that used
by Canale *et al.* (1974). With the simplified stoichiometry discussed above,
the mechanisms that govern each of the three nutrient cycles can be formulated; Figure 2.6 is provided as an explanation of these cycles.
Appendix A   contains a detailed listing of the model equations. The
definition and units of each term not discussed in the text are given in
Appendix B.   Also included in Appendix B   are the coefficient values
used during the calibration to be discussed in a later section. The source
of the model of these coefficients is described in detail by Canale *et al.*
(1975).

The bacterial and physical breakdown of detrital nutrients obviously
encompasses a myriad of complex reactions, most of which have been
investigated very little to date. The chain of reactions proposed in the
model hopefully depicts to some degree the mechanisms of recycling observed in lakes. The dynamics of the nitrogen subsystem seem to proceed
at rates comparable to the biological rates of the system, so the detail
incorporated into modeling this subsystem is justified. The Lake Michigan
food web model being discussed considers detrital nitrogen in addition to
the three nitrogen forms considered by Hydroscience (1973). The reaction
rate for the breakdown of detrital forms into dissolved organic forms has
not been found in the literature. Presently for nitrogen, the rate has
been set equal to the rate used for the conversion of organic nitrogen
to ammonia.

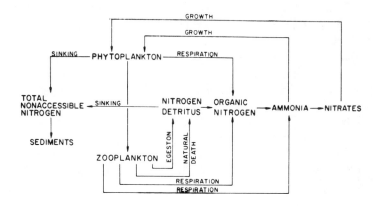

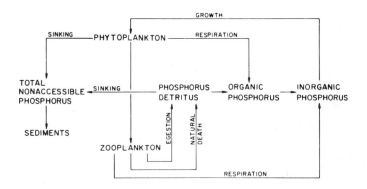

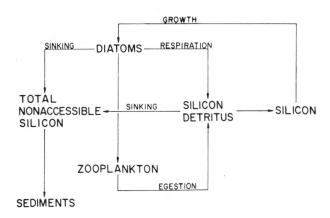

Figure 2.6   Nutrient cycle mechanisms.

Hydroscience (1973) and Canale *et al.* (1974) have attempted to partition phosphorus into components. The reaction rates governing detrital-to-organic phosphorus transfer and organic-to-inorganic phosphorus transfer far exceed the fastest biological rate observed in the system. Since the time constant for these phosphorus reactions is less than a week for most of the production season, the phosphorus subsystem might be more conveniently modeled by employing a quasi-equilibrium assumption.

Besides the pathways from nitrogen detritus to organic nitrogen and from phosphorus detritus to organic phosphorus, organic forms are also generated during phytoplankton respiration. Approximately 70% of the nitrogen respired by zooplankton enters the water in organic form, with the remainder entering as ammonia. Research indicates that most of the zooplankton-respired phosphorus emerges in an inorganic form (Raymont 1963). The form of these source terms in the nutrient equations follows directly from mass balance concepts.

The final products of the recycle reactions, *i.e.,* the inorganic forms, provide the essential nutrients for primary production. Phytoplankton utilize both inorganic nitrogen forms, ammonia and nitrate. Previous models, Hydroscience (1973) and Canale *et al.* (1974), have distinguished between algal preferences for the two nutrients. Auer (1974), in an extensive literature survey, has reported on nitrogen utilization by algae. The uptake of the inorganic nitrogen forms seems to be controlled by complex environmental and intercellular conditions including trace element availability, light and carbon dioxide levels, internal and external pH, and cell age. Although some reports of ammonia preference have been found, no consensus has been reached among researchers on the subject. Presently, the uptake mechanism is modeled so that ammonia is preferred 95% to 5% over nitrate when the two forms are in equal abundance. Regardless of which form the algae prefer, the inorganic nitrogen forms may be pooled into a single state with no change in plankton behavior. The zero degree bias in nitrification rates reported by Hydroscience (1973) has not been included.

MODEL FORCING FUNCTIONS

Inherent in the equations for the dynamics of the system are certain forcing functions, or exogenous variables, that describe environmental or other conditions assumed to be unaffected by the internal conditions of the system. Four of these variables—temperature, light intensity, photoperiod, and extinction coefficient—directly control phytoplankton production and are described in Canale *et al.* (1975). Light extinction has been included as a forcing function rather than as a calculated function of the plankton and other dissolved and particulate matter. Thus, the predictive value of the model is limited. If standing stocks of detritus or plankton deviate significantly from present conditions, a more appropriate technique for estimation of light extinction is necessary.

A sinking rate function simulates the limited data on seasonal variations in phytoplankton and detritus sinking. The alewife-eating rates and concentrations are necessary to approximate the effect of the upper food chain on the lower trophic levels; they would be unnecessary if the alewives were calculated as a state variable internal to the model. The detailed behavior of these functions is also described in Canale *et al.* (1975). The forcing functions represent mean conditions observed in the lake and cannot be expected to duplicate a particular yearly cycle. The amount of deviation from these conditions encountered from year to year dictates the scatter observed in the biological and chemical data from year to year.

COMPUTATIONAL TOPICS

The highly nonlinear nature of the differential equations in this system makes it desirable to use a numerical integration technique capable of adapting quickly to changing modes of behavior. DVDQ, a variable-order variable-stepsize scheme developed by Krogh (1969), meets the demands of such a complex system and was used for all calculations described in this paper. Because the model forcing-functions must be developed and stored using a finite number of time points, interpolation must be used. It has been observed that DVDQ can be sensitive to the interpolation scheme chosen. Some investigations have focused on a comparison of simple linear interpolation with the piecewise cubic spline technique.

The latter approach interpolates between data points with a cubic polynomial with continuity in the interpolation function and its first and second derivatives enforced at the data points. When linear interpolation is applied to the forcing functions of the model, discontinuities are introduced into the derivative of $\dot{x}(t)$ at the discreet times where the forcing functions are defined. These discontinuities in $\dot{x}(t)$ provide no insurmountable numerical difficulties for DVDQ, but do require a reduction in stepsize and a corresponding increase in computation time. It has been found that the spline interpolation scheme can cut computation time by a factor of two to three. A disadvantage of the splines procedure, however, appears to be that it exhibits unusual behavior between data points when the partitioning of the time axis is coarse in comparison to the variation in the data. This enigma has been circumvented by avoiding large gaps between the points that define the forcing functions.

RESULTS

Model Calibration

Data collected from Grand Traverse Bay stations (Figure 2.2) during 1971, 1972, and 1973 have been compared with model calculations. The water quality at these stations is similar to that in northern Lake Michigan and is superior to that in lower Lake Michigan and Green Bay. Biological data are available as species counts for both the phytoplankton and the zooplankton. These data have been converted to model units using carbon content per cell conversion factors (see Canale *et al.* 1975). With the exception of a few inconsistencies, these years indicate patterns of variation that are generally reproducible by the model (see Figures 2.7 and 2.8). When investigations reported in the literature have identified the value of a coefficient within a narrow range, this range has not been violated during calibration of the model. However, when the reported range of a coefficient indicates substantial scientific uncertainty, the coefficient has been freely manipulated to achieve desired results. In a sense, then, the coefficient values listed in Appendix B, especially those associated with zooplankton production rates, are a hypothesis requiring substantial experimental validation. Without this validation the model can be considered calibrated but cannot be considered completely verified.

In general the seasonal variation of the zooplankton is controlled by food availability and predation. The annual cycles of the two herbivorous cladoceran components (*Daphnia,* and *Bosmina* and *Holopedium*) in both model and data, show two bimodal peaks, one in the summer and one in the fall. The bimodal shape of the herbivorous cladoceran cycles is apparently due to the decline of their summer food (small diatoms) and the increase of *Leptodora*. For both states, recovery in the fall results from an increase in green algae.

Phosphorus controls the diatom fluctuations in the model, although slow silicon turnover keeps diatoms low during late summer and early fall. Large amounts of silicon locked in the detrital state and the presence of regenerated phosphorus enable the greens to bloom in the fall. However, in the model the green algal peak is higher than that represented in Grand Traverse Bay data. Model calculations for the blue-greens follow the data closely, both showing a small mid-September bloom. Since Lake Michigan is never warm enough for the phytoplankton taxa to attain the maximum growth rates, nutrient-independent growth rates are dominated by the annual temperature cycle, with diatoms having an advantage over greens and blue-greens. Although the greens are more temperature-controlled than diatoms, their competitive position in the food web is critically affected by light. During high light periods, greens are more inhibited than diatoms because their optimum light intensity is lower. According to the model, the greens gain a competitive advantage over diatoms during low light periods. Because of high light and temperature requirements, blue-greens never dominate.

Although nitrogen is not limiting in Lake Michigan, phosphorus limitation reduces the growth rates significantly. The diatom growth rates are further reduced by silicon limitation. These computed cycles can be indirectly justified by comparing model phytoplankton trajectories with the observed data (Figure 2.8). However, extensive primary productivity data are not available for direct comparison. In the computer runs silicon starts high and declines slowly during winter, then declines rapidly during spring and early summer (Figure 2.8). Recovery takes place after fall overturn and is complete by winter. Recovery in the model is mainly due to the fall overturn; however, according to the data, there is some regeneration of silicon from diatom decomposition prior to overturn. Thus, during the summer, the data are higher than the model calculates.

Dissolved-organic and dissolved-inorganic phosphorus are assigned separate states in the model. However, only total-dissolved phosphorus data are available for Grand Traverse Bay. Thus, the overall results of this approach can be checked only by adding the two model states and comparing the results against data for total-dissolved phosphorus. Since phosphorus appears to be the limiting nutrient in Grand Traverse Bay, it is critical that the model duplicate the observed phosphorus cycles. The data, as seen in Figure 2.8 (although erratic because of sampling and laboratory limitations), and the computed curve for total-dissolved phosphorus agree. It is noted that the model-calculated total-dissolved phosphorus levels are higher than spring data. This suggests that some spring diatom growth may be cells with excess phosphorus. An internal reservoir need not be large to account for the extra dissolved phosphorus in the model. For example, at day 60 the concentration of algae is about 100 μg/l carbon. These algae have a requirement of about 3.6 μg/l of phosphorus. On day 60 the excess total dissolved phosphorus calculated by the model is 1.6 μg/l. The difference could be explained by phosphorus storage of about 50%. This phosphorus would be in the algae and not soluble, though usable for growth. Azad and Borchardt (1970) have shown that 50% excess phosphorus is easily possible.

Model Calculated Energetics

In order to track the flow of mass through such a complex system model as the Lake Michigan food web model, information beyond the

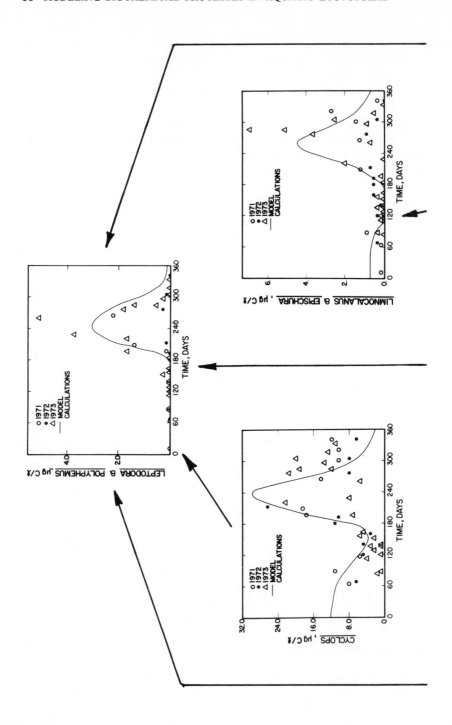

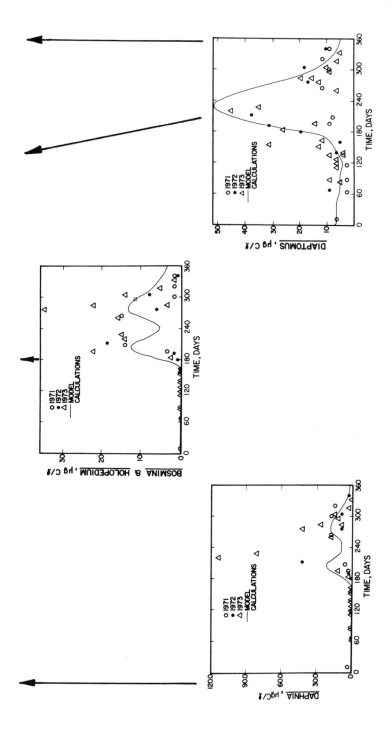

Figure 2.7   Model calibration for zooplankton.

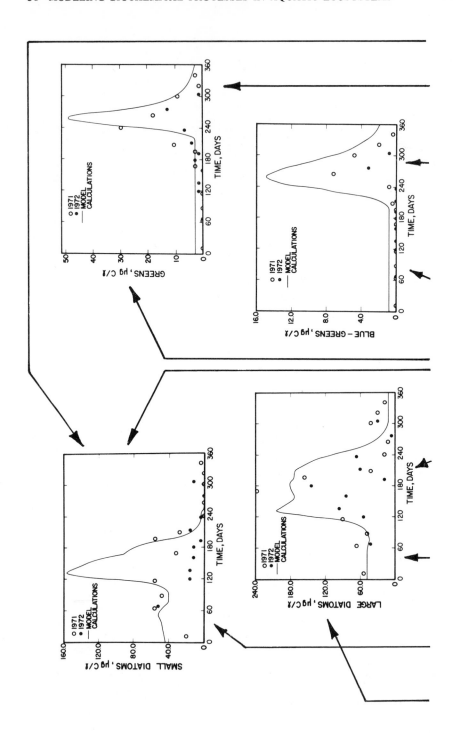

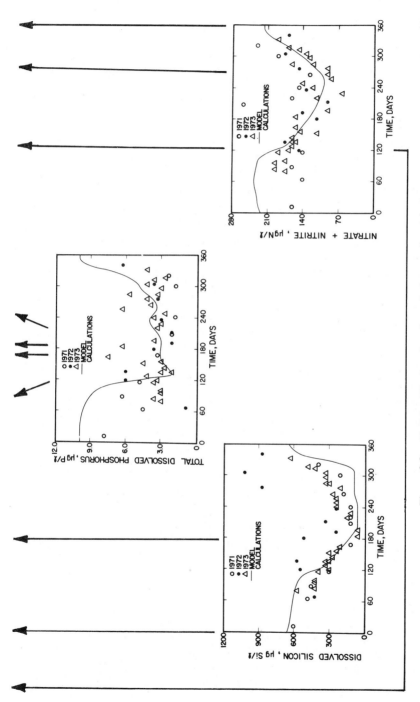

Figure 2.8   Model calibration for phytoplankton and nutrients.

integrated state responses is essential. The dynamic behavior of the states only suggests some of the pathways critical to the behavior of the model, such as the relationship between the sharp rise in phytoplankton and the decline of phosphorus. Other interactions can occur unrecognized. Phytoplankton loss due to a zooplankter might be disguised by excessive zooplankton respiration or predation on the zooplankter from the higher levels of the food chain. One way to study the interactions is to plot the various contributions to the derivative of each state as a function of time. However, it is difficult to extract useful quantitative information from such plots. For example, one would find it very laborious to compare the consumption of various food sources for a zooplankter.

What is needed is the integral of these derivative contributions. It is known that $X_i(t) - X_i(o) = \int_o^t \dot{X}_i \, dt = \sum_j \int_o^t f_{ij} \, (\overline{X},t)dt$, where the summation is over the various mechanisms in the ith differential equation. Since the equations are coupled (f_{ij} may depend on $\overline{X}$, not just X_i) the system must first be integrated using a numerical integrator as discussed earlier. Such integrals of the individual components of the derivative $\int_o^t f_{ij} \, (\overline{X},t)dt$ can be evaluated without difficulty.

Canale et al. (1975) generated plots for the phytoplankton and adult zooplankton states and for the alewife. For ease of comparison, the cumulative contributions $\int_o^t f_{ij}(\overline{X},t)dt$ are stacked one upon another. Therefore, the uppermost curve is generated by integrating the positive (growth) term of the equation from the initial condition. The curve just below this is attained by subtracting off the first negative contribution, and so on, until the lowermost curve represents the total of all contributions and coincides with the state trajectory. It is observed that the width of a band is nondecreasing from day 0 to day 360 since by construction $f_{ij} \leqslant 0$ or $f_{ij} \geqslant 0$ holds for all time. It should be noted that the procedure used to generate these plots is not equivalent to that of integrating the system without a mechanism to determine the contribution of all other mechanisms. Thus for example, the contribution of respiration to small diatom biomass loss is not that which occurs without other phytoplankton effects in the system. Figure 2.9 shows the various components that contribute to the loss of small diatoms and Bosmina and Holopedium.

Model Simulations

Simulations were conducted with the calibrated model for the purpose of examining the behavior of the model when phosphorus and alewife concentrations are higher than present conditions. The first simulation assumes that the initial level of phosphorus is 3.0 times the present initial

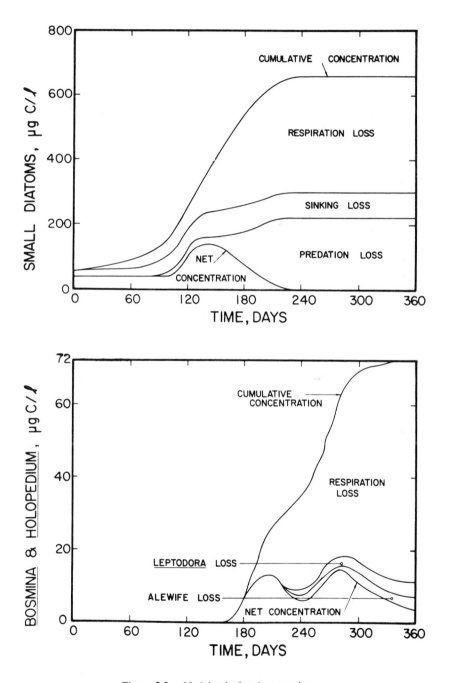

Figure 2.9   Model calculated energetics.

conditions; the second simulation assumes that alewife predation is 5 times present conditions.

Figure 2.10 shows excess nitrate under existing phosphorus conditions. However, as phosphorus is increased by a factor of 3, nitrogen becomes limiting. It is expected that nitrogen-fixing blue-green algae would develop under such circumstances, and indeed nitrogen limitation and large populations of blue-green algae are known to occur in other parts of the Great Lakes, such as Green Bay. At present the model does not consider nitrogen-fixing blue-greens, although the theoretical framework for including this effect has been considered by Bierman (1973). It has been found that nitrate, as well as the other dissolved oxidized nutrient states, is not sensitive to increases in alewife predation. Figure 2.10 shows that high levels of inorganic phosphorus cause complete dissolved silicon utilization (down to the unavailable level) approximately 60 days earlier than under normal circumstances. This earlier utilization also results in an earlier midsummer regeneration of silicon.

The above changes in nutrient patterns are filtered up the food chain, causing changes in both the phytoplankton and the zooplankton. In the simulation where the initial phosphorus was increased by a factor of 3, diatoms became less dominant and bloomed earlier in the year (see Figure 2.11). This has been reported in southern Lake Michigan by Schelske and Stoermer (1972). The blue-green population increases by an order of magnitude, and the green pattern becomes bimodal. The grazing of the herbivorous cladocerans on the greens allows the blue-greens to become more dominant. However, because cladocerans are low during the last part of the year, the greens are able to bloom a second time, thus halting the blue-greens. The first bloom of greens allows herbivorous zooplankton such as *Bosmina* to peak higher than normal (see Figure 2.12). The disappearance of the second *Bosmina* peak is probably caused by the early collapse of the small diatoms. Increases in the predatory zooplankton states, such as *Leptodora*, demonstrate a generally greater increase than their prey (see Figure 2.12) due to increased phosphorus. Thus new productivity resulting from increased phosphorus is transferred up the food web and accumulates in the top levels. This suggests that the increase in alewives that occurred in Lake Michigan during the 1960s was facilitated by increasing eutrophication of the lake.

The second simulation assumes that phosphorus levels are normal and that alewife predation increases by a factor of 5, which corresponds roughly to the peak abundance of this fish observed in Lake Michigan around 1966. Following the increase of alewives the abundance of predatory forms such as *Leptodora* decrease markedly (Figure 2.12). As a result of reduced predation, small herbivorous zooplankton such as *Bosmina*

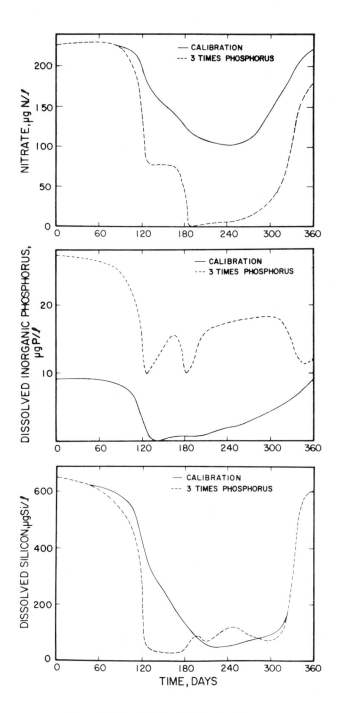

Figure 2.10 Model nutrient simulations.

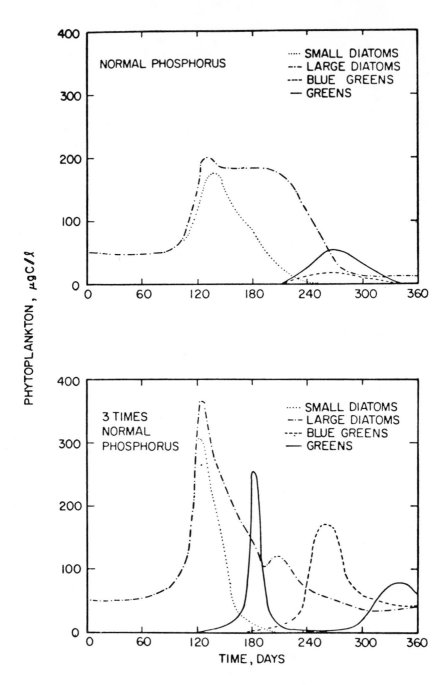

Figure 2.11   Model phytoplankton simulations.

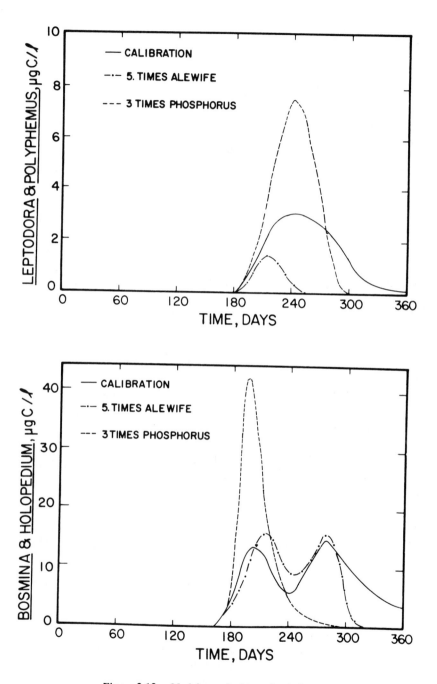

Figure 2.12 Model zooplankton simulations.

are enhanced slightly during certain parts of the year. These types of change in the zooplankton composition in Lake Michigan have been noted by Wells (1970).

SUMMARY AND CONCLUSIONS

Previous plankton models for the Great Lakes have considered the interactions among dissolved nutrients, chlorophyll *a*, and total zooplankton. These models have been developed primarily for the purpose of predicting the gross effects of nutrient input control and urban growth on the phytoplankton standing crop. However, more complex ecological problems, such as the effect of eutrophication on the forage fish and the effect of foreign marine fish species on the plankton composition, require complex food web models. The purpose of this paper has been to justify, develop, and verify a preliminary lower food web model for Lake Michigan.

The model has no horizontal definition and two vertical layers. However, it has a complex representation of the food web interactions and consists of 12 nutrient equations, 4 phytoplankton equations, and 9 zooplankton equations. The nutrients considered in the model are nitrogen, phosphorus, and silicon. Nitrogen and phosphorus can appear in detrital, dissolved organic, or dissolved inorganic form, whereas silicon can be detrital or dissolved. Diatom, green, and blue-green components of the algal community are considered. Furthermore, the diatom population is divided on the basis of size to accommodate feeding preferences in the zooplankton. The zooplankton in the model are divided on the basis of size, feeding mechanism, and taxonomic grouping. A crude representation of age structure within the zooplankton community is included by placing the nauplii in a separate compartment. Alewife predation upon the zooplankton is included through use of time variable forcing-functions that represent the feeding rate of the adult and juvenile fish.

The model calculations have been compared with field data from Grand Traverse Bay with encouraging results. The calibrated model has been used to make some preliminary estimates of the effects of eutrophication and higher forage fish levels on the plankton and nutrient distributions in the lake. Simulations have been made in which the initial phosphorus levels were increased to the level currently found in Lake Erie. Another set of simulations has examined the effect of forage fish levels at high values that approximate the alewife population explosion that occurred in Lake Michigan in 1966. The results suggest nutrient limitation by silicon and nitrogen following phosphorus increases and subsequent phytoplankton shifts from diatoms to green and blue-green forms. Alewife predation causes a decline of large predator zooplankton forms and enhancement of small herbivorous taxa.

It is expected that the model will provide guidance on the requirements of additional scientific research on food web transfer mechanisms and on the requirements of a comprehensive field monitoring program. Ultimately the model may be useful to decision-makers concerned with the control of eutrophication and the management of the Lake Michigan fishery.

ACKNOWLEDGMENTS

The authors are indebted to the following University of Michigan graduate students who have participated in this study: Janice Boyd (Natural Resources) who formulated many of the model mechanisms, contributing much to the conceptual framework on which the zooplankton-eating mechanisms are based; Richard Bortins (Computer, Information and Control Engineering) who spent many hours calibrating the model to Grand Traverse Bay data and assisted in the evolution of the zooplankton life cycle mechanisms and the nutrient regeneration mechanisms; and Barry R. Bochner (Bioengineering) who developed much of the computer software necessary to integrate the differential equations.

This work is a result of research sponsored by NOAA office of Sea Grant, Department of Commerce, under Grant No. 04-4-158-23. The U.S. Government is authorized to produce and distribute reprints for governmental purposes notwithstanding any copyright notation that may appear hereon.

REFERENCES

Arnold, D. E. "Ingestion, Assimilation, Survival, and Reproduction by *Daphnia pulex* Fed Seven species of Blue-Green Algae," *Limnol. Oceanogr.* **16**, 906 (1971).

Auer, N. A. *Nitrogen Utilization by the Algae: A Review.* Unpublished report to University of Michigan Sea Grant Program. (1974) 30 pages.

Azad, H. S. and J. S. Borchardt. "Variations in Phosphorus Uptake by Algae," *Environ. Sci. Technol.* **4**, 737 (1970).

Bierman, V. J., Jr., F. H. Verhoff, T. L. Paulson, and M. W. Tenney. "Multi-Nutrient Dynamics Models of Algal Growth and Species Competition in Eutrophic Lakes," In *Modeling the Eutrophication Process*, E. J. Middlebrooks, D. H. Falkenborg and T. E. Maloney, Eds. (1973), pp. 89-109.

Burns, N. W. and A. E. Pashley. "*In Situ* Measurement of the Settling Velocity Profile of Particulate Organic Carbon in Lake Ontario," *J. Fish. Res. Bd. Canada* **31**, 291 (1974).

Canale, R. P., D. J. Hineman, and S. Nachiappan. *A Biological Production Model for Grand Traverse Bay*, Sea Grant Technical Report No. 37 (Ann Arbor, Mich.: The University of Michigan, 1974).

Canale, R. P., L. M. DePalma, and A. H. Vogel. *A Food Web Model for Lake Michigan, Part 2—Model Formulation and Preliminary Verification*, Sea Grant Technical Report No. 43 (Ann Arbor, Mich.: The University of Michigan, 1975), 150 pages.

Canale, R. P. and A. H. Vogel. "The Effects of Temperature on Phytoplankton Growth," *ASCE J. Environ. Eng. Div.* **100**(EE1); 231 (1974).

DiToro, D. M., D. J. O'Connor, and R. P. Thomann. "A Dynamic Model of the Phytoplankton Population in the Sacramento-San Joaquin Delta," *Advances in Chemistry Series 106: Nonequilibrium Systems in Natural Water Chemistry* (1971), p. 131.

Dugdale, R. C. "Nutrient Limitation in the Sea: Dynamics, Identification and Significance," *Limnol. Oceanogr.* **12**, 685 (1967).

Eppley, R. W., J. N. Rogers, and J. J. McCarthy, Jr. "Half-Saturation Constant for Uptake of Nitrate and Ammonium by Marine Phytoplankton," *Limnol. Oceanogr.* **14**, 912 (1969).

Frost, B. W. "Effects of Size and Concentration of Food Particles on the Feeding Behavior of the Marine Planktonic Copepod *Calanus Pacificus*," *Limnol. Oceanogr.* **17**, 805 (1972).

Gause, G. F. *The Struggle for Existence*. (Baltimore: Williams and Wilkins, 1934).

Herbes, S. E. "Biological Utilizability of Dissolved Organic Phosphorus in Natural Waters," Ph.D. dissertation, The University of Michigan, Ann Arbor (1974).

Hydroscience, Inc. *Limnological Systems Analysis for Great Lakes. Phase I: Preliminary Model Design*, Great Lakes Basin Commission, Ann Arbor, Mich., DACW-35-71-30030 (1973), 473 pp.

Kitchell, J. F., J. F. Koonce, R. V. O'Neill, H. H. Shugart, Jr., J. J. Magruson, and R. S. Booth. *Implementation of a Predator-Prey Biomass Model for Fishes*, Eastern Deciduous Forest Biome, Memo Report 72-118 (Madison: The University of Wisconsin, 1973).

Krough, F. T. "VODQ/SVDQ,DVDQ-variable Order Integrators for the Numerical Solution of Ordinary Differential Equations," TV Document No. CP-2308, NOP-11643 *JPL*, Pasadena, California (1969).

Main, R. A. "The Life History and Food Relations of *Epischura lacustris Forbes (Copepoda: Calanoida)*," Ph.D. dissertation, The University of Michigan, Ann Arbor (1962).

O'Neill, R. V. "Indirect Estimation of Energy Fluxes in Animal Food Web," *J. Theoret. Biol.* **22**, 284 (1969)'

Paasche, E. "Silicon and the Ecology of Marine Plankton Diatoms. I. *Thalassiosira pseudonana* Grown in a Chemostat with Silicate as Limiting Nutrient," *Marine Biol.* **19**, 117 (1973a).

Paasche, E. "Silicon and the Ecology of Marine Plankton Diatoms. II. Silicate-Uptake Kinetics in Five Diatom Species," *Marine Biol.* **19**, 262 (1973b).

Raymont, J. E. G. *Plankton and Productivity in the Oceans* (Oxford: Pergamon Press, 1963).

Riley, G. A. "A Mathematical Model of Regional Variations in Plankton," *Limnol. Oceanogr.* **10** (Suppl.) 202 (1965).

Ryther, J. H. "Photosynthesis in the Ocean as a Function of Light Intensity," *Limnol. Oceanogr.* **1**, 61 (1956).

Schelske, C. L. and E. F. Stoermer. "Phosphorus Silica and Eutrophication in Lake Michigan," in *Nutrients and Eutrophication*, G. D. Likens, Ed., *Am. Soc. Limnol. Oceanog.* special symp. 1, 157 (1971).

Steele, J. H. "Notes on Some Theoretical Problems in Production Ecology," in *Primary Production in Aquatic Environments*, C. R. Goldman, Ed. *Mem. Inst. Idrobiol.* 18 Suppl., University of California, Berkeley, (1965), p. 383.

Strickland, J. D. H. "Production of Organic Matter in the Preliminary States of Marine Food Chain," in *Chemical Oceanography*, J. P. Riley and G. Skirrow, Eds. (New York: Academic Press, 1965), p. 478.

Vanderhoef, L. N., C.-Y. Huang, R. Musil, and J. Williams. "Nitrogen Fixation (Acetylene Reduction) by Phytoplankton in Green Bay, Lake Michigan, in Relation to Nutrient Concentrations," *Limnol. Oceanogr.* 19, 126 (1974).

Warren, C. E. *Biology and Water Pollution Control* (Philadelphia: Saunders, 1971).

Wells, L. "Effects of Alewife Predation on Zooplankton Populations in Lake Michigan," *Limnol. Oceanogr.* 15, 556 (1970).

APPENDIX A
SUMMARY OF THE MODEL EQUATIONS

$$\dot{c}_z = \left[\text{growth}\right]_z - \begin{bmatrix}\text{predation} \\ \text{by other} \\ \text{zooplankton}\end{bmatrix}_z - \begin{bmatrix}\text{predation} \\ \text{by alewife}\end{bmatrix}_z - \begin{bmatrix}\text{respiration} \\ \text{loss}\end{bmatrix}_z$$

z = 1,2,...,9

$$- A14_z(t)*c_z$$

natural death
for copepods

$$\dot{c}_p = \left[\text{growth}\right]_p - \begin{bmatrix}\text{predation} \\ \text{loss}\end{bmatrix}_p - \begin{bmatrix}\text{respiration} \\ \text{loss}\end{bmatrix}_p - \begin{bmatrix}\text{sinking} \\ \text{loss}\end{bmatrix}_p$$

p = 10,11,12,13

(1) Hatching-maturation mechanisms not written into zooplankton equations.

(2) Turnover mechanism not written into the nutrient equations.

$$\dot{c}_{14} = NCR * \sum_z \left(1.- \begin{bmatrix} assimilation \\ efficiency \end{bmatrix}_z \right) * \begin{bmatrix} eating \\ rate \end{bmatrix}_z * c_z + NCR * \sum_z \begin{bmatrix} natural \\ death \end{bmatrix}_z$$

$$- A18 * T * c_{14} - A23 * SINK(t) * c_{14}$$

$$\dot{c}_{15} = NCR * \sum_p \begin{bmatrix} respiration \\ loss \end{bmatrix}_p + NCR * A21 * \sum_z \begin{bmatrix} respiration \\ loss \end{bmatrix}_z + A18 * T * c_{14}$$

$$- A20 * T * c_{15} + LOAD_{15}$$

$$\dot{c}_{16} = NCR * (1.-A21) * \sum_z \begin{bmatrix} respiration \\ loss \end{bmatrix}_z + A20 * T * c_{15} - A22 * T * c_{16}$$

$$- NCR * \left(\frac{ANH3*c_{16}}{ANH3*c_{16} + (1.-ANH3)*c_{17}} \right) * \sum_p \begin{bmatrix} growth \end{bmatrix}_p + LOAD_{16}$$

$$\dot{c}_{17} = A22 * T * c_{16} - NCR * \left(\frac{(1.-ANH3)*c_{17}}{ANH3*c_{16} + (1.-ANH3)*c_{17}} \right) * \sum_p \begin{bmatrix} growth \end{bmatrix}_p$$

$$+ LOAD_{17}$$

$$\dot{c}_{23} = A23 * SINK(t) * c_{14} + NCR * \sum_p \begin{bmatrix} sinking \\ loss \end{bmatrix}_p * VOLEP$$

$$+ [LOAD_{15} + LOAD_{16} + LOAD_{17}] * VOLHY$$

$$\dot{c}_{18} = PCR * \sum_z \left(1.- \begin{bmatrix} assimilation \\ efficiency \end{bmatrix}_z \right) * \begin{bmatrix} eating \\ rate \end{bmatrix}_z * c_z + PCR * \sum_z \begin{bmatrix} natural \\ death \end{bmatrix}_z$$

$$- A17 * T * c_{18} - A23 * SINK(t) * c_{18}$$

$$\dot{c}_{19} = PCR * \sum_p \begin{bmatrix} respiration \\ loss \end{bmatrix}_p + A17 * T * c_{18} - A19 * T * c_{19} + LOAD_{19}$$

$$\dot{c}_{20} = PCR * \sum_{z} \begin{bmatrix} \text{respiration} \\ \text{loss} \end{bmatrix}_{z} + A19 * T * c_{19} - PCR * \sum_{p} [\text{growth}]_{p} + LOAD_{20}$$

$$\dot{c}_{24} = A23 * SINK(t) * c_{18} + PCR * \sum_{p} \begin{bmatrix} \text{sinking} \\ \text{loss} \end{bmatrix}_{p} * VOLEP$$

$$+ [LOAD_{19} + LOAD_{20}] * VOLHY$$

$$\dot{c}_{21} = SCR * \sum_{\text{diatoms}} \begin{bmatrix} \text{respiration} \\ \text{loss} \end{bmatrix}_{p} + \begin{bmatrix} \text{predation} \\ \text{loss} \end{bmatrix}_{p} - A16 * T * c_{21}$$

$$- A23 * SINK(t) * c_{21}$$

$$\dot{c}_{22} = A16 * T * c_{21} - SCR * \sum_{\text{diatoms}} [\text{growth}]_{p} + LOAD_{22}$$

$$\dot{c}_{25} = A23 * SINK(t) * c_{21} + SCR * \sum_{\text{diatoms}} \begin{bmatrix} \text{sinking} \\ \text{loss} \end{bmatrix}_{p} * VOLEP$$

$$+ LOAD_{22} * VOLHY$$

APPENDIX B
MODEL COEFFICIENTS

Zooplankton Related Coefficients

Symbol	Definition	Units	Value Used
$A7_{\text{Lept.-Poly.}}$	maximum snatching rates @ $20°$ C	$\dfrac{\text{mg food c}}{\text{mg z c·day}}$	0.70
$A7_{\text{Cyc.}}$			0.43
$A7_{\text{Cyc. n}}$	maximum filtering rates @ $20°$C	$\dfrac{1}{\text{mg z c·day}}$	2.6
$A7_{\text{Diap. n}}$			6.5
$A7_{\text{Lim.-Ep. n}}$			5.2
$A7_{\text{Diap.}}$			1.0
$A7_{\text{Lim.-Ep.}}$			1.25
$A7_{\text{Daph.}}$	filtering rates @ $20°$ C		4.0
$A7_{\text{Bosm.-Holo.}}$			3.5
$KFOOD_{\text{Lept.-Poly.}}$	half-saturation food level for raptors	$\dfrac{\text{mg food c}}{1}$	0.2
$KFOOD_{\text{Cyc.}}$			0.2
A9	minimum filtering rate multiplier	unitless	0.1
A10	food level where multiplier is ½ (1+A9)	$\dfrac{\text{mg food c}}{1}$	0.2
A11R	assimil. eff. of raptors	$\dfrac{\text{mg z c}}{\text{mg food c}}$	0.4
A11S	assimil. eff. of selectives		0.7
A11N	max. assimil. eff. of nonselectives	$\dfrac{\text{mg food c}}{1}$	0.8
A24	half eff. food level for nonselectives		0.2
$A12_{\text{adults}}$	respiration rates @ $20°$ C	day^{-1}	0.06
$A12_{\text{nauplii}}$			0.04
$A14_{\text{Cyc.}}$	natural death rates for the copepods	day^{-1}	0.005
$A14_{\text{Diap.}}$			0.005
$A14_{\text{Limn.-Ep.}}$			0.003

Phytoplankton Related Coefficients

Symbol	Definition	Units	Value Used
$Al_{sm.\ diatoms}$			2.1
$Al_{lg.\ diatoms}$	maximum 20°C growth rate	day^{-1}	2.0
$Al_{blue-greens}$			1.6
Al_{greens}			1.9
$IS_{sm.\ diatoms}$			225
$IS_{lg.\ diatoms}$	optimum light intensities	$\dfrac{langleys}{day}$	225
$IS_{blue-greens}$			600
IS_{greens}			160
$A6_{sm.\ diatoms}$			0.05
$A6_{lg.\ diatoms}$	maximum sinking rates	day^{-1}	0.03
$A6_{blue-greens}$			0
$A6_{greens}$			0.02
KN			0.015
KP	Michaelis constants for phytoplankton growth	$\dfrac{mg\ nutrient}{l}$	0.0025
KS			0.030
DEPTH	depth of euphotic zone	m	20
A3	respiration rate @ 20°C for phytoplankton	day^{-1}	0.08

Nutrient Related Coefficients

Symbol	Definition	Units	Value Used
NCR			0.2
PCR	nutrient to carbon ratios	$\dfrac{mg\ nutrient}{mg\ c}$	
SCR			
A18	detrital nitrogen→organic nitrogen		0.001
A20	organic nitrogen→ammonia		0.0012
A22	ammonia→nitrate	$\dfrac{day^{-1}}{°C}$	0.008
A17	detrital phosphorus→organic phosphorus		0.01
A19	organic phosphorus→inorganic phosphorus		0.01
A16	detrital silicon→dissolved silicon		0.0015

Symbol	Definition	Units	Value Used
A23	maximum detrital sinking rate	day^{-1}	0.05
A21	fraction of zooplankton respired nitrogen that is organic	unitless	0.7
Q	net flow through Lake Michigan	l/day	1.37×10^{11}
VOLEP	volume of Lake Michigan epilimnion	l	1.218×10^{15}
VOLHY	volume of the hypolimnion	l	3.654×10^{15}

3

A LONG-TERM PHOSPHORUS MODEL FOR LAKES: APPLICATION TO LAKE WASHINGTON

M. W. Lorenzen, D. J. Smith, and L. V. Kimmel[1]

The effects of decreasing or eliminating phosphorus inputs to lakes have not been clearly established. The diversion of sewage from Lake Washington was dramatically successful in reducing lake phosphorus and algal biomass concentrations (Edmondson 1972). However, diversion of approximately 30% of the external phosphorus loading to Lake Sammamish has had no detectable effect on phosphorus or algal concentrations (Emery et al. 1973).

Predictive models are in the early stages of development, and none have been verified to the extent that the results of nutrient diversion can be reliably projected. Various model formulations have been shown to yield quite different results (Lorenzen 1974). In order to develop and refine reliable predictive tools, models must be applied to a number of lakes where adequate historical data are available. Only through repeated demonstrations of model applicability on diverse lakes will a methodology become acceptable.

It is the purpose of this paper to develop a simple phosphorus budget model and to demonstrate its application to Lake Washington. Previous models are reviewed briefly and a detailed derivation of the proposed model is given. Methods to estimate values for model constants are discussed. The model is then applied to Lake Washington and results compared to historical records. Implications of model formulations and sensitivity of results to rate constants are discussed.

---

[1] Tetra Tech, Inc., Lafayette, California; Tetra Tech, Inc., Lafayette, California; and Battelle-Northwest, Richland, Washington, respectively.

PREVIOUS MODELS

Nutrient budget models have been based on mass balance equations applied to one or two compartment representations of lakes. Vollenweider (1969) suggested a mass balance that considered input, output, net loss to the sediments, and a correction factor for stratified lakes. The mass balance was applied to a single, completely mixed system. Megard (1971) used essentially the same formulation as Vollenweider to calculate steady-state phosphorus concentrations after a change in phosphorus income. In addition, he estimated the time required to attain steady-state.

More recently, Imboden (1974) suggested a two compartment model for phosphorus budgets in lakes. The model considers a stratified lake and includes input, output, exchange between hypolimnion and epilimnion, as well as sediment exchange. The model formulation leads to a set of four coupled differential equations for dissolved and particulate phosphorus. Imboden compared actual loading rates, calculated "tolerable" rates, and trophic state of several lakes. Results were in general agreement with the guidelines presented by Vollenweider (1970).

O'Melia (1972) and Snodgrass and O'Melia (1974) have presented a phosphorus model very similar to Imboden's. However, no release of sediment phosphorus was included, but depth-dependent rates of turbulent diffusion were considered. Lorenzen (1974) evaluated three different nutrient budget formulations and demonstrated that each formulation led to significant differences in results for well-defined but hypothetical lakes.

None of these models have been used to simulate long term observed phosphorus concentrations in lakes that have undergone significant changes in loading rates. The model developed here is intended to be as simple as possible yet realistic enough to simulate long term changes resulting from modified loading rates.

MODEL DERIVATION

The model developed here considers a completely mixed system with phosphorus input, outflow, loss to sediments, and release from sediments. The model also accounts for the fact that some of the phosphorus input to the sediments is not available for release. A diagrammatic representation of this system is shown in Figure 3.1. The sediment phosphorus concentration, C_s, can be considered the "exchangeable or releasable" amount.

This representation is based on the assumption that for long-term calculations the complex processes of algal growth and sedimentation with subsequent decay and release of phosphorus from the sediments can be

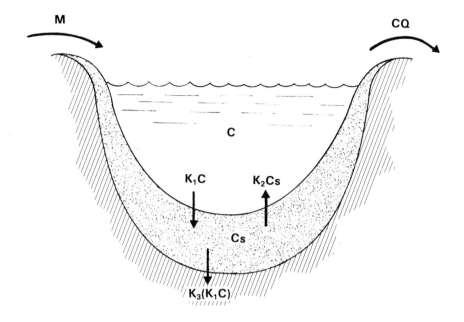

Figure 3.1 Diagrammatic representation of phosphorus budget model.

represented by simple first order reactions. This assumption may not be unreasonable. However, it cannot be rigorously defended based on theoretical principles. The utility of such simplifications remains to be evaluated by comparing calculations to observations for a number of lakes.

A mass balance on both sediment and water phosphorus concentration yields the following coupled differential equations:

$$\frac{dC}{dt} = \frac{M}{V} + \frac{K_2 A C_s}{V} - \frac{K_1 A C}{V} - \frac{CQ}{V} \tag{3.1}$$

$$\frac{dC_s}{dt} = \frac{K_1 A C}{V_s} - \frac{K_2 A C_s}{V_s} - \frac{K_1 K_3 A C}{V_s} \tag{3.2}$$

where

   C = average annual total phosphorus concentration in water column (g/m^3)

   C_s = total exchangeable phosphorus concentration in the sediments (g/m^3)

M = total annual phosphorus loading (g/yr)

V = lake volume (m^3)

V_s = sediment volume (m^3)

A = lake surface area (m^2) = sediment area (m^2)

Q = annual outflow (m^3/yr)

K_1 = specific rate of phosphorus transfer to the sediments (m/yr)

K_2 = specific rate of phosphorus transfer from the sediments (m/yr)

K_3 = fraction of total phosphorus input to sediment that is unavailable for the exchange process

These equations can be solved numerically or analytically. The analytical solution for C, the phosphorus concentration in the water, as a function of time when M, Q, V, V_s, A, K_1, K_2, and K_3 are considered constant is given by

$$
C = \left[\frac{\alpha + X_5}{\alpha - \beta} e^{-\alpha t} + \frac{\beta + X_5}{\beta - \alpha} e^{-\beta t} \right] C_o
$$

$$
- \left[\frac{X_3}{\alpha - \beta} e^{-\alpha t} + \frac{X_3}{\beta - \alpha} e^{-\beta t} \right] C_{s,o}
$$

$$
+ \frac{X_1}{X_2 X_5 - X_3 X_4} \left[\left(\frac{\alpha + X_5}{\alpha - \beta} e^{-\alpha t} + \frac{\beta + X_5}{\beta - \alpha} e^{-\beta t} - 1 \right) X_5 \right.
$$

$$
\left. + \left(\frac{X_3 X_4}{\alpha - \beta} e^{-\alpha t} + \frac{X_3 X_4}{\beta - \alpha} e^{-\beta t} \right) \right]
\tag{3.3}
$$

where C_o and $C_{s,o}$ are the concentrations of total phosphorus in the water and in the sediment at time equal zero, and where

$$
X_1 = \frac{M}{V}
\tag{3.4}
$$

$$
X_2 = \frac{(K_1 A + Q)}{V}
\tag{3.5}
$$

$$
X_3 = \frac{K_2 A}{V}
\tag{3.6}
$$

$$
X_4 = (1 - K_3) \frac{K_1 A}{V_s}
\tag{3.7}
$$

$$X_5 = \frac{K_2 A}{V_s} \tag{3.8}$$

$$\alpha + \beta = -(X_2 + X_5) \tag{3.9}$$

$$\alpha\beta = (X_2 X_5 - X_3 X_4) \tag{3.10}$$

The sediment phosphorus concentration as a function of time for constant parameter values is given by

$$C_s = -\left[\frac{X_4}{\alpha-\beta} e^{-\alpha t} + \frac{X_4}{\beta-\alpha} e^{-\beta t}\right] C_o$$

$$+ \left[\frac{\alpha+X_2}{\alpha-\beta} e^{-\alpha t} + \frac{\beta+X_2}{\beta-\alpha} e^{-\beta t}\right] C_{s,o}$$

$$- \frac{X_1}{\alpha\beta} \left[\left(\frac{X_4}{\alpha-\beta} e^{-\alpha t} + \frac{X_4}{\beta-\alpha} e^{-\beta t}\right) X_5\right.$$

$$\left. + \left(\frac{\alpha+X_2}{\alpha-\beta} e^{-\alpha t} + \frac{\beta+X_2}{\beta-\alpha} e^{-\beta t} - 1\right)\right] X_4 \tag{3.11}$$

If loading rates are held constant for a period of time, an equilibrium or steady-state concentration will be reached. The steady-state concentration of phosphorus in the water, C_∞, can be found by evaluating Equation (3.3) at $t = \infty$. The result is

$$C_\infty = \frac{-X_1 X_5}{X_2 X_5 - X_3 X_4} \tag{3.12}$$

$$= \frac{M}{Q + K_1 K_3 A} \tag{3.13}$$

or in terms of the average influent phosphorus concentration, C_{in} ($C_{in} = M/Q$).

$$C_\infty = \frac{C_{in}}{1 + \frac{K_1 K_3 A}{Q}} \tag{3.14}$$

The steady-state value of the sediment phosphorus concentration $C_{S\infty}$ is found by a similar solution to Equation (3.11) at t = ∞.

$$C_{S\infty} = \frac{X_4 X_1}{X_2 X_5 - X_3 X_4} \tag{3.15}$$

$$= \frac{M K_1 (1-K_3)}{K_2 (Q+K_1 K_3 A)} \tag{3.16}$$

In terms of the average influent phosphorus concentration, the steady-state sediment value is

$$C_{S\infty} = \frac{(1-K_3) C_{in} K_1}{\dfrac{K_2 (1+K_1 K_3 A)}{Q}} \tag{3.17}$$

The ratio of steady-state water to sediment phosphorus is given by

$$\frac{C_\infty}{C_{S\infty}} = \frac{K_2}{K_1} \frac{1}{(1-K_3)} \tag{3.18}$$

Although these steady-state solutions are not likely to apply to many lakes, the solutions yield some insight into the relative importance of various terms and may be useful in estimating parameter values. The steady-state phosphorus concentration in the water as given by Equation (3.14) is independent of the specific release rate from the sediments (K_2) and depends only on the influent concentration, hydraulic flow rate, area, and amount of phosphorus input to sediments that is not available for exchange.

PARAMETER ESTIMATION

In order to apply the time dependent model a number of parameters must be evaluated or estimated. An accurate estimate of total phosphorus loading rate is essential, and physical parameters including lake volume, depth, and area must be known. An estimate of effective sediment volume, tributary and effluent hydraulic flows on an annual average basis and the rate constants K_1 and K_2 as well as K_3, the fraction of phosphorus input to sediments that is unavailable for release, must also be determined.

The physical lake parameters are generally known or easily determined and hydrologic data are available from the U.S. Geological Survey. However, nutrient loading rates are sometimes difficult to determine. Point source monitoring programs are useful, but consideration must be given to precipitation and ground water inflow. The estimation of rate constants and percent of phosphorus input to sediments that is unavailable are the most tenuous. It should be kept in mind that K_1, K_2, and K_3 are used to represent annual averages and not instantaneous rates. These parameters are therefore not amenable to direct measurement. Reasonable estimates based on reported experimental studies would provide initial values.

The following procedure to estimate constants can be used when reasonably good estimates of loading rates, average water and sediment concentrations and exchangeable sediment phosphorus concentrations are known for an historical period. The period should span three to four detention times when loading rates were fairly constant and observed concentrations of phosphorus in the water did not change significantly.

The first step is to estimate the value of $K_1 K_3$ from Equation (3.13).

$$K_1 K_3 = \frac{M - CQ}{CA} \qquad (3.19)$$

If an estimate of K_3, the fraction of phosphorus input to the sediment that is not exchangeable, is available, then K_1 and K_3 are determined. The value of K_2 can then be calculated from Equation (3.18).

$$K_2 = \frac{C_\infty}{C_{S\infty}} K_1 (1 - K_3) \qquad (3.20)$$

It is thus apparent that the values of the model constants are not independent and a consistent set that satisfies the constraints of Equations (3.12) through (3.18) should be chosen. This constraint places considerable limits on the ability to arbitrarily calibrate the model.

LAKE WASHINGTON

In order to apply this model to Lake Washington, phosphorus loading rates must be estimated, the lake area and volume determined, and model constants must be established. In order to test the predictions, historical phosphorus concentrations are needed.

Phosphorus loading rates for the period 1931 to 1967 were estimated from information presented by Edmondson *et al.* (1972) and Brown and Caldwell (1953). The values used are summarized in Table 3.1. Physical

Table 3.1 Phosphorus Loading to Lake Washington

Year(s)	Kg-P/year
1931-1940	45,000
1941-1950	61,000
1951-1956	81,000
1957	89,000
1958	140,000
1959	190,000
1960	240,000
1961	280,000
1962	288,000
1963	257,000
1964	257,000
1965	196,000
1966	196,000
1967	80,000
1968-2000[a]	80,000

[a]Projected for calculations.

characteristics of the lake are summarized in Table 3.2. The lake volume and area were estimated from values reported by Water Resources Engineers (1972) and are not in precise agreement with values reported by Edmondson (1969). However, the discrepancy makes a negligable difference in calculated results.

Table 3.2 Physical Characteristics of Lake Washington

Flow, Q	9×10^8 m^3/yr
Area, A	1×10^8 m^2
Volume, V	3.8×10^9 m^3
Sediment volume, V_s [a]	1×10^7 m^3
Detention time	4.2 years

[a]Assumed 10 cm thick.

Shapiro *et al.* (1971) reported total sediment phosphorus concentrations in Lake Washington cores to vary from one to three micrograms phosphorus per milligram of dry sediment. Water content of the cores ranged from 75% to 90%. These values indicate that total phosphorus concentrations may have ranged from approximately 700 g/m^3 of wet sediment near the surface to 300 g/m^3 in deeper layers. Acid extractable phosphorus in the top ten centimeters of sediment was reported to range from 50% to 80% of total phosphorus. Reductant extractable phosphorus was also reported to range from 50% to 80% in the top ten centimeters.

There is no clear cut method for determining the fraction of sediment phosphorus that will be unavailable for exchange. However, it appears that a large percentage (50% to 80%) of the sediment phosphorus is associated with iron compounds. Due to the oxidizing condition of bulk Lake Washington water it was assumed that 60% of the phosphorus input to the sediments would not be releasable. The value of K_3 was therefore assumed to be 0.6. Due to the tenuous nature of this selection, calculations were performed for K_3 values ranging from 0.3 to 0.7. These results are discussed later under sensitivity analysis.

In order to calculate a consistent set of model constants, the period 1941 to 1950 was assumed to represent relatively steady conditions. Edmondson's (1972) data indicate the average lake phosphorus concentration at the end of that period was approximately 0.02 mg/l. The procedure outlined previously was used to calculate a consistent set of constants based on the K_3 value of 0.6, a total sediment phosphorus concentration of 600 g/m^3, a constant loading rate of 61,000 Kg P/year, and a steady-state concentration of 0.02 mg/l. The resulting values are given in Table 3.3.

Table 3.3   Parameter Values for Lake Washington

K_1	36 m/yr
K_2	0.0012 m/yr
K_3	0.6
$C_{s,o}$	240 mg/l
C_o	0.015 mg/l

These values were used to calculate the phosphorus concentration in the water as a function of time according to Equation (3.3). The loading rate, M, was input as a constant value for each annual time step. The results are compared to Edmondson's (1972) reported values in Figure 3.2, which also shows the calculated phosphorus concentration that would have occurred if the 1962 maximum loading rate had been continued (dashed line). For the conditions assumed, the approach to steady-state is very slow.

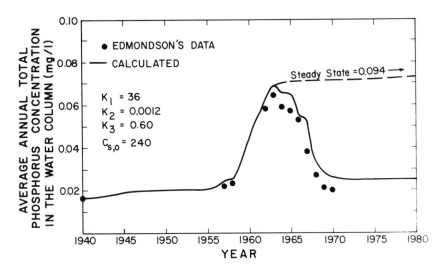

Figure 3.2   Calculated and observed annual average total phosphorus in Lake Washington.

The importance of the time-dependent solution as opposed to a steady-state model (Equation 3.14) is shown in Figure 3.3. The steady-state solutions indicate the phosphorus concentrations that would occur if loading rates were held constant for long periods of time. The calculated steady-state concentrations are plotted as a dashed line in comparison to the model solution shown as a solid line. The contrast between the steady-state and time-dependent solutions is more pronounced for the sediment phosphorus than for the overlying waters. As shown in Figure 3.4, the sediments never approach steady-state and would probably not reflect the brief period of high loading rates.

The fact that model calculations indicate that the sediments approach steady-state very slowly may have significant implications in regard to rates of lake recovery following nutrient diversion. Lake Washington received high phosphorus inputs for only a brief time. Had these high loading rates continued for a sufficient period for sediments to come to equilibrium, the rate of recovery may have been slower when a nutrient

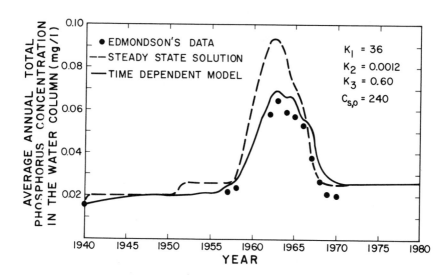

Figure 3.3   Comparison of time-dependent and steady-state model solutions
for phosphorus in the water column.

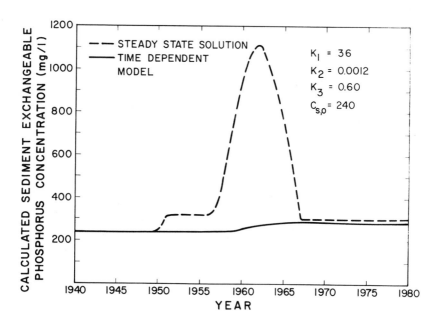

Figure 3.4   Comparison of time-dependent and steady-state model solutions
for phosphorus in the sediment.

diversion program was undertaken. In order to illustrate this possibility, calculations were performed for the same nutrient reduction program, but the initial conditions (water and sediment phosphorus concentrations) were assumed to be those calculated for steady-state conditions coexisting with the high loading rates of 1962. The results shown in Figure 3.5 indicate an initial rapid rate of recovery followed by a much slower approach to steady-state. An intermediate case, with initial conditions corresponding to 50 years of high loading rates, is also shown in Figure 3.5.

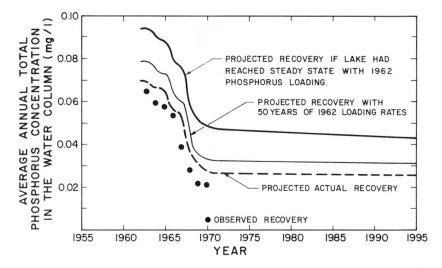

Figure 3.5   Effect of initial conditions on rate of recovery.

These calculations demonstrate the importance of past history and the proper determination of the initial state of a lake. For example, lakes whose sediments have not equilibrated with high loading rates would be expected to recover quickly in response to nutrient diversion. Lakes whose sediments have equilibrated may require much longer times following nutrient diversion in order to reach new steady-state phosphorus concentrations.

It should be noted that the calculated rates of recovery are dependent on the assumed depth of sediment. The importance of the assumed sediment thickness is shown graphically in Figure 3.6. These calculations were performed for two sets of initial phosphorus concentrations and a range of assumed sediment thickness values. For the case when the lake and sediment phosphorus concentrations started at equilibrium with 1962

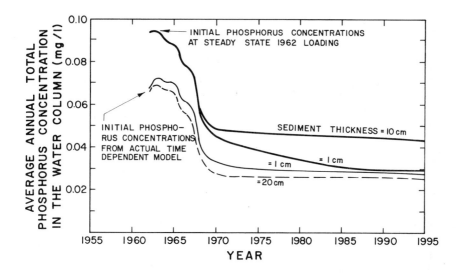

Figure 3.6   Effect of assumed sediment thickness on rate of recovery.

phosphorus loading rates, the effect of assumed sediment thickness is negligible during the first five years following diversion. However, during the following 20 years the calculated concentrations approach steady-state more rapidly when the sediment is assumed to be 1 cm thick.

For the case when initial phosphorus concentrations were calculated by simulating Lake Washington with a range of sediment thickness values, the resulting recovery is very insensitive to the assumed sediment thickness. The 1 cm-thick sediment case predicts a slightly slower recovery because the sediment concentration reached a much higher level (nearer equilibrium) during the period of high loading. The high concentrations in the thin sediment layer were then released more rapidly and slowed recovery.

SENSITIVITY ANALYSES

In order to evaluate the sensitivity of model predictions to the choice of constants a number of runs with alternative values were made.

Sensitivity analyses are often performed by varying one parameter value while all others are held constant. However, it has been shown that certain relationships exist between the values of K_1, K_2, K_3, and C_s. Therefore a sensitivity analysis should allow one parameter to vary and still maintain the proper relationships between the parameters.

Calculations were performed by assuming a range of values for K_3, the fraction of phosphorus input to the sediments that was not exchangeable. Values of 30%, 40%, 50%, 60%, and 70% were assumed and the other parameter values calculated to be consistent with the loading rates and observed concentrations for the period 1940-1950. The results are shown in Figure 3.7. These results indicate that in general the model predictions

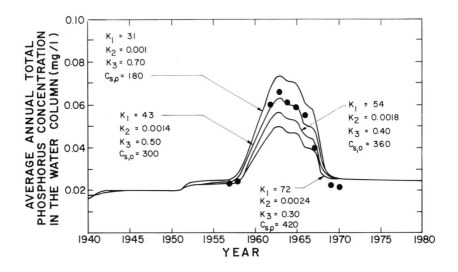

Figure 3.7   Sensitivity of predictions to model constants.

are not very sensitive to a wide range of model constants. It is interesting to note that high values of K_3 (less exchangeable phosphorus in sediment) and lower sediment release rates ($K_2 C_s$) result in *higher* phosphorus concentrations in the water during the transient response period. The long term response is the same for all combinations of constants. The reason for higher transient phosphorus concentrations with low sediment concentrations and release rates is that when K_3 is large, K_1 must be correspondingly small in order to satisfy the initial steady-state assumption. Because the loss rate of phosphorus from the water is small, the water concentration increases more rapidly during periods of increased loading, and during periods of decreased loading the water concentration decreases more slowly.

The rate of loss from water to sediment dominates the transient response because small changes in the mass of phosphorus significantly affect the concentration in the water and thus the rate ($K_1 C$), whereas large changes in the mass of phosphorus in the sediment are required to change the release rate ($K_2 C_s$).

These relationships should not be used to draw inferences about sediment buffer capacities for different lakes. The results are based on knowing a set of steady-state conditions for a lake. The implications are related to the results of varying parameter values for that lake.

The procedure described here should not be followed when a lake changes from being substantially oligotrophic to eutrophic and has periods of anaerobiosis in the hypolimnion. Under such conditions the releasable phosphorus fraction can change substantially. The same relationships between parameter values should exist. However, the steady-state values of C and C_s may be quite different and a new set of parameter values must be determined. Although the parameter values for aerobic and anaerobic periods will not necessarily be the same, it is possible that average values could be used to represent annual conditions.

One of the most difficult problems with nutrient budget determinations is associated with establishing loading rates. Considerable expenditures of time and funds are required to quantify tributary and other sources of nutrients. In order to estimate the required accuracy of such determinations, model runs were made with loading rates increased and decreased by 10% and 25%. The results are shown in Figure 3.8. The case when loading rates were reduced by 10% actually gives a slightly better fit than the original run when a best estimate of parameters was used. These results indicate that in general loading rates plus or minus 10% will not greatly affect model predictions. However, a 25% variation in loading rates can significantly change model predictions.

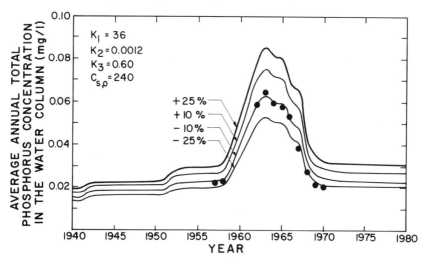

Figure 3.8 Sensitivity of model predictions to phosphorus loading rates.

DISCUSSION

A simple mass balance model has been shown to be capable of simulating long-term phosphorus concentrations in Lake Washington. The calculated results are in remarkable agreement with the reported concentrations. It should be emphasized that the model constants were chosen to agree with observations for the period 1940-1950, and were not changed for the remaining time period, which included large changes in loading rates.

The implications of slow rates of sediment phosphorus equilibration indicate that the times required for lake recovery may be highly variable depending on the period of time over which the lake received high nutrient loads. A theoretical and quantitative basis to evaluate the importance of the length of time a lake has been eutrophic is provided. The results indicate that the importance of the effective sediment volume depends on the length of time a lake has received high loading rates and how near to steady-state the phosphorus concentrations have become.

Preliminary sensitivity analysis indicates that the percentage of phosphorus input to the sediments that is not available for exchange is an important factor controlling phosphorus concentrations in the overlying water. The percentage of unavailable phosphorus may depend on the type of algae grown, the presence and chemical composition of clay particles, the amount of iron available, and the redox potential of the sediments and overlying water.

Significant further research on a number of lakes will be required to determine whether this model represents a generally adequate approach to predicting long term total phosphorus concentrations in lakes. The application to Lake Washington has provided encouraging results.

REFERENCES

Brown and Caldwell Engineers. "Metropolitan Seattle Sewage and Drainage Survey," (1953).

Edmondson, W. T. "Eutrophication in North America," in *Eutrophication: Causes, Consequences, Correctives* (Washington, D.C.: National Academy of Science, 1969).

Edmondson, W. T. "Nutrients and Phytoplankton in Lake Washington," in *Nutrients and Eutrophication*, G. E. Likens, Ed. (Lawrence, Kansas: The American Society of Limnology and Oceanography, 1972).

Emery, R. M., C. E. Moon, and E. B. Welch. "Delayed Response of a Mesotrophic Lake after Nutrient Diversion," *J. Water Poll. Control Fed.* **45**, 913 (1973).

Imboden, D. M. "Phosphorus Models of Lake Eutrophication," *Limnol. Oceanog.* **119**, 297 (1974).

Lorenzen, M. W. "Predicting the Effects of Nutrient Diversion on Lake Recovery," in *Modeling the Eutrophication Process*, E. J. Middlebrooks, D. H. Falkenborg, and T. E. Maloney, Ed. (Ann Arbor, Mich.: Ann Arbor Science Publishers, 1974).

Megard, R. O. "Eutrophication and the Phosphorus Balance of Lakes," presented at the 1971 Winter Meeting of the American Society of Agricultural Engineering, Chicago, Illinois (December 1971).

O'Melia, C. R. "An Approach to the Modeling of Lakes," *Swiss J. Hydrol.* **34**, 1 (1972).

Shapiro, J., W. T. Edmondson, and D. E. Allison. "Changes in the Chemical Composition of Sediments in Lake Washington, 1958-1970," *Limnol. Oceanog.* **16**, 437 (1971).

Snodgrass, W. T. and C. R. O'Melia. "A Predictive Model for Phosphorus in Lakes," draft manuscript (August 1974).

Vollenweider, R. A. "Possibilities and Limits of Elementary Models Concerning the Budget of Substances in Lakes," *Arch. Hydrobiology* **66**, 1 (1969).

Vollenweider, R. A. "Scientific Fundamentals of the Eutrophication of Lakes and Flowing Waters with Particular Reference to Nitrogen and Phosphorus as Factors in Eutrophication," Organization for Economic Cooperation and Development, Paris (September 30, 1970).

Water Resources Engineers. *Ecologic Simulation for Aquatic Environments*, Office of Water Resources Research (December 1972).

4

THE DISPERSION OF CONTAMINANTS
IN THE NEAR-SHORE REGION

W. Lick, J. Paul, and Y. P. Sheng[1]

INTRODUCTION

A contaminant (either heat or a physical substance) disperses through a lake by convection and turbulent diffusion. As the contaminant moves from its source to a sink, it may be degraded or converted to other forms by chemical, biological, or physical processes. An understanding of these dispersion and transformation processes is essential in understanding aquatic ecosystems.

In particular, a knowledge of these processes in near-shore regions is essential since (1) the near-shore regions are where contaminants are generally introduced and therefore their concentrations and effects are generally greater than in the off-shore regions, and (2) the near-shore regions are of more particular interest to us for such uses as recreation, water supplies, and fishing. Recent studies have demonstrated significant differences between near-shore and open-lake waters.

A large amount of work has been and is being done to understand the transformation processes of importance in aquatic systems. At the same time, it has been recognized that the dispersion process must also be understood. In addition to field work, considerable theoretical work has been done to predict the circulation in lakes. Much of this is reported

[1] Case Western Reserve University, Cleveland, Ohio 44106

in the *Proceedings of the Great Lakes Conferences.* Until recently, most of this work has dealt with the overall circulation in lakes and not specifically with the near-shore. For example, most numerical models of the circulation in large lakes employ a numerical grid of 3-10 km, a grid far too large to treat the near-shore in adequate detail.

At Case Western Reserve University as part of a research effort sponsored by the United States Environmental Protection Agency, we have developed numerical models capable of realistically describing the currents and the dispersion of contaminants throughout large lakes, especially in the near-shore. A summary of the hydrodynamic modeling is reported by Lick (1975). In addition to the general formulation and development of near-shore hydrodynamic and dispersion models, which necessarily include procedures for coupling the near-shore and overall lake circulation models, specific applications of these models have been made for general understanding of the numerical model and of the physical consequences of contaminant dispersion.

In the present report, three specific problems have been chosen to illustrate quantitatively the dispersion of a contaminant. These are: (1) the dispersion of heat in the discharge from the Point Beach power plant on Lake Michigan, (2) a detailed description of flow and dispersion near the mouth of a river entering a lake, and (3) on a larger scale, a description of the wind-driven flow and dispersion from a point source in the near-shore region of a large lake (Lake Erie) under present conditions and as modified by large man-made structures or islands, such as a jetport in the lake. These problems have been investigated by means of numerical models.

The basic equation describing the dispersion of a contaminant is

$$\frac{\partial C}{\partial t} + \frac{\partial (Cu)}{\partial x} + \frac{\partial (Cv)}{\partial y} + \frac{\partial (Cw)}{\partial z} = \frac{\partial}{\partial x}\left(D_H \frac{\partial C}{\partial x}\right) + \frac{\partial}{\partial y}\left(D_H \frac{\partial C}{\partial y}\right)$$

$$+ \frac{\partial}{\partial z}\left(D_v \frac{\partial C}{\partial z}\right) + S \qquad (4.1)$$

where

C = concentration (or temperature if the contaminant being considered is heat)

x,y = horizontal coordinates

z = vertical coordinate

t = time

u,v = particle (or fluid) velocities in the x and y directions, respectively w   =   particle velocity (the sum of the fluid velocity and the settling velocity w_s of the contaminant relative to the fluid) in the z direction

D_H =   horizontal eddy diffusivity

D_v =   vertical eddy diffusivity

S   =   source term.

In order to solve this equation for the generally time-dependent concentration, the velocities u, v, and w must be known. In the present study, these velocities were determined by means of numerical hydrodynamic models. In problems (1) and (2), the effects of buoyancy due to variable temperatures and hence densities were significant. Therefore, the above dispersion equation for temperature (or energy) must be, and was, coupled with the fluid momentum equations for a valid solution for the velocities and the temperature. However, in problems (2) and (3), when considering the dispersion of a physical substance, it was assumed that the concentrations were small enough that the effect of the substance on the flow could be neglected. In the latter case, the dispersion equation above can be uncoupled from the fluid flow equations. The fluid flow and the dispersion of a substance can then be solved as separate problems.

The hydrodynamic models used here have previously been used for overall lake circulation calculations and have been modified so as to give more detail and accuracy in the near-shore regions. These models are completely three-dimensional and generally time-dependent. For simplicity and ease of understanding the results, only the steady-state velocities are used in the present calculations although the use of time-dependent velocities is a straight-forward procedure.

In the dispersion of heat, the boundary conditions considered are either the specification of a temperature or the specification of a heat flux according to

$$- D_v \frac{\partial T}{\partial z} = H(T - T_{eq}) \tag{4.2}$$

where H is a surface heat transfer coefficient and T_{eq} is the equilibrium temperature.

In the dispersion of a physical substance, a boundary condition sufficiently general to describe practically all situations encountered can be written as (Monin and Yaglom 1971)

$$w_s C - D_v \frac{\partial C}{\partial z} = - \beta C + E$$

$$= - \beta (C - C_{eq}) \tag{4.3}$$

On the left-hand side of this equation, the first term represents the flux to the boundary due to gravitational settling while the second term represents the flux to the boundary due to vertical turbulent diffusion. On the right-hand side, the first term depends on the porosity or stickiness of the wall. The case $\beta=0$ corresponds to perfect reflection of the substance from the boundary while the case $\beta \to \infty$ corresponds to perfect absorption at the boundary. For $0 < \beta < \infty$, partial reflection and absorption occurs. The second term on the right-hand side is due to entrainment. In the second form of the equation, C_{eq} represents an equilibrium concentration that in general depends on the shear stress and sediment composition. When the concentration is greater than C_{eq}, net deposition results, while when the concentration is less than C_{eq}, net erosion results.

The processes of deposition and entrainment that determine β and E (or C_{eq}) are extremely complex. Although a considerable amount of work has been done, mainly by civil engineers, on entrainment processes for noncohesive sediments, little is known about the quantity E for cohesive sediments or about β. Little investigation has been done to analyze the effects of β or E on contaminant dispersion.

The above dispersion equation and the appropriate boundary conditions were solved numerically by means of finite difference schemes. In all cases, a strictly conservative numerical scheme was used. In addition, a stretching of the vertical coordinate proportional to the local depth was used. With this transformation, the same number of vertical grid points are present in the shallow as in the deeper parts of the basin. This ensures that in the shallow areas there is no loss of accuracy in the computations due to lack of vertical resolution, a significant factor in near-shore calculations.

THERMAL PLUMES FROM A POWER PLANT

A time-dependent, variable density model (Paul and Lick 1973, 1974; Lick 1975) was used to calculate the flow field and temperature distribution in the discharge from the Point Beach power plant on Lake Michigan. The calculations include realistic geometry, buoyancy effects, wind stresses, and cross-flows in the lake. The basic numerical procedure of solution and details of the calculations are presented in the references mentioned above; therefore, except for a brief statement of the assumptions used in the model, the model will not be discussed here. However, as examples of the dispersion of heat, the results of two calculations and their comparison with field observations will be presented.

The basic assumptions used in the hydrodynamic model are: (a) The Boussinesq approximation is valid. This assumes that density variations are small and can be neglected in the equations of motion except in the gravity term. The coupling between the energy and momentum equations is retained. (b) Eddy coefficients are used to account for turbulent diffusion effects in both the momentum and energy equations. For each flow situation, the horizontal eddy coefficients are assumed constant but the vertical eddy coefficients may vary depending on the temperature and velocity gradients. (c) The pressure is assumed to vary hydrostatically. (d) The rigid-lid approximately is valid, *i.e.*, $w(z = 0) = 0$. This approximation is used to eliminate surface gravity waves and the small time scales associated with them, greatly increasing the maximum time step possible in the numerical computations. In this approximation, only the high frequency surface variations associated with gravity waves are neglected while the steady-state results, with which we are here solely concerned, are calculated correctly and are the same as for the free-surface case.

Calculations for two cases of the discharge from the Point Beach power plant have been made: the discharge into an almost quiescent lake and the discharge into a lake with a cross-flow and a surface wind present. The relevant parameters, based on field measurements taken at the site of the discharge (Frigo, *et al.* 1974) are listed in Table 4.1.

Table 4.1   Parameters for Point Beach Power Plant Discharge Calculations

Flow rate	24.7 m^3/sec
Outfall width	10.8 m
Outfall depth	4.2 m
Ambient lake temperature	9.5°C
Vertical eddy coefficient	(50 - 200 $\frac{\partial T}{\partial z}$) cm^2/sec
Horizontal eddy coefficient	1000 cm^2/sec
Surface heat transfer coefficient	30 watt/m^2 - °C
Temperature variation	8.5°C
Maximum velocity	0.9 m/sec
Densimetric Froude number	4.2
Equation of state	$\Delta\rho = -1.25 \times 10^{-3} \Delta T$

The bottom topography is shown in Figure 4.1. The outfall extends into the lake and the discharge forms a 60° angle with the shore. In the calculations, a variable grid was used. In the horizontal, grid sizes varied from 3 m near the outfall to 80 m far from the outfall. Six grid points in the vertical were used.

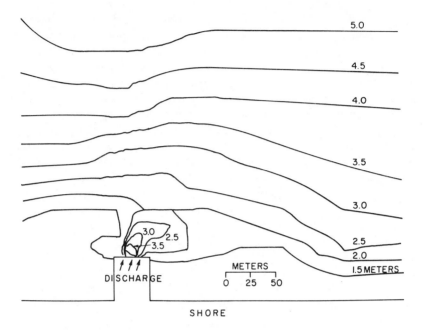

Figure 4.1   Bottom topography for the Point Beach power plant.

Boundary Conditions

The inlet velocity profile was specified as a smoothed average of that measured at the outfall. The inlet temperature was taken as the constant value measured. A surface heat transfer proportional to the difference in temperature between the surface water and the air was assumed, with the heat transfer coefficient determined from the work of Edinger and Geyer (1965). The stress acting on the water surface due to the wind (measured) is calculated by the formulas developed by Wilson (1960). The outer x and y boundaries for the numerical calculations must be taken at some finite distance. In the present calculations, a region appropriately 1000m by 1500 m was used. Roache (1972) discusses outer boundary conditions and concludes that the actual boundary conditions are not that important as long as the boundary conditions used do not severely restrict the flow. The boundary conditions used here are that the normal derivatives of the velocities and the temperatures are zero.

The eddy viscosities and eddy conductivities are taken to be identical in magnitude but dependent on the local vertical temperature gradients; they are assumed to be given by

$$A_v = \alpha - \beta \frac{\partial T}{\partial z} \qquad (4.4)$$

where α and β are constants depending on local conditions. The constant α is equal to the vertical eddy diffusivity under vertically stable conditions.

For the above boundary conditions and parameters, time-dependent calculations were made starting with an initial guess of the flow and proceeding until a steady-state had been reached. The first case was that of the discharge into a quiescent lake. No winds or cross-flows in the basin were present. From the calculations, one can show that as the flow is discharged from the outfall, it is forced towards the surface by the decreasing depth (see Figure 4.1) as well as by buoyancy. The large densimetric Froude number [defined by $u/\Delta\rho gh/\rho)^{1/2}$, where u is the outfall velocity, $\Delta\rho$ is the density difference between the discharge and lake waters, and h is the outfall depth] of 4.2 indicates that at least initially buoyancy effects are not dominant. About 75 m from the outfall, the depth begins to increase. After this point, the discharge tends to remain very near the surface. The resulting surface temperature field is shown in Figure 4.2.

A second calculation was made for the case when a cross-flow (current of 9.2 cm/sec at an angle of 125°) and a wind (approximately 5 m/sec

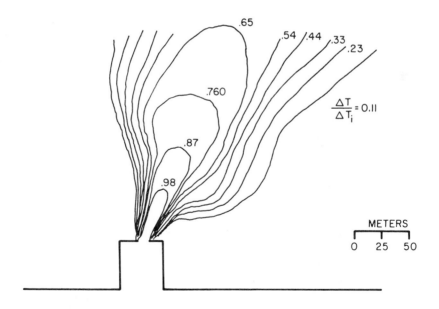

Figure 4.2   Surface temperature distribution, no wind, no cross-flow.

at an angle of 270°) were present. In this case, the discharge is physically swept in the direction of the cross-flow. The temperatures (see Figure 4.3) are displaced in the direction of the cross-flow and towards the shore.

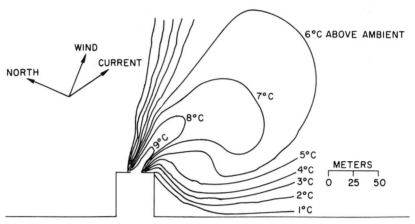

Figure 4.3 Surface temperature distribution with wind and cross-flow. ΔT_i is the temperature difference between the lake and outfall waters while ΔT is the temperature difference between the surface plume waters and outfall.

For this second case, a comparison of the calculated results with field observations (Frigo, *et al.* 1974) is shown in Figures 4.4 and 4.5. The former shows the temperature decay along the centerline, while the latter shows the isotherm areas for various temperatures. It can be seen that there is good agreement between the predicted results and field observations. Similar agreement was obtained between predicted results and field observations for the first case above. The calculations are presently being extended to include more of the flow field.

Additional field data are available (Frigo, *et al.* 1974) from which one can determine more details of the flow field. However, the flow, due to its turbulent nature (see Csanady 1973 for a general description of the turbulent diffusion and nature of plumes), is highly variable both in space and time. Continuous field measurements over a sufficiently long period of time to average out these variations must be made before more general comparisons can be made between our calculations and observations. This has not been done yet. However, the above calculated results, although limited, seem more than adequate at this point and do give one confidence in the numerical model.

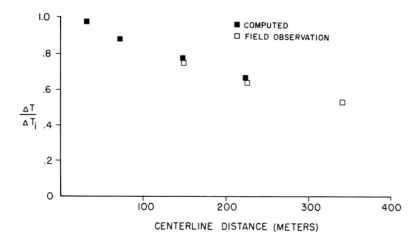

Figure 4.4   Surface temperature decay along centerline for case with wind and cross-flow.

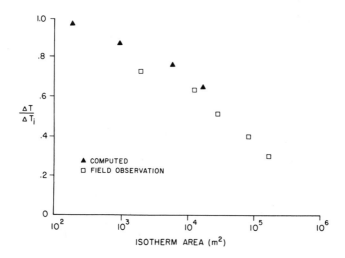

Figure 4.5   Surface isotherm areas for case with wind and cross-flow.

CONTAMINANT DISPERSION NEAR
A RIVER MOUTH

A somewhat idealized model of the flow from a river into a lake has been used to investigate the dispersion of contaminants near a river mouth. The parameters needed to describe the flow are identical to those typical of the Cuyahoga River during the summer. However, it is assumed that in the offshore direction the basin depth doubles in 2.5 river widths and then remains constant. This depth variation is not correct for the Cuyahoga River but is used to indicate the effects of a variable-depth basin. Previous calculations of the flow field have been made for a constant depth basin (Paul and Lick 1974).

For this calculation, the parameters were assumed to be: river width, 76.2 m; river depth, 8.5 m; ambient lake temperature, 22.2°C; vertical eddy coefficient 10 cm^2/sec; horizontal eddy coefficient, 26 cm^2/sec; maximum river velocity, 1 cm/sec; Froude number, 1.1×10^{-3}; densimetric Froude number, 0.41; and settling velocity of particulate matter, 0.01 cm/sec $\doteq$ 10 m/day. The numerical model is the same as that used to describe the flow in the previous section and the same basic assumptions and approximations apply. A variable grid size was also used with horizontal grid sizes varying from 8 m near the river mouth to 350 m far from the mouth. Eight grid points in the vertical were used.

Results of the calculations for the velocities and temperatures are shown in Figures 4.6 and 4.7 which show the velocities and temperatures in a vertical plane along the centerline of the discharge. Vertical distances are nondimensionalized with respect to the river depth, while horizontal distances are nondimensionalized with respect to the river width.

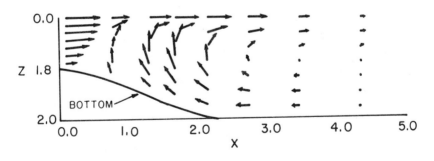

Figure 4.6 Velocity distribution in a vertical plane along the centerline of the discharge. Current magnitude → 1 cm/sec.

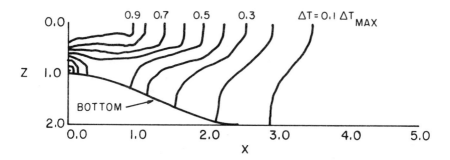

Figure 4.7   Temperature distribution in a vertical plane along the centerline
of the discharge.

The low densimetric Froude number of 0.41 indicates that buoyancy
is important. The effects of this buoyancy can be seen in Figure 4.6,
which shows large vertical velocities near the mouth of the river. Due
to its buoyancy, the river water rises as it enters the lake and spreads
out over the surface. Water is entrained from below with lake water
flowing toward the river mouth near the bottom. It can be shown that
the velocities are larger at the surface than in the constant depth case,
in which the entrainment is considerably reduced due to frictional drag
at the bottom, thus reducing the amount of fluid reaching the upper
layer. Figure 4.7 shows the effect of entrainment and vertical mixing
on the temperature field, *i.e.,* cooler water is brought in along the
bottom and warmer moved out along the surface, with vertical exchange
of heat between the two.

Assuming that the solids do not modify the flow, the dispersion of
dissolved and suspended solids has also been investigated. To investigate
the effects of settling and the bottom boundary condition on the concentrations, four examples were calculated. These were: (1) no settling,
zero flux at the bottom; (2) settling, zero flux at the bottom; (3) no
settling, zero concentration at the bottom; and (4) settling, zero concentration at the bottom. The boundary conditions of zero flux and
zero concentration are two easily calculable and important limits of
Equation (4.3). Zero flux corresponds to setting the right-hand side
of Equation (4.3) to zero, *i.e.,* no entrainment and a perfectly reflecting
boundary, while zero concentration corresponds to zero entrainment and
letting $\beta \rightarrow \infty$, *i.e.,* a perfectly absorbing boundary.

The contaminant was assumed to have a concentration of one in the
river with the contaminant discharge starting at zero time and constant

thereafter. For case (1), the steady-state would be a constant concentration of one throughout the basin, but this takes time to develop. Graphic representation of the concentrations in a vertical plane along the centerline of the discharge after approximately 10 days of steady discharge are shown in Figure 4.8. Far from the river mouth, the concentration is fairly constant in the vertical direction due to vertical mixing. However, near the river mouth, entrainment is important and strong vertical gradients of concentration exist.

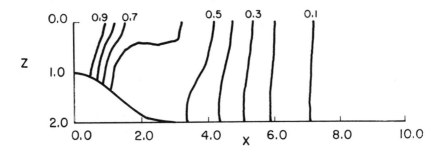

Figure 4.8 Concentration distribution in a vertical plane along the centerline of the discharge. No settling and a zero flux condition at the bottom.

For case (2), a zero flux bottom but with a particle settling velocity of 10^{-2} cm/sec, the results for the concentration after 10 days discharge in a vertical plane along the centerline of the discharge are shown in Figure 4.9. At this time, the concentration had approximately reached a steady state. The distribution is decidedly different from case (1).

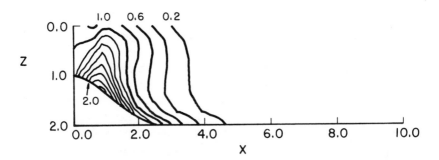

Figure 4.9 Concentration distribution in a vertical plane along the centerline of the discharge. Settling velocity of 10^{-2} cm/sec and a zero flux condition at the bottom.

Of interest is the horizontal distribution of concentration. At the surface
with increasing distance from the river mouth, the concentration first decreases to about 0.7 then increases to almost 1.0 before decreasing again.
Near the bottom, the concentration increases from 1.0 at the river mouth
to approximately 2.6 near x = 1.0 before decreasing to lower levels farther
out in the basin. This locally high concentration, reflected as high turbidity in the field, is due to settling of the particles as they leave the
river, entrainment of the particles by the near-bottom currents as the
particles settle, and subsequent recirculation of the contaminant. This
phenomenon generally will be present at a river mouth whenever settling
and entrainment occur (Postma 1964). However, the observability of a
concentration maximum and its magnitude depend very strongly on the
fluid velocities, settling velocities, the bottom boundary condition (see
below) and other parameters such as the local geometry.

The steady-state concentrations in a vertical plane along the centerline
for case (3), zero concentration at the bottom with no settling, are shown
in Figure 4.10. In this case, the steady state was approached in approximately five days, a result of the strong flux out of the water through
the bottom. Again, the effects of entrainment are noticeable. The concentration distribution is quite different from case (1), indicating the
importance of the bottom boundary condition on the concentration
distribution.

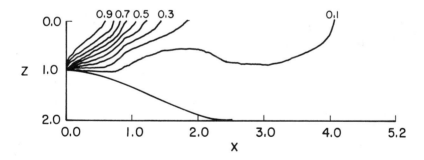

Figure 4.10   Concentration distribution in a vertical plane along the centerline
of the discharge. No settling and a zero concentration condition at the bottom.

Figure 4.11 shows the steady-state distribution for case (4), zero concentration at the bottom with a settling velocity of 10^{-2} cm/sec. Again
there are significant differences from the other three cases. The steady
state was reached even more rapidly (approximately 2½ days) than in
any of the other cases due to the combined effects of settling and the

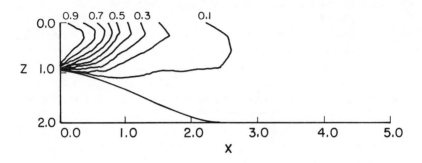

Figure 4.11   Concentration distribution in a vertical plane along the centerline of the discharge.   Settling velocity of 10^{-2} cm/sec and a zero concentration condition at the bottom.

zero concentration bottom boundary condition. In the vertical direction, maximum concentrations occur at an intermediate depth due to the combined effects of settling and vertical diffusion.

In general, a contaminant can be transported to a boundary by both settling and turbulent mixing. An effective time for settling can be defined as

$$t_s = \frac{h}{w_s}$$

where h is the average thickness of the water layer (one-half of the depth). An effective time for turbulent mixing can be defined as

$$t_D = \frac{h^2}{D_v}$$

In the present example, if we take h to be 10 m and $D_v = 10$ cm^2/sec, then $t_s = t_D = 10^5$ secs $\doteq$ 1 day. Since

$$\frac{t_s}{t_D} = \frac{D_v}{hw_s}$$

it is quite apparent that for realistic values of D_v, h, and w_s (for clay-size particles, $w_s \leqslant 10^{-2}$ cm/sec), the contaminant may be transported to the bottom, and hence taken out of the system, much more rapidly by diffusion than by settling.

DISPERSION BY WIND-DRIVEN CURRENTS

The problem considered here is the long term and large scale dispersion of contaminants after entering Lake Erie from the Cuyahoga River and the effect of large man-made structures on this dispersion. In the previous section, currents driven by the wind were neglected. However, in the present problem, the effects of the wind-driven currents on dispersion are significant and are specifically investigated. The river itself is treated as a point source of pollution and the effect of the river on the flow in the lake is neglected. The contaminant is assumed to be discharged from the river at a constant rate starting at some initial time.

The specific conditions assumed for the calculation were: (a) a steady-state velocity field caused by a W 32°S wind with a velocity of 5.2 m/sec (the dominant wind in the area) and a wind stress of 0.9 dynes/cm^2, (b) negligible variable density effects, (c) a concentration in the river of 10 units, (d) mass flux of the river was 27 m^3/sec, (e) vertical eddy diffusivity was 17 cm^2/sec while horizontal mixing was neglected, and (f) the boundary condition at the bottom was that of zero flux.

For these conditions, the velocities were calculated using a numerical model previously used (Gedney and Lick 1972; Sheng and Lick 1975) for both offshore and near-shore hydrodynamic calculations in Lake Erie. Throughout the lake, a two-mile horizontal grid and seven grid points in the vertical were used. The surface velocities are shown in Figure 4.12, while the near-bottom (1/6 of the depth from the bottom) velocities are shown in Figure 4.13. Strong currents in the along-shore direction are apparent at the surface, and it can be shown that they extend almost to the bottom. However, the near-bottom currents, especially in the vicinity of Cleveland, are directed away from shore. From the calculations, it can be shown that downwelling is present throughout the water column over the entire near-shore area shown in the figures. Due to the sharper bottom slopes near shore, the vertical velocities in regions near shore are larger than those in regions far from the shore.

Once the velocities are known, the dispersion of a contaminant can be calculated by means of Equation (4.1). In the present calculation, the dispersion of a conservative, dissolved substance was considered. Relative concentrations at the lake surface at 128 hours are shown in Figure 4.14, which shows that most of the contaminant is transported along shore toward the East but some is transported in the offshore direction. This offshore transport is caused mainly by the bottom currents in the offshore direction followed by vertical mixing. Including a horizontal diffusivity of 2 x 10^5 cm^2/sec made little difference in the results.

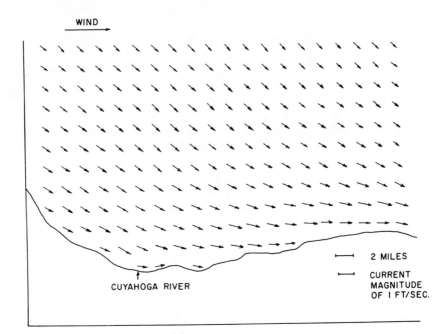

Figure 4.12 Surface velocities for a wind-driven current of 5.2 m/sec from W 32°S.

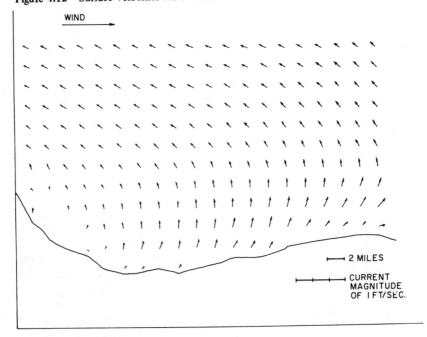

Figure 4.13 Velocities at 5/6 of the depth for a wind-driven current of 5.2 m/sec from W 32°S.

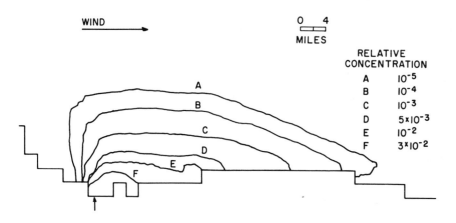

Figure 4.14   Surface concentrations at 128 hours.

To show qualitatively the effects of an offshore Jetport on the currents and the dispersion of a contaminant, two additional cases were calculated under the same conditions as above but with two different proposed configurations of a Jetport. These configurations were (1) an island 4 miles square approximately 6 miles offshore of Cleveland, and (2) a landfill extension of this island to shore. The results of the calculations for the surface concentrations at 128 hours are shown in Figures 4.15 and 4.16. The island configuration, as seen in Figure 4.15, does not affect the contaminant transport appreciably, except in the immediate vicinity of the island because the flow over most of the near-shore region is not appreciably modified by the presence of the island.   In the case of the island

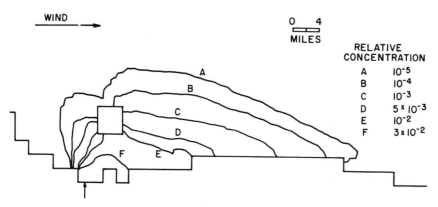

Figure 4.15   Surface concentrations at 128 hours; island Jetport.

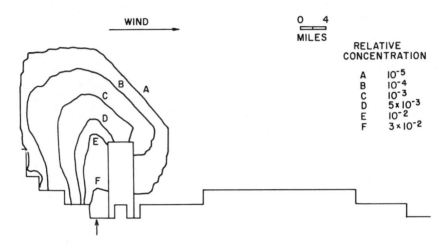

Figure 4.16 Surface concentrations at 128 hours; island with extension to shore.

with an extension to shore, the contaminant is transported much further offshore than in the previous two cases. The extension blocks the flow in the long-shore direction, and very little contaminant is found at the down-wind side of the Jetport. It can be shown for this case that there are strong subsurface, offshore currents in regions near the river mouth and upwind of the Jetport. A report (Sheng and Lick 1975) gives more details of the calculations and presents additional examples.

SUMMARY AND CONCLUSIONS

The dispersion of contaminants, especially in near-shore regions, is an important, essential, and complex problem. Practical applications require a knowledge of physical, chemical, and biological processes. A few examples have been given to illustrate quantitatively the dispersion of contaminants as affected by physical processes.

The first example, concerned with the dispersion of heat from the dis-charge from a power plant, is a realistic, quantitative calculation. The cal-culations included realistic geometry and topography, buoyancy effects, wind stresses, and cross-flows in the lake, all of which were important in their effect on the flow field. Limited comparison with field data showed good agreement and gives one confidence in the basic procedure and in the results of the calculations.

The second example was concerned with the dispersion of contaminants near a river mouth but was a more idealized case than the first example.

The effect of temperature and density differences were significant in modifying the flow field and caused large vertical velocities and entrainment near the mouth of the river. This in turn modified considerably the contaminant dispersion. When settling was considered (with a zero flux bottom condition), a local maximum in the concentration was observed caused by the recirculation of the contaminant near the river mouth. Changing the bottom boundary condition to a zero concentration condition considerably modified the concentration, indicating the importance of knowing what this boundary condition is in the real situation. The flux from the system depends not only on this boundary condition but also on the rate at which contaminants are transported to the boundary, which occurs by a combination of settling, turbulent diffusion, and vertical convection, any one of which may be dominant in an actual case.

The third example illustrated the long-time dispersion near shore due to wind-driven currents and the modification of this dispersion by large man-made structures. The dispersion depended on the three-dimensionality of the flow with surface currents moving along shore while bottom currents moved offshore, introducing an effective horizontal diffusion. The island structure does not significantly modify the flow field or contaminant dispersion, but the island with extension to shore does. These results of course depend on the particular wind stress and may vary as the wind stress varies.

ACKNOWLEDGMENT

The authors wish to acknowledge with appreciation the support of this research by the U.S. Environmental Protection Agency. Mr. William L. Richardson served as Grant Project Officer.

REFERENCES

Csanady, G. T. *Turbulent Diffusion in the Environment.* (Boston: D. Reidel Publ. Co., 1973).

Edinger, J. E. and J. C. Geyer. "Heat Exchange in the Environment," Publication No. 65-902, Edison Electric Institute, New York (1965).

Frigo, A. A., N. Frye, and J. V. Tokar. "Field Investigations of Heated Discharges from Nuclear Power Plants on Lake Michigan," Argonne National Laboratories ANL/ES-32 (1974).

Gedney, R. and W. Lick. "Wind-Driven Currents in Lake Erie," *J. Geophys. Res.* **77**, 2714 (1972).

Lick, W. "Numerical Models of Lake Currents," U.S. Environmental Protection Agency Report.

Monin, A. S. and A. M. Yaglom. *Statistical Fluid Mechanics*, Vol. 1 (Cambridge, Mass.: The MIT Press, 1973).

Paul, J. F. and W. Lick. "A Numerical Model for a Three-Dimensional, Variable-Density Jet," Technical Report, Case Western Reserve University (1973).

Paul, J. F. and W. Lick. "A Numerical Model for Thermal Plumes and River Discharges," *Proceedings 17th Conference on Great Lakes Resear Research* (Hamilton, Ontario, 1974).

Postma, H. "Marine Pollution and Sedimentology," in *Pollution and Marine Ecology*, Olson and Burgess, Eds. (New York: Interscience, 1964).

Roache, P. J. *Computational Fluid Dynamics*. (Albuquerque, N.M.: Hermosa Publ., 1972).

Sheng, Y. P. and W. Lick. "The Wind-Driven Currents and Contaminant Dispersion in the Near-Shore," Case Western Reserve University Report (1975).

Wilson, B. W. "Note on Surface Wind Stress Over Water at Low and High Speeds," *J. Geophys. Res.,* 65 (1960).

5

AN EVALUATION OF THE TRANSPORT CHARACTERISTICS OF SAGINAW BAY USING A MATHEMATICAL MODEL OF CHLORIDE

William L. Richardson[1]

INTRODUCTION

The U. S. Environmental Protection Agency, Grosse Ile Laboratory, has undertaken a study of Saginaw Bay to quantify water quality processes. The study, which is being conducted in support of the U.S.-Canada Great Lakes Agreement through the International Joint Commission's Upper Lakes Reference Study, includes development of water quality models and field investigations for model calibration and verification.

The program is designed to produce as its final product a water quality model capable of simulating the effect of nutrients on the growth, composition, and distribution of phytoplankton biomass. This management model will eventually be used to predict the effect of a series of probable nutrient control programs.

The model consists of two major submodels: (1) physical circulation of water (*i.e.,* material transport) and (2) chemical-biological processes. This chapter presents the results of the submodeling effort devoted to quantify Saginaw Bay circulation using a simple coarse-grid, mass balance formulation that traces the transport of chlorides. First, a steady-state, average annual circulation pattern previously calibrated

[1] U.S. Environmental Protection Agency, Grosse Ile Laboratory, Grosse Ile, Michigan.

with 1965 data is reviewed (Richardson 1974), and the attempt to verify this pattern using 1974 data is presented. Also, a time-dependent approach is presented that is used to calibrate seasonal circulation patterns for 1974.

SAGINAW BAY WATER QUALITY

Saginaw Bay is a shallow (mean depth 5 meters) extension of Lake Huron located on the eastern side of the State of Michigan as shown in Figure 5.1. The bay is 21-42 km in width, approximately 82 km in length and receives runoff from a 21000 km^2 drainage basin. The basin supports a population of over one million (1960 census), and a variety of land uses including large industrial-urban, as well as agricultural, recreational, and undeveloped areas. The water quality of the bay has been the concern of many resource managers because of the bay's value for recreation, fishing, and water supply (USDI, 1969).

A review of historical water quality conditions in Saginaw Bay has been prepared by Freedman (1974), and results of water quality surveys conducted in 1974 at the stations indicated in Figure 5.2, in support of the IJC Upper Lakes Reference Study, have been summarized by Smith (1975). Water quality is characterized by high concentrations of dissolved solids, particularly chloride, (2-39 mg/l in 1974) and excessive nutrient concentrations that support prolific growths of algae. Chlorophyll *a* concentrations (an indirect indication of phytoplankton biomass) have been recorded in the range of 0.3-74 μg/l measured in 1974. The primary source of material inputs is the Saginaw River, which, as shown in Figure 5.1, drains the major portion of the total bay drainage basin. The concentration levels of most materials in the bay are highest near the Saginaw River and lowest near Lake Huron. There is also a general gradient of low to high concentration from the northwest to southeast side of the bay.

MODELING PROCESS

The two primary components of the final water quality management model shown in Figure 5.3 are (1) the physical transport or circulation component and (2) the chemical-biological process component. Bierman (1975) has developed a chemical-biological submodel for Saginaw Bay. His research has focused primarily on the nutrient-phytoplankton growth and composition problems, and he has investigated growth kinetics in detail. However, there is no spatial detail in the initial model for the inner bay, which is considered as a single, completely mixed reactor.

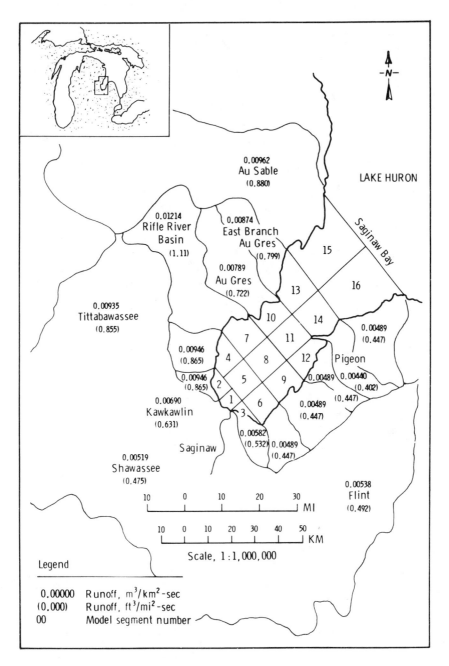

Figure 5.1 Saginaw Bay Drainage Basin indicating average 1965 runoff for model segment subbasins (Richardson 1974).

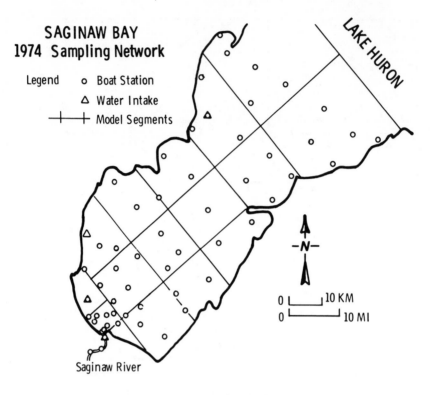

Figure 5.2 Saginaw Bay 1974 sampling network.

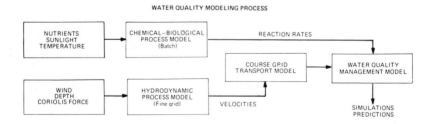

Figure 5.3 Water quality modeling process.

The spatial resolution will be obtained by the superposition of the chemical-biological batch model with the circulation regime provided by the coarse-grid transport model.

SAGINAW BAY CIRCULATION

Although there have been several field investigations of Saginaw Bay circulation, most of the results have been qualitative (Ayers 1956; USDI 1956). These studies involved releasing drift bottles and determining their direction and rate of travel. Ayers (1956) concluded from the analysis of drift bottle information and a qualitative assessment of material distributions that the bay acts like a simple estuary; he thus arrived at "probable" circulation patterns in the bay. He deduced that prevailing west winds and the Coriolis effect tend to deflect the outflow of the Saginaw River along the southeast shore and the same forces tend to hold it there during its course to Lake Huron. The Fish and Wildlife (1956) study concluded that no one stable surface current pattern exists in the bay. Their drift bottle results revealed more variability that was closely related to meteorological conditions. These studies are not only limited by their lack of quantification, but are also limited by the short time period covered by the study. To model the temporal and spatial transport of materials and the nutrient-phytoplankton dynamics, a quantified circulation scheme is needed over the entire annual growth cycle of phytoplankton.

ALTERNATIVE CIRCULATION MODELING APPROACHES

The temporal requirements and inadequacy of field measurements necessitates the use of a mathematical determination of circulation. Two basic circulation modeling approaches have been identified.

Fine Grid Hydrodynamic Approach

The first approach involves a complex mathematical formulation of the physical forces that influence circulation. Hutchinson (1957) has listed these forces, which include wind, temperature (density), river influent and effluent, atmospheric pressure, Coriolis force (*i.e.*, rotation of the earth) and friction (bathymetry). The hydrodynamic formulations involve rather complex partial differential equations for the conservation of mass, momentum, and energy written for a relatively fine spatial grid (Lick 1975; Simons 1973). The model inputs include the above factors, and the outputs are the current velocities in three dimensions.

There are several practical problems arising with this approach. First, the computer programs for solving the equations require considerable computer resources primarily because of their fine time and space resolutions. It has been estimated, for example, that a nine-month simulation of two dimensional currents in Saginaw Bay would require about 40 hours of large-scale computer time. Second, the verificiation of such a model would require an extensive current metering and meteorological measurement program.

Coarse-Grid, Mass Balance Approach

A more economical approach to quantifying circulation has been used by Thomann (1972). This approach utilizes one of the key hydrodynamic principles, mass balance, to trace a material through a system that has been segmented on a relatively coarse grid. The material can be anything for which most reactions are known; some investigators have traced the transport of dye released into the system. For Lake Ontario, Thomann (1975) used temperature as a transport tracer, in estuaries chloride intrusion from the ocean has been used (Thomann 1972), and in Saginaw Bay, the most appropriate tracer was the chloride inflow from the Saginaw River. Chloride provides a continuous tracer over the entire year, is nonreactive, and can be conveniently measured at its sources and in the bay.

The coarse-grid approach provides sufficient detail to define problems within large time and space scales. This is adequate for the water quality problems being considered by this study, which require definition over time periods of weeks to months and space scale of 5 to 50 kilometers.

With the chloride mass load known along with the resulting chloride concentrations in the bay, the circulation parameters are the unknowns. These parameters are determined through a simple iterative calibration and verification process using a computer solution to the mass balance formulation, which requires less than a minute of computer time per run.

BASIC MASS BALANCE FORMULATION

The basic mass balance equation is formulated as follows:

$$\frac{\partial c}{\partial t} = 0 = -\frac{\partial}{\partial x}(uc) - \frac{\partial}{\partial y}(vc) + \frac{\partial}{\partial x}(E_x \frac{\partial c}{\partial x}) + (E_y \frac{\partial c}{\partial y}) - K(x,y)c \qquad (5.1)$$

where

u = velocity in x direction (L)

v = velocity in y direction (L)
E_x = dispersion coefficients in x direction (L^2/T)
E_y = dispersion coefficients in y direction (L^2/T)
c = concentration of substance (M/L^3)
K = decay rate of substance c (l/T)

The approach used to solve Equation (5.1) is described by O'Connor and Thomann (1971). In effect, the water body is divided into an array of completely mixed segments, and a mass balance is defined about each by converting Equation (5.1) to the form

$$V_k \frac{dc}{dt} = 0 = \Sigma_j \ [-Q_{kj} \ (\alpha_{kj}c_k + \beta_{kj}c_j) + E'_{kj} \ (c_j - c_k)] - V_k K_k c_k + W_k$$

(5.2)

where

V_k = volume of segment K (L^3)
E_{kj} = bulk dispersion coefficient = $E_{kj}A_{kj}/\bar{L}$ (L^3/T)
L = average length of adjacent sections (L)
A_{kj} = cross-sectional area between segments k and j (L^2)
α_{kj} = dimensionless mixing coefficient; $\beta_{kj} = 1-\alpha_{kj}$
Q_{kj} = advective transport parameter (L^3/T)
W_k = source of c in mass units per time (M/T)

Units:

L = length
T = time
M   mass

The parameters α and β are weighting factors that correct for cases of adjacent segments of unequal length.

$$\alpha_{kj} = \frac{\bar{L}_j}{\bar{L}_j + \bar{L}_k}$$

(5.3)

where positive solutions are maintained by the stability criteria

$$\alpha_{kj} > 1 - \frac{E'_{kj}}{Q_{kj}}$$

(5.4)

Alternative Solutions

The system of simultaneous equations (one equation for each model segment) can be solved in two ways, steady-state ($dc/dt = 0$) or time-variable ($dc/dt \neq 0$). The steady-state approach is used to investigate average conditions over some defined time interval. All model inputs must be defined in terms of averages over this period; model output (the concentrations in each model segment) is also in terms of a time average. A steady-state solution to the system of equations (Equation 5.2) has been documented by Chaptra (1973).

A computer program for the time-dependent solution of Equation (5.1) has been developed at the Grosse Ile Laboratory, which uses a fourth order Runga-Kutta integration technique. The output from such a solution includes material concentrations in each model segment for any time interval required. The time resolution of the output, however, is only as fine as that of the input data.

MODEL APPLICATION TO SAGINAW BAY

Whether the steady-state or time-variable solution is used, the final circulation scheme is determined using the same iterative calibration and final verification process. As shown in Figure 5.4, the model formulation requires:

(1) segmentation of the water body
(2) determination of segment physical characteristics
(3) measurement of chloride loads
(4) measurement of chloride concentration in most segments
(5) specification of the circulation parameters.

Segmentation

Specification of a particular segmentation scheme depends on several factors: (1) the spatial resolution desired, (2) water quality gradients, (3) research time available, and (4) computer time available. Based on the these considerations, Saginaw Bay was divided into 16 segments as shown in Figure 5.1.

Chloride Loads

Chloride loads are obtained by field measurement of chloride concentration and flows at all major sources. For Saginaw Bay the primary source is the Saginaw River; however, loads from all tributaries and major point sources have been accounted for and included as input to the appropriate model segment. The average chloride load in 1965 was 2.8 million kg/day (Richardson 1974). In 1974, the average annual load had been reduced to 1.2 million kg/day, as shown in Figure 5.5.

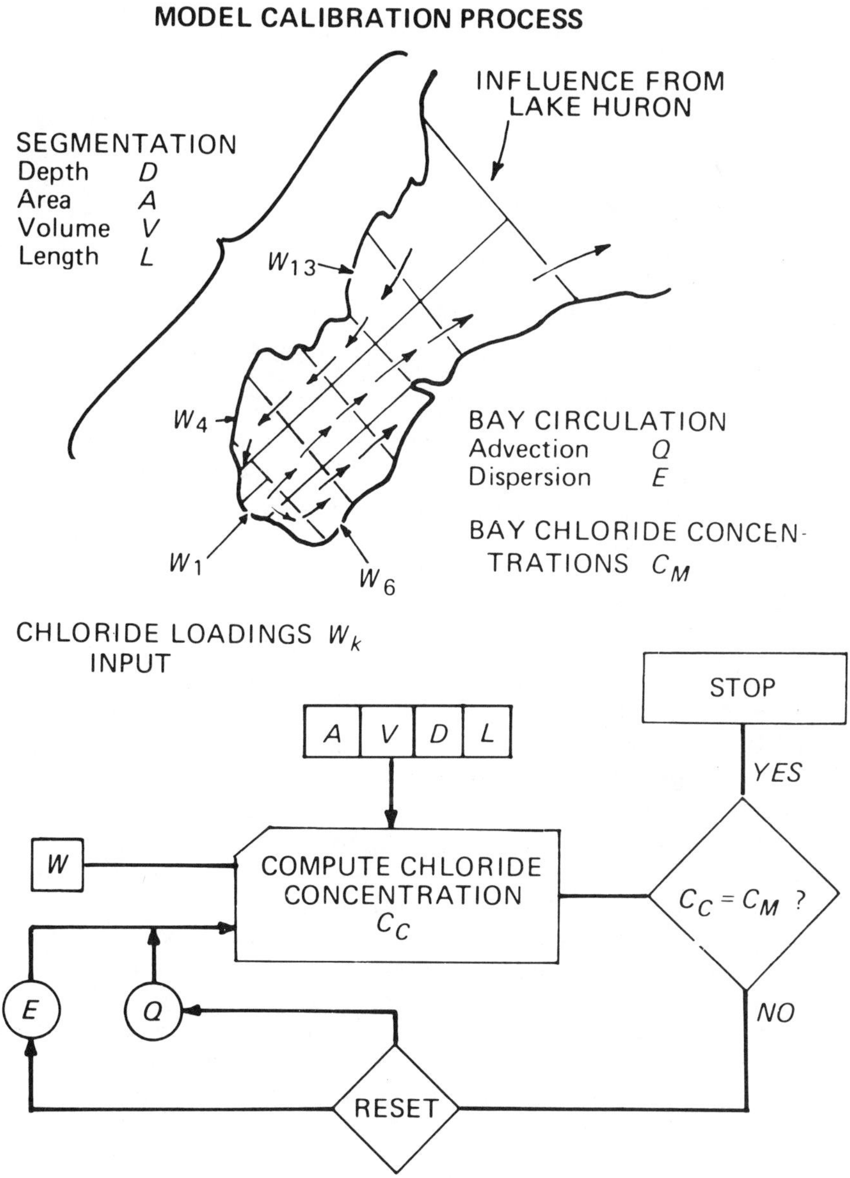

Figure 5.4 Model calibration process.

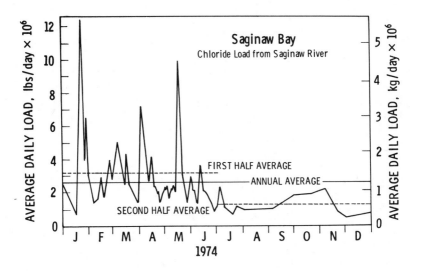

Figure 5.5 1974 chloride load for Saginaw Bay.

Segmental Chloride Concentrations

Results of historical chloride data have been reviewed by Freedman (1974). The specific data sources used for this investigation include the 1965 FWPCA chloride data measured at about 20 sampling stations (USDI 1969) and the 1974 Cranbrook Institute of Science data measured at the 59 stations shown in Figure 5.2 (Smith 1975). In addition to the monthly to bimonthly cruise stations in 1974, four stations were sampled at water intakes on a one- to three-day frequency. The cruise data are averaged on a segment basis for comparison to calculated concentrations.

The calibration process proceeds with initial estimates of the transport parameters E_{kj} and Q_{kj}, and an initial calculation of segment concentrations is made. The results are compared to the measured segment concentrations. Adjustments of E_{kj} and Q_{kj} are made and the process repeated until an acceptable comparison is obtained. Verification of the previously calibrated circulation scheme is attained when a new set of load and concentration data is used, as above, and an acceptable comparison obtained without adjustments.

STEADY-STATE CALIBRATION

This process was first used to test the hypotheses that the circulation in Saginaw Bay could be described by one average annual pattern

(Richardson 1974). The rationale was that circulation is highly random and closely related to meteorological events. These events, however, should average out over a period of time, within the time and space scales of interest to this study (*i.e.*, weeks to months and 5 to 50 km). The highly fluctuating and random events would be included in the dispersive transport parameters, E_{kj}. Any prevailing pattern (*i.e.*, net circulation) would be included in the advective parameter Q_{kj}.

The calibration process was performed over a range of values for the transport parameters, and the initial advective pattern used was that suggested by Ayers (1956). The results are presented in the form of a sensitivity analysis in Figure 5.6. With the 1965 average chloride concentrations indicated for each segment, a valid range for the parameters was obtained. It appeared, for example, that dispersion must range between 2.5 and 16 km^2/day with valid solutions for advection in the entire range of values investigated, zero to 25500 m^3/sec net inflow from Lake Huron to segment 15. The final advective scheme (Figure 5.7) was determined by matching computed to measured concentrations as shown in Figure 5.8. For this final calibrated scheme, inflow from Lake Huron was 6400 m^3/sec, and dispersion was 5 km^2/day, set at all interfaces except those in the outer bay which were set at 15.6 km^2/day.

STEADY-STATE VERIFICATION

The verification process simply involves one calculation using the final advective and dispersive scheme from the 1965 calibration and the 1974 average chloride load of 1.2 million kg/day. The computed chloride concentrations in each segment were then compared to those measured in 1975 and the results plotted in Figure 5.9.

These results indicate that the model does predict the general decrease in chloride concentration one would intuitively expect as a result of the load reduction; however, the spatial comparison is significantly different, most distinctly in the inner bay. At this point, it is concluded that the steady-state (annual average model) can only provide a gross approximation of observed data.

The rate poor verification of the annual steady-state model led to a more detailed analysis of the data for both 1965 and 1974. In comparing the two data sets, it was noted that many more surveys were conducted during the spring in 1974 than were conducted in 1965. In fact, there was only one survey prior to June 1, in 1965, whereas in 1974 there were 5 between March and June. Therefore, another verification attempt was made for comparable data periods. The portion of the 1974 data related to the same data period in 1965 was used along

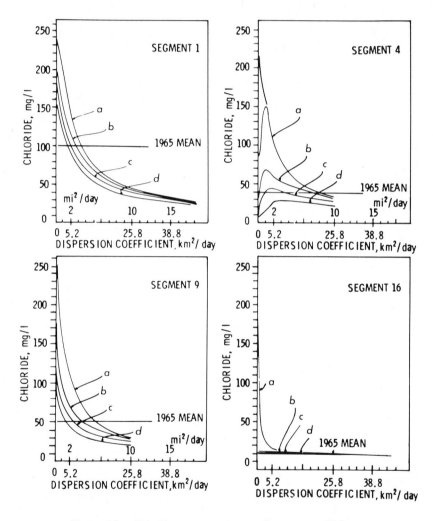

Figure 5.6 Chloride concentration *vs.* disperson coefficient
for inflows of (a) o, (b) 6370, (c) 12700, and (d) 25485 cubic meters per second
(Richardson 1974).

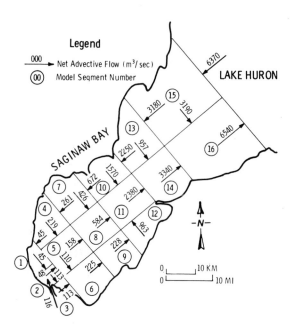

Figure 5.7 Calibrated average advective transport scheme for Saginaw Bay.

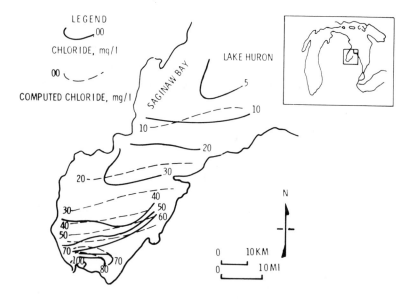

Figure 5.8 Saginaw Bay calibrated *vs.* measured 1965 average chloride isopleths.

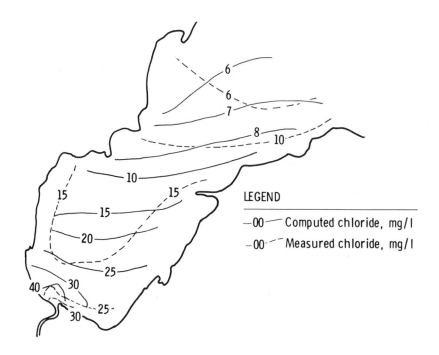

Figure 5.9   Computed *vs.* measured 1974 average chloride isopleths for for Saginaw Bay.

with the corresponding chloride load. The remaining 1974 data (March through June) were also grouped with its corresponding loading, and verification was then attempted for these two data sets with results given in Figure 5.10.

For the March through June period, verification failed again. However, for the July through November period, Figure 5.10b, the comparison was excellent and as good as the fit obtained for the 1965 data. The conclusion drawn here is that the circulation pattern calibrated for the 1965 data and verified for the same data period in 1974 is a valid average pattern for summer through fall conditions. Additional phenomena must be present during the spring period, which invalidates the use of one circulation pattern for the entire year.

TIME VARIABLE SOLUTION

To investigate the circulation pattern (*i.e.*, chloride transport) in more detail, especially during the first part of 1974, the system of mass balance

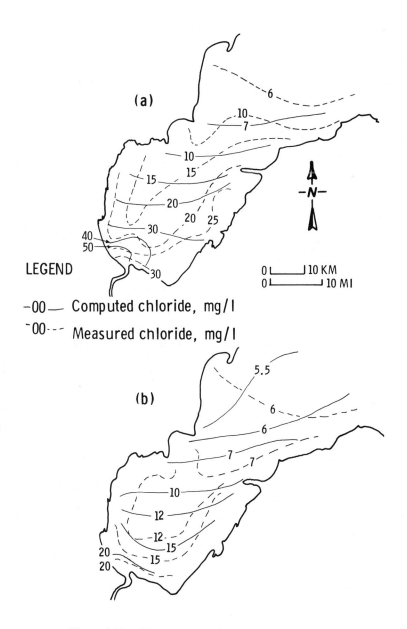

Figure 5.10 Computed *vs.* measured chloride isopleths
(a) March through June 1974, (b) July through November 1974.

equations is integrated with respect to time to obtain a computed chloride time series.

Time-Variable Data

The chloride data are presented for selected segments in Figure 5.11. Chloride concentrations, particularly those measured at the water intakes, show considerable variation. In general, the time-series data for most segments show an increase in chloride in the spring, starting near the end of February, with a dramatic decrease at the end of May and early June, another increase in the fall and a decline during winter months.

TIME-VARIABLE MODEL RESULTS

For the initial time-variable solution, the circulation parameters determined from the steady-state analysis were used for the entire year along with the 1974 time-variable loads. An intuitive conclusion is that the spring concentration rise is due to the high spring load. However, the initial model solution did not substantiate this hypothesis. The model solution with the time variable loading shown in Figure 5.5 fails to reproduce this spring increase, but fairly well describes the chloride concentration time series in the bay for the remainder of the year (Figures 5.12 and 5.13). The same conclusion was drawn from the previous steady-state investigation. It is necessary to explain this spring peak and to decide on a rationale for adjustments of the model parameters, E_{kj} and Q_{kj}, to obtain a reasonable fit.

It is possible to deduce from previous sensitivity analyses that during spring a circulation regime must exist that reduces transport within the bay. Therefore, for the next model computation, the dispersion transport parameter, E_{kj}, was reduced at all interfaces by 50%, or to about 2 km^2/day during the spring from March 1 through June 1. The rationale for this change was that ice cover and floes would tend to dampen the dispersion transport.

This solution shows little change from the initial result for the outer bay segments. However, quite a substantial reduction in chloride concentration occurred in segments near the Saginaw River while concentration in segments along the southeast shore (6, 9, and 12) increased substantially. In the segments along the northwest shore (4, 7, and 10) concentration actually decreased, primarily during the spring when the dispersion coefficient had been reduced but when loads are high. This was contrary to the effect first expected. It was apparent that a change was necessary in the advective scheme. However, some rationale or underlying physical reason for this recalibration was sought.

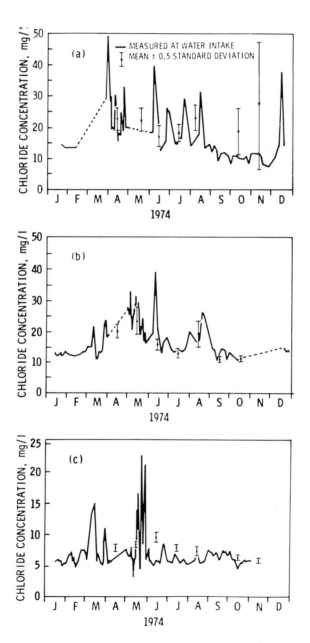

Figure 5.11   1974 measured chloride concentrations in Saginaw Bay model segments (a) segment 2, (b) segment 4, and (c) segment 13.

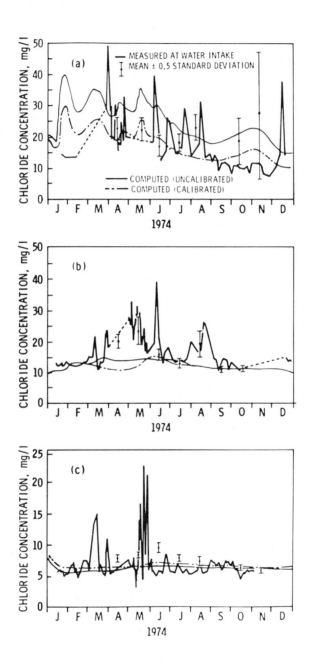

Figure 5.12 1974 measured *vs.* computed chloride concentrations for (a) segment 2, (b) segment 4, and (c) segment 13.

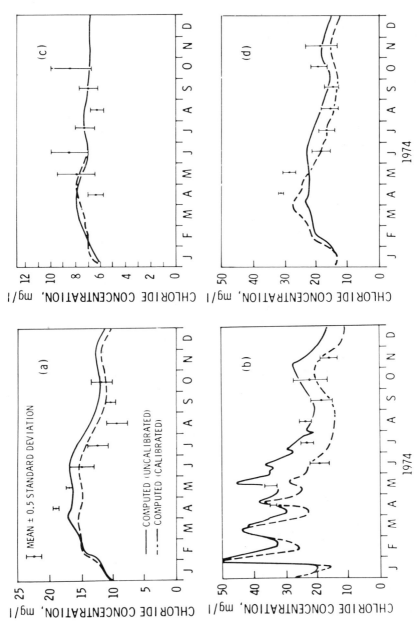

Figure 5.13 1974 measured *vs.* computed chloride concentrations for (a) segment 8, (b) segment 3, (c) segment 16, (d) segment 9.

THERMAL BAR EFFECT

The physical factors that influence lake circulation (wind, Coriolis force, and bathymetry) would not create any regular seasonal alteration to circulation. One phenomenon, which highly influences transport in the Great Lakes on a seasonal basis, is the temperature or density structure. The well-known limnological phenomenon common to many temperate lakes is the thermocline effect (Hutchinson 1957). Saginaw Bay is too turbulent and shallow for a persistent thermocline to exist (USDI 1956). However, a related phenomenon, the "thermal bar" could have a significant effect. This phenomenon occurs in the spring and precedes the formation of the thermocline. The thermal bar has been observed and documented by several Great Lakes investigators (Rodgers 1965 and 1971; Nobel 1968; Elliott 1971; and Hubbard 1973). Rodgers (1965) describes the thermal bar as the mixing zone between waters less than $4°C$ and greater than $4°C$ that constitutes a vertical barrier to extensive offshore movement of the warmer water near the shore. His description of the thermal bar formation is as follows:

> As spring heating progresses, nearshore waters heat to $4°C$ before the deeper central regions. Warm runoff water probably forms near the shore as a boundary between the midlake waters less than $4°C$ and warm water inshore. During the following weeks, as heating progresses, the thermal bar moves toward the middle of the lake like a contracting cylinder. As the bar moves offshore, the central region maintains a near-uniform vertical homogeneity, and in the region around the shore a thermocline develops. . . . The thermal bar, when it is close to shore in spring, appears to impound runoff and probably limits the dissipation of "pollution" injected into the zone between the shore and thermal bar. . . .

Although the 1974 survey of Saginaw Bay (Smith 1974) was not conducted for the purpose of defining its fine structure, enough temperature measurements were made to obtain a somewhat detailed picture of the thermal bar. The thermal bar was present as indicated by the vertical temperature profiles (Figure 5.14) along a transect of sampling points as indicated for surveys on April 29, May 13, June 14, and July 9. For the April and May profiles, the isotherms follow Rodgers' classical description of the thermal bar. In June, the isotherms are horizontal and the thermal bar is no longer apparent. This rather qualitative analysis provides a good rationale for altering the circulation pattern for the period from March 1 when chloride levels begin to rise to May 20 when they rapidly decrease and the thermal bar disappears.

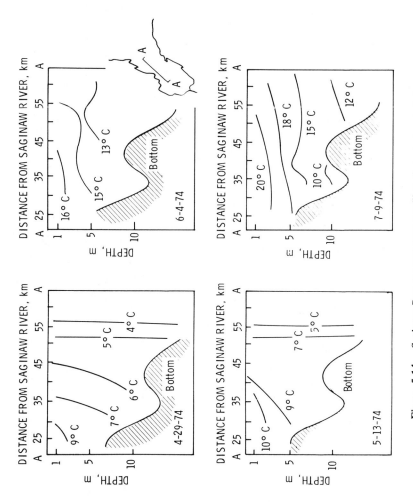

Figure 5.14 Saginaw Bay temperature profiles along transect A-A.

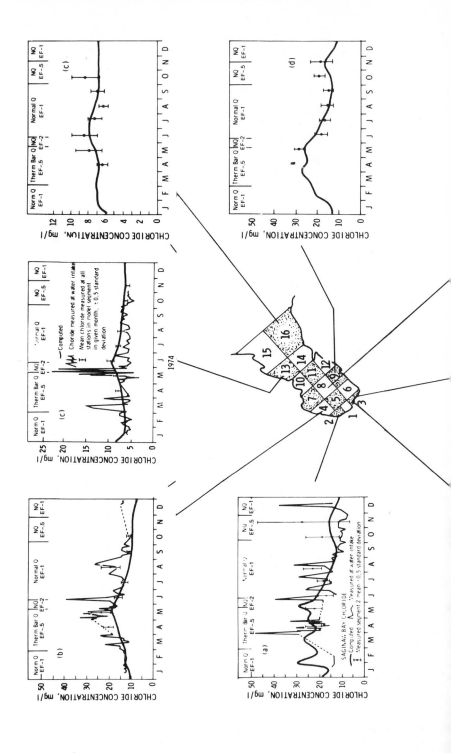

CONCLUSIONS

The use of a simple coarse-grid, mass balance model is an economical and invaluable tool for deducing quantified circulation characteristics of Saginaw Bay, as well as a unique methodology for interpreting other limnological data (*i.e.*, thermal bar effect). The average circulation scheme calibrated using 1965 data (Richardson 1974) was verified as valid for most of the year, except for the spring. Using a time-variable solution to the model equations it was determined that during the spring a 75% reduction of advective transport from outer to inner bay must be included, as well as a 50% reduction in dispersion, to obtain a reasonable fit to measured spring chloride levels. This reduction in transport is indicated by the development of a thermal bar structure, which can form a vertical barrier to mass exchange between the nearshore and open lake. A reasonable fit of computed *vs.* measured chloride concentrations was obtained for most segments throughout the year. The circulation scheme is considered to be calibrated and the necessary verification will be the subject of future research. Further research will also be conducted using a fine-grid hydrodynamic model, which should provide much insight into the finer physical processes involved. However, it is concluded that the presently calibrated circulation scheme is adequate to combine with the biological process model to form the first spatially segmented version of a nutrient-phytoplankton water quality management model.

ACKNOWLEDGMENTS

The author wishes to acknowledge the contributions made by several of his colleagues. First, the modeling effort could not have been done without the data provided by the Cranbrook Institute of Science study on Saginaw Bay chemistry. Particular thanks are due Elliott Smith, Principal Investigator, and his staff who provided open access to unpublished data and to Michael M. Mullin, who expedited the chemical analysis of the samples. Appreciation is also given to Victor J. Bierman and David M. Dolan who developed the time-variable computer program.

The author also wishes to acknowledge the encouragement given during this research by Tudor T. Davies and Nelson A. Thomas.

REFERENCES

Ayers, J. C., D. V. Anderson, D. C. Chandler, G. H. Lauff. "Currents and Water Masses of Lake Huron (1954) Synoptic Surveys," Ontario Department of Lands and Forests, Division of Research and University of Michigan, Great Lakes Research Institute (1956).

Bierman, V. J., Jr. "Mathematical Model of the Enhancement of the Blue-Green Algae by Nutrient Enrichment," presented at American Chemical Society, Division of Environmental Chemistry, 169th National Meeting, Philadelphia, Pennsylvania (April, 1975).

Chapra, S. "Interim Documentation for HARO1, a Steady-State Estuarine Water Quality Computer Model," U.S. Environmental Protection Agency, Region II, Data Systems Branch (1973).

Elliott, G. H. "A Mathematical Study of the Thermal Bar," in *Proceedings, Fourteenth Conference on Great Lakes Research* (International Association for Great Lakes Resarch, 1971).

Freedman, P. L. "Saginaw Bay: An Evaluation of Existing and Historical Conditions," The University of Michigan, Environmental Protection Agency, Region V, Enforcement Division (1974).

Hubbard, D. W., and F. C. Elder. "The Fine Structure of the Early Spring Thermal Bar in Lake Superior," in *Proceedings, Sixteenth Conference on Great Lakes Research* (International Association for Great Lakes Research, 1973).

Hutchinson, G. E. *A Treatise on Limnology*. (New York: J. Wiley and Sons, Inc., 1957).

Lick, W. "Numerical Models of Lake Currents," Case Western Reserve University, Department of Earth Sciences, for U.S. EPA, Grosse Ile Laboratory (in preparation).

Noble, V. E. and R. F. Anderson. "Temperature and Current in the Grand Haven, Michigan, Vicinity During Thermal Bar Conditions," *Proceedings, Eleventh Conference on Great Lakes Research* (International Association for Great Lakes Research, 1968).

O'Connor, D. J. and R. V. Thomann. "Water Quality Models: Chemical, Physical and Biological Constituents," in *Estuarine Modeling: An Assessment*, George H. Ward, Ed. U.S. EPA Research Series, 16070 DZU (1971).

Richardson, W. L. "Modeling Chloride Distribution in Saginaw Bay," *Proceedings Seventeenth Conference on Great Lakes Research* (International Association of Great Lakes Research, 1974).

Rodgers, G. K. "The Thermal Bar in the Laurentian Great Lakes," *Proceedings, Eighth Conference on Great Lakes Research,"* The University of Michigan, Great Lakes Research Division, Publication No. 13.

Rodgers, G. K. "Field Investigation of the Thermal Bar in Lake Ontario: Precition Temperature Measurements," in *Proceedings, Fourteenth Conference on Great Lakes Research* (International Association for Great Lakes Research, 1971).

Simons, T. J. "Comparison of Observed and Computed Currents in Lake Ontario During Hurricane Agnes (June 1972)," *Proceedings, Sixteenth Conference on Great Lakes Research* (International Association of Great Lakes Research, 1973).

Smith, V. E. *Annual Report—Upper Lakes Reference Study: A Survey of Chemical and Biological Factors in Saginaw Bay (Lake Huron)*. (Bloomfield Hills, Michigan: Cranbrook Institute of Science, for the U.S. Environmental Protection Agency, Grosse Ile Laboratory, 1975).

Thomann, R. V. *Systems Analysis and Water Quality Management*. (New York: Environmental Science Services Division, 1972).

Thomann, R. V., D. M. DiToro, R. P. Winfield, D. J. O'Connor. "Mathematical Modeling of Phytoplankton in Lake Ontario, 1. Model Development and Verification," U.S. EPA Publication 660/3-75-005 (1975).

U.S. Department of the Interior, Fish and Wildlife Service. "Surface Current Studies of Saginaw Bay and Lake Huron," (1956).

U.S. Department of the Interior, Fish and Wildlife Service. "Lake Huron—Michigan, Water Quality Data; 1965 Data," Lake Huron Basin Office (1969).

U.S. Department of the Interior. "Water Quality Control Plan, Lake Huron," Federal Water Pollution Control Administration, unpublished report (April, 1969).

6

SEASONAL PHYTOPLANKTON SUCCESSION AS A FUNCTION OF SPECIES COMPETITION FOR PHOSPHORUS AND NITROGEN

Joseph V. DePinto, Victor J. Bierman, Jr., and Francis H. Verhoff[1]

INTRODUCTION

Excessive nutrient input and the resulting nuisance algal blooms in freshwater lakes have recently become a major topic of limnological investigation. From the viewpoint of lake management and/or restoration it is desirable to be able to predict *a priori* if a given set of environmental conditions will tend to favor a stable (nonblooming) or an unstable (frequent algal blooms) system. Furthermore, since cyanophyte blooms appear to be the most objectionable type, knowledge of the pattern of seasonal phytoplankton succession in a lake would be useful. It is well known that phytoplankton succession is affected by many factors including temperature, light, turbulence, inorganic nutrients, accessory organic materials, antibiotics, parasitism, predation, and all forms of competition (Hutchinson 1967). Attempts are being made to determine what variations in the above factors tend to bring about the dominance of unwanted blue-green algal blooms.

Among the earliest proponents of the theory that seasonal variations in chemical composition of lake water are to a large extent

---

[1] Civil and Environmental Engineering Department, Clarkson College of Technology, Potsdam, New York; E.P.A. Grosse Ile Laboratory, Grosse Ile, Michigan; and Department of Chemical Engineering, University of Notre Dame, Notre Dame, Indiana, respectively.

responsible for periodicity of phytoplankton was Pearsall (1932). In his studies of the net plankton of English lakes, he found that high nutrient levels in the spring caused a vernal diatom maximum that was often replaced by a chrysophycean when the silicate concentration fell and the nitrogen/phosphorus ratio rose. Later Hutchinson (1944) stressed that competitive interactions for nutrients between various species could be important factors in seasonal succession in lakes. As an example, he cited data from Linsley Pond, indicating that competition for phosphorus between *Fragilaria* and *Anabaena* was controlled by nitrate availability, with *Fragilaria* favored at high nitrate levels and *Anabaena* favored at low nitrogen levels.

More recently a number of authors (Bush and Welch 1972; Fitzgerald 1969; Shapiro 1973; and Stoermer 1973), have noted that variations in phosphorus and nitrogen levels, due to inputs and/or cycling, can have a causative effect on changes in species dominance. In this regard, a field study was undertaken to observe any correlations between nutrient levels and seasonal phytoplankton succession in a hypereutrophic lake (Stone Lake, Cassopolis, Michigan). The mechanisms behind the observed successional pattern can be qualitatively explained in terms of phosphorus and nitrogen levels in the epilimnion of the lake and a two-step process model of the kinetics of algal growth in eutrophic water bodies developed by Bierman, *et al.* (1973a, 1973b) and Verhoff, *et al.* (1972, 1973).

MODEL DESCRIPTION

The model utilizes separate nutrient transport and cell synthesis mechanisms to enable it to account for differences in nutrient uptake and storage and in intrinsic growth rate among phytoplankton species. Much experimental evidence (*e.g.,* Fuhs 1969, 1971; Droop 1973; Azad and Borchardt 1970; Caperon 1972, 1972b; and Eppley and Thomas 1969) indicates that the mechanisms of nutrient uptake and cell growth are quite distinct and that specific cell growth rates are dependent on intracellular levels of the limiting nutrients. This mechanism is in contrast to the classical Michaelis-Menten approach relating growth rates directly to medium nutrient concentrations.

The model components utilized in the application to Stone Lake are illustrated in Figure 6.1, which is a schematic of the nutrient compartments used in this case and the potential pathways of phosphorus and nitrogen between compartments. For the growth simulations employed herein, the epilimnion of Stone Lake during summer stratification was considered to be a completely mixed batch reactor (4 meters depth),

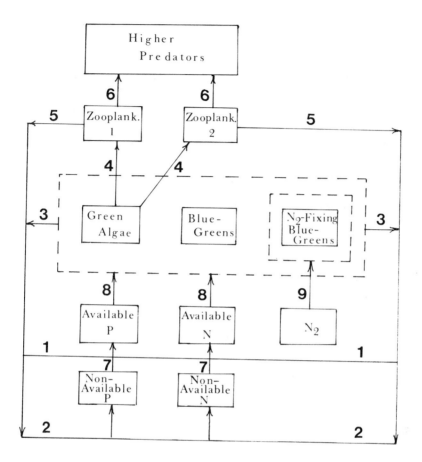

Figure 6.1 Schematic of the nutrient compartments used in the algal growth model for this study and the potential pathways of phosphorus and nitrogen between compartments.

with no external nutrient inputs and the only nutrient output being sinking of biomass into the hypolimnion. Note in Figure 6.1 that three important functional groups of phytoplankton are considered for this application: green algae and two blue-green algal groups, one of which can fix atmospheric nitrogen and the other of which must use dissolved, combined nitrogen. Different physiological input parameters and requirements (some based on species-specific data and others on estimations) have been assigned to the different groups, and along with other factors can explain the group successions.

The major working assumptions and specific coefficients used for the model applications in this paper are identical to those used by Bierman (1975) for Saginaw Bay, Lake Huron with the exception of those coefficients in Table 6.1. Since many of the coefficients in complex multiclass models must be estimated anyway, the different physical and hydrological characteristics between Stone Lake and Saginaw Bay naturally led to further ambiguities. Stone Lake has a very long hydraulic retention time (~6 years) and no significant external sources of nutrients, while Saginaw Bay has a retention time of only two months with heavy waste loadings from domestic as well as potentially inhibitory industrial sources. Also, it is unreasonable to assume that different algal species within a group would all have the same growth rate and sinking rate. Since the species list for Stone Lake is not the same as for Saginaw Bay, these parameters could easily vary from one application to another.

Table 6.1 Phytoplankton Parameters Unique to Stone Lake Application[a]

Parameter	Phytoplankton Group		
	Greens	Blue-Greens (non-N_2-fixing)	Blue-Greens (N_2-fixing)
Maximum growth rate (25°C) (day)$^{-1}$	1.6	1.1	1.1
Sinking rate (m/day)	0.3	0.1	0.1

[a]Remaining parameters used in this model application are detailed in Bierman (1975).

The model, of course, contains available phosphorus and nitrogen pools from which the algal groups can draw these nutrients. All other nutrients are considered to be in excess and never limiting. The box labeled N_2 indicates that when the concentration of available nitrogen falls below 300 μg N/l, the N_2-fixers gain a competitive advantage by fixing atmospheric nitrogen. This is a simplifying assumption based primarily on the work of Horne and Fogg (1970). For operation of the model, nitrogen fixation (pathway 9) is simulated by ignoring nitrogen transport from the medium by the N_2-fixers and arbitrarily setting the intracellular N at its maximum level. The specific growth rate as a function of nitrogen is therefore at its maximum level for the prevailing temperature and light attenuation conditions (Bierman 1975). This step has important repercussions on the nitrogen budget when these cells are lysed and decomposed with subsequent nutrient recycle.

The model also includes nonavailable nutrient pools for phosphorus and nitrogen. Lines 1 and 7 represent inputs directly to the available pools, while line 2 represents the input to the nonavailable pools. Both pools are fed by the recycle of nutrients from the lysing and decomposition of algae (line 3) and from the excretion of zooplankton (line 5). Note that the zooplankton only graze the green algal group. The fraction of pathways 3 and 5, which goes into pathway 1, represents the recycle of excess phosphorus and nitrogen over the minimum cell stoichiometric requirements. DePinto (1975) found that *excess* intracellular phosphorus (present as inorganic phosphates) was rapidly released at the beginning of aerobic bacterial decomposition of laboratory cultures of *Chlorella vulgaris*. On the other hand, regeneration of organic cellular phosphorus was proportional to the degree of decomposition of the cellular organic matter. In line with this additional finding, the model places the minimum stoichiometric levels of phosphorus and nitrogen into the nonavailable pools (pathway 2). The resulting detrital material in these pools is then bacterially decomposed. In this way, phosphorus and nitrogen are slowly released back into the available pools via pathway 7.

The nutrient uptake step in this model involves a carrier-mediated transport of phosphorus and nitrogen using a reaction-diffusion mechanism presented by Verhoff, *et al.* (1973). The importance of this mechanism with respect to phytoplankton succession is that it allows the potential intermediate nutrient storage in excess of the cell's immediate metabolic needs, as well as species differences in nutrient uptake rates. In general, non-N_2-fixing blue-green algae appear to have the highest affinity for phosphorus (most efficient phosphorus transport system) followed by the N_2-fixing group, and finally green algae. With regard to nitrogen, all three groups are considered to have the same kinetics and stoichiometries and, therefore, compete on an equal basis for dissolved, combined nitrogen (except, of course, when N_2-fixation occurs). Due to the sparsity of data on nitrogen uptake, this assumption was necessary; however, the nitrogen uptake data of Fitzgerald (1968) seems to support this premise.

The second step in this model is cell synthesis, which operates independently of nutrient transport. Mass growth rate is related to the intracellular phosphorus levels according to the formalism of Fuhs (1969, 1971) and to the intracellular nitrogen levels according to the formalism of Caperon (1972). Whichever nutrient yields the lowest specific growth rate is considered to be limiting, and that limiting growth rate is the one used for the synthesis period under consideration. Temperature and light limitation are also considered in this model, where deviations from the optimum values cause reductions in specific growth rates as well as nutrient uptake rates (Bierman 1975). As far as algal group differences in growth rates are concerned, green algae have intrinsically higher maximum phosphorus-dependent growth rates than blue-greens (Morton, *et al.* 1971; Payne 1973). Whether or not the specific growth rate of one algal group leads to its dominance over the others, however, depends on a great many factors other than its ability to grow under ideal conditions.

EXPERIMENTAL METHODS

The lake studied in this investigation is Stone Lake, located in the southern part of the State of Michigan (North 41° 45', West 86°00). It is a dimictic, seepage lake with a surface area of approximately 61 hectares (150 acres). The maximum depth is 18.5 meters and the mean depth is 6 meters. Figure 6.2 is a hydrographic map of the lake charted by the University of Notre Dame (Tenney, *et al.* 1970).

Since Stone Lake is located immediately adjacent to the city of Cassopolis (population ∼ 5000), beginning in 1939 it was used as the receiving water body for the wastewater effluent from the village. The lake received secondary effluent for approximately 27 years until, in 1965, a new wastewater treatment facility was built outside the drainage basin. Despite the almost complete curtailment of domestic pollution the lake has failed to show any significant improvement over the past 10 years. The continued hypereutrophic state of the lake is attributable to the 27 years of excessive inorganic nutrient inputs and a long hydraulic residence time (∼ 6 years) which prevents flushing of these nutrients. Also, since the drainage basin of this lake (87 hectares or 215 acres) is relatively small in comparison with the surface area, the allochthonous nutrient inputs are quite small in comparison with the already high levels in the lake.

Based on rainfall data and analysis of total phosphorus concentrations in the runoff, it was estimated that the increase in total phosphorus concentration in the epilimnion over the six month period from mid-April to mid-October was approximately $17 \mu g$ P/l (Bierman 1973). This increase, due to external sources, amounts to less than 3% of the soluble orthophosphate concentration in the lake in January, 1973 (720 μg P/l). It is likely, therefore, that any fluctuations in nutrients in Stone Lake during summer stratification are due to internal cycling.

For this study, Stone Lake was monitored on a biweekly basis during the growing season of 1973. The following physical and chemical parameters were monitored: temperature, dissolved oxygen, pH, suspended solids, soluble orthophosphate, total phosphorus, soluble ammonia, nitrate and organic nitrogen. A quantitative and qualitative analysis of the phytoplankton community was made at the same time. All sampling was

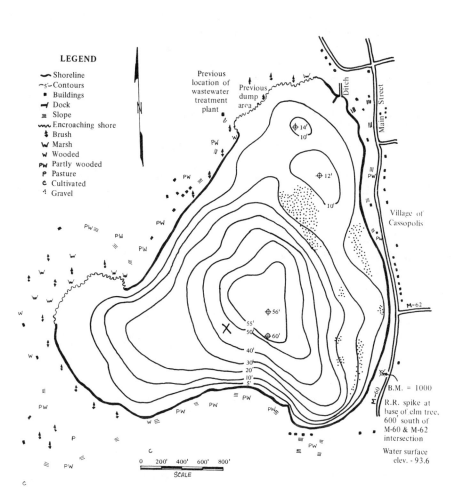

Figure 6.2 Hydrographic map of Stone Lake showing sampling station
at point marked with X.

conducted at a sampling station located roughly in the center of the lake
over the deepest point. The sampling station for this study, as indicated
in Figure 6.2, was chosen to allow a study of the representative phytoplankton in the lake as well as a complete water column profile. A
depth profile of all parameters was obtained by sampling at 1 meter
intervals through the first 4 meters and then at 2 meter intervals from
6 through 14 meters.

Temperature and dissolved oxygen were measured in the field with an oxygen meter and probe. The remaining chemical parameters were determined by procedures described in *Standard Methods for the Examination of Water and Wastewater* (APHA 1971). Orthophosphate was determined by the stannous chloride method.

Phytoplankton samples were taken from the various depths in the lake with a plastic Kemmerer Sampler and preserved in formalin. Quantitative estimates were obtained by passing a suitable portion of the sample through a 1.2 μm membrane filter. The algae collected on the filter were stained with eosin Y and aniline blue (deNoyelles 1968), and the filters were dried and mounted in Permount to clear the filter for microscopic examination (McNabb 1960). Phytoplankton densities were obtained on the basis of the average number of cells (or groups of cells) per microscopic field (100x and 430x magnification) and the volume of water passing through the area of a given field.

Although the dominant phytoplankton found in the lake were keyed to the genus level, for the purpose of comparison of successional patterns with the model predictions the phytoplankton were categorized into three functional groups: (1) green algae, (2) non-N_2-fixing blue-green algae, and (3) N_2-fixing blue-green algae. A fourth group, diatoms, was not observed in significant numbers during the period of investigation (May-October, 1973). Cell counts were converted to dry weight biomass for the model application by applying the following conversion factors: 0.27×10^{-7} mg dry wt/cell (Barber 1968) for greens, 0.25×10^{-7} mg dry wt/cell for non-N_2-fixing blue-greens (AAP 1971), and 0.41×10^{-7} mg dry wt/cell for N_2-fixers (AAP 1971).

RESULTS OF LAKE MONITORING

For the purposes of this study we were interested in observing the nutrient-phytoplankton relationships in the epilimnion of a summer stratified lake. Figure 6.3 is a temperature isopleth for Stone Lake in which the two turnover periods and the thermal stratification period are graphically illustrated. The spring turnover occurred in late March with mixing conditions present until the middle of May. By the end of May a fairly stable thermal stratification had developed, lasting until the fall turnover at the end of October. Based on this temperature contour an average epilimnion depth of 4 meters was chosen for this study. Averages of the orthophosphate and inorganic nitrogen levels and phytoplankton biomass through the epilimnion for the stratification period from the end of May through the end of October are presented for analysis.

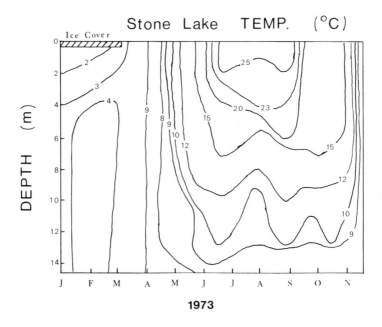

Stone Lake TEMP. ($^{\circ}$C)

Figure 6.3 The temperature isopleth for Stone Lake (1973) indicates a fairly stable thermal stratification from the beginning of June through October.

Phytoplankton Succession

Like many eutrophic lakes, Stone Lake is a highly unstable system, and its phytoplankton assemblage is, therefore, prone to domination by only one or two species at any given time. This tendency is illustrated by noting the composition of the major blooms occurring in the lake. The relative magnitude and timing of the blooms are shown in Figure 6.4, which is a plot of the average phytoplankton biomass (dry weight) in the upper 4 meters of Stone Lake during the summer stratification. The graph illustrates the variation not only of the total crop but also of the three functional groups discussed above. The total crop exhibited two prominent peaks, one in late July and the other in October. The first peak was composed of two somewhat overlapping blooms, first a N_2-fixing bloom of *Aphonizomenon* and *Anabaena* followed closely by a non-nitrogen-fixing bloom consisting primarily of *Microcystis*. Although the green algae group had been represented by a small but diverse population throughout the summer, it was not until late September that this group attained any appreciable biomass. This second major peak in the

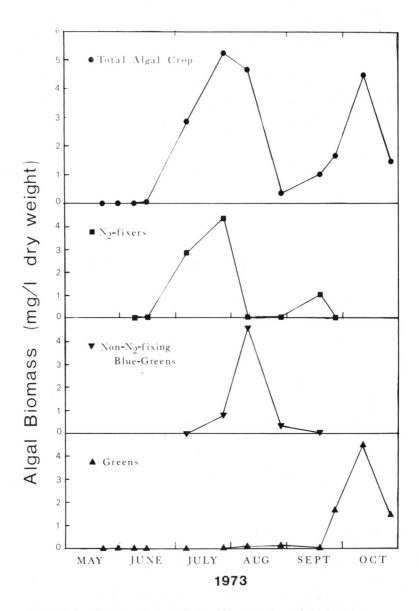

Figure 6.4   The average phytoplankton biomass (dry weight) in the upper
four meters of Stone Lake during the summer stratification illustrates
the unstable nature of the system.

total crop can be attributed mainly to the bloom-forming green algae
Microspora, although other green algae contributed to the production
during this period.

It should be mentioned at this point that the authors are aware that
perhaps two weeks between sampling dates (see Figure 6.4) is probably
too long for a phytoplankton successional study.   There is no question
that more frequent sampling would have provided a better curve and
maximum biomass for the algal blooms;   nevertheless, it can be stated
that phytoplankton concentrations are probably changing at least as fast
as indicated in Figure 6.4.   Also, the rapid increase and decrease of algae
blooms has been evident in suspended solids data throughout the previous
seven years of observation of Stone Lake (Theis, *et al.* 1975).   It is
believed, therefore, that these data are sufficient for the purposes of this
study.

The successional pattern of the three phytoplankton groups is more
easily visualized in Figure 6.5, which is a graphic illustration of the percentage of the total phytoplankton biomass occupied by each group.

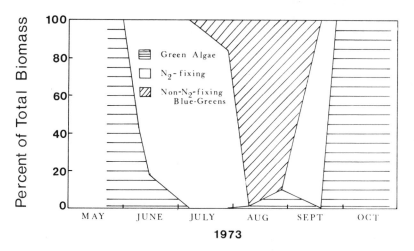

Figure 6.5   The phytoplankton succession in Stone Lake during the summer of 1973
is summarized as a cyclic pattern from non-N_2-fixing algae to N_2-fixing algae
and back again, a cycle that is repeated twice during this time period.

The rapid and almost complete shifts in group dominance can be fully
appreciated with this type of figure.   In the early stages of stratification
a small (less than 1 mg/l dry weight) but relatively diverse group of green
algae were the only phytoplankton present.   By the beginning of July,
however, the N_2-fixing group accounted for virtually 100% of the algal

biomass. The next shift came towards the end of July and favored the non-N_2-fixing blue-green *Microcystis* in combination with some greens. This bloom preceded another N_2-fixing peak (exclusively *Anabaena*) in September. Finally, towards the end of summer stratification, the green algae group regained the majority of the total biomass. The phytoplankton succession in Stone Lake during the summer of 1973 can therefore be summarized as a cyclic pattern from non-N_2-fixing algae (greens and non-N_2-fixing blue-greens) to N_2-fixing blue-greens and back again. This cycle repeated itself twice from May through October.

Nutrient Variations

As indicated earlier Stone Lake has an extremely high level of phosphorus throughout the year (Figure 6.6). At no point in time or depth did the soluble orthophosphate concentration go below 1.0 mg/l as PO_4 (326 μg P/l). This concentration is extremely high even in comparison with other eutrophic lakes. For example, the soluble reactive phosphorus at 5-meter depth in Lake Erie varies from 1 to 11 μg P/l, and the total phosphorus at the same depth ranges from 15 to 50 μg P/l (Great Lakes Water Quality Board 1973). It is therefore highly unlikely that phosphorus is limiting at any time in Stone Lake.

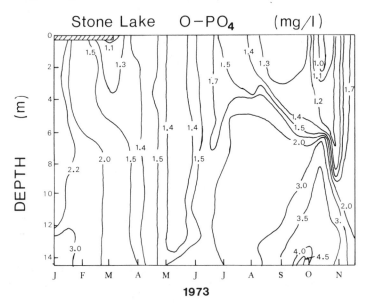

Figure 6.6 The soluble orthophosphate isopleth for Stone Lake (1973) graphically points out the constant high levels of this nutrient during the period of study.

A plot of the average soluble orthophosphate concentration in the epilimnion through the summer stratification period is presented in Figure 6.7. With the exception of a sharp rise in early June and a sharp decline in mid-October, the phosphorus level in the epilimnion remains relatively

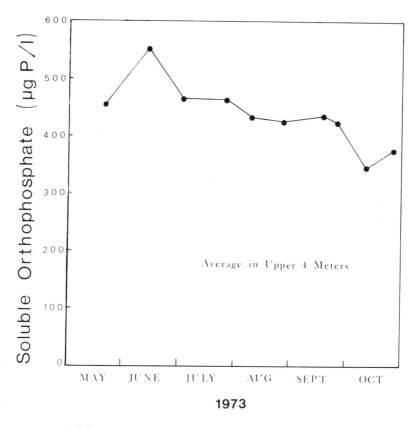

Figure 6.7   A relatively small percentage of the available phosphorus in the epilimnion of Stone Lake was utilized for algal growth.

unchanged. It is worthy of note that the 100 μg P/l increase from May 21 through June 14 coincided with the die-off of a large-rooted macrophyte growth that covered approximately one-third of the lake bottom. The subsequent decrease in phosphorus back to the previous level coincides with the onset of the large nitrogen-fixing cyanophyte bloom in the last half of June. Subsequent algal growth did little to further reduce the phosphate until the large green algae bloom in October apparently caused about an 80 μg P/l decrease in soluble orthophosphate. This apparent additional phosphorus requirement may be explained by the fact that some green algae that tend to develop in hypereutrophic environments have higher minimum intracellular phosphorus requirements than blue-greens (Soeder, *et al.* 1971; Uhlmann 1971).

It is obvious from the above discussion that some physical or chemical parameter besides phosphorus is limiting the size of the algal community in Stone Lake. On the other hand, combined inorganic nitrogen appears to be an important factor in regulating the seasonal succession of the three phytoplankton groups in question. In contrast to phosphorus, the nitrogen levels in the epilimnion of Stone Lake at the beginning of this study are quite low for a eutrophic system (Figures 6.8 and 6.9). A plot of the combined inorganic nitrogen in the epilimnion of the lake reveals that the average nitrogen level was about 100 μg N/l at the onset of stratification (Figure 6.10). These low nitrogen levels coincided with the onset of the nitrogen-fixing blue-green bloom (*Aphanizomenon* and *Anabaena*) in the last two weeks of June. It is also noteworthy that the apparent high rate of nitrogen fixation during late June and early July corresponded to a rather dramatic increase in the soluble inorganic nitrogen in the euphotic zone. Although N_2-fixation itself was never actually measured (*e.g.,* acetylene reduction), heterocysts were observed during this bloom. Also, the model simulations suggest that it is not likely that this rise in the inorganic nitrogen concentration was entirely due to external inputs.

A decrease in nitrogen through August was closely correlated with the *Microcystis* bloom during that month. Again, immediately following the nitrogen dip, a bloom of N_2-fixing *Anabaena* became predominant. Inorganic nitrogen levels rose sharply in the lake through October. Based on the magnitude of this increase (Figure 6.10), it is highly unlikely that this rise in nitrogen levels was due entirely to nitrogen regeneration in the upper waters of the lake. Based on Figures 6.8 and 6.9, it is apparent that the majority of the combined nitrogen increase in the lake was the result of ammonia increase. The gradual build-up of ammonia in the hypolimnion of the lake (probably due to decomposition of falling plankton and deamination of organic nitrogen compounds in the sediments) through the summer (Figure 6.8) and the decreased thermal stability undoubtedly contributed to the rise in surface values through a vertical transport mechanism. Regardless of its source, this large increase in inorganic nitrogen may well have been the stimulus for the substantial green algae bloom in October.

One could characterize the nitrogen-phytoplankton relationship in this particular lake by saying that given the high available phosphorus levels,

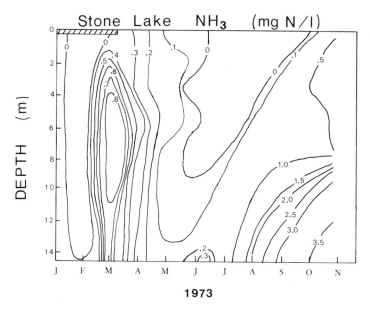

Figure 6.8   Very low NH$_3$-N levels were noted in the surface waters of Stone Lake until the concentrations began to build up in the hypolimnion and the thermal stratification began to break down.

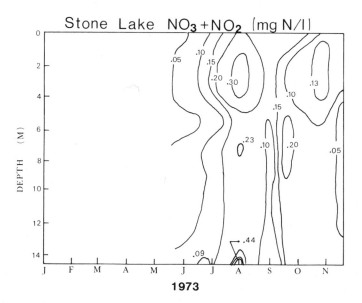

Figure 6.9   Nitrate regeneration occurs through July, during the period of N$_2$-fixing blue-green algae dominance.

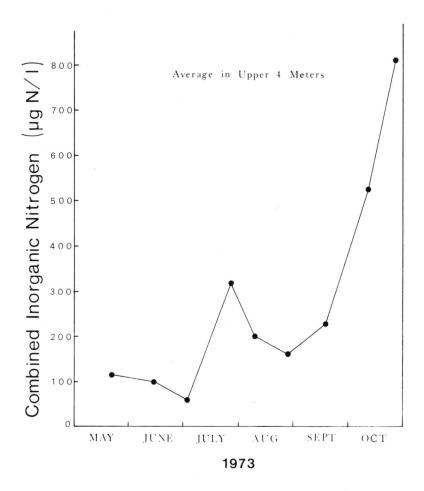

Figure 6.10 At minimum values of combined inorganic nitrogen in the upper four meters of Stone Lake, the N_2-fixing algae were dominant. Their activity apparently recharged the surface waters with available nitrogen.

the system favors a succession towards N_2-fixing blue-green algae when combined inorganic nitrogen concentrations in the euphotic zone drop below about 200 μg N/l. This tendency for nitrogen-deficient conditions to provide a competitive advantage to N_2-fixing algae has been observed by others (Ogawa and Carr 1969; Vanderhoef, *et al.* 1974).

COMPARISON WITH MODEL SIMULATIONS

The application of the algal growth model to the Stone Lake data reveals some very interesting insights into the mechanisms involved in phytoplankton growth and succession. Since phosphorus is present in vast excess it appears that the nitrogen dynamics, in particular, N_2-fixation and nitrogen recycle, play a large role in determining phytoplankton group dominance. Also of significance is the means with which the model handles the death rate (decay rate) of the algal blooms. It will become obvious that some death mechanism other than nutrient or light limitation must be important; from the studies of DePinto (1975) the mechanism appears to be heterotrophic activity. The following model simulations will concentrate on illustrating the importance of the mechanisms of N_2-fixation, nitrogen regeneration, and algal death rates in regulating the Stone Lake phytoplankton periodicity. It should be understood that these simulations are not predictive in nature but are only attempts to describe a given data set and to ask research questions.

The best model simulation of the Stone Lake data was obtained when atmospheric nitrogen was allowed as a source of available nitrogen for the N_2-fixing group (providing an unlimited supply of available nitrogen), nitrogen regeneration from all three algal groups contributed to the available nitrogen pool, and a stringent mechanism for the decline of algal group blooms was applied (Figure 6.11). Of these three mechanisms the most difficult one to include in this model was the decay of the phytoplankton blooms. Since the data of DePinto (1975) indicates that most of the nutrients contained within the algae can be released within 30 days of microbial degradation in the dark, the algal decay mechanism must reflect this finding. To model this phenomenon properly, however, a component of heterotrophic microbes containing various degradation stoichiometries should be included. But the interest of this report was to examine the effect of differences in algal groups, and hence, insufficient information on the heterotrophic activity made it impossible to incorporate this component on anything but an arbitrary basis.

A simple description of the decay process might clarify it. As the algal crop increases, more carbonaceous materials become available for microbiological decay; the microbe population then also increases but with some lag behind the algae. Because of the intrinsically more rapid growth rate of the heterotrophic bacteria, however, their ability to degrade the algae eventually surpasses the rate of algal growth. As a result a rapid decrease in the algal population ensues and is then followed by a decline in heterotrophic activity.

There are several ways in which this phenomenon of microbial decay can arbitrarily be included in the model. Generally we can say that this

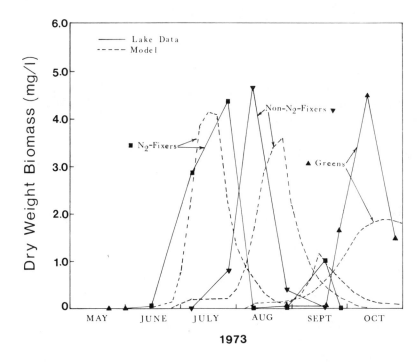

Figure 6.11   The best model simulation was obtained by allowing nitrogen
fixation and regeneration and by employing a second-order decay term for
each algal group based on its biomass and the biomass of the total crop.
In addition the specific growth rate of each group was constrained
to zero as that particular group reached its peak.

decay rate not only depends upon the algal concentration at a given instant but also upon the previous 30 days. One way of including this
factor is to employ, as a variable, the average algal concentration over
some previous period or possibly the integral of the concentration minus
some steady-state value. The easiest way, however, is to adjust the growth
and decay rates as algal concentrations get high. This is essentially the
procedure used here. For each algal group the death mechanism includes
a second order algal decay term that is proportional to the product of
the dry weight biomass of the group in question and the biomass of the
entire phytoplankton assemblage. This term represents an algal death
(or lysing) rate, which is at its largest during the peak of any bloom
but which also depends on the abundance of other algal groups, representing a competitive inhibition. This decay term takes the form

$$r_i = -k \cdot T \cdot A_i \cdot A_t$$

where

k = decay constant $[\text{day} \cdot {}^\circ C \cdot \text{mg/l}]^{-1}$

T = water temperature $[{}^\circ C]$

r_i = change in biomass of ith algal group due to lysing $[\text{mg/l} \cdot \text{day}]$

A_i = biomass of ith algal group $[\text{mg/l}]$

A_t = biomass of total algal crop $[\text{mg/l}]$

It can be used to adjust the biomass obtained from the growth rate term over any given time period. The value for k used in this model application is 0.0015 $(\text{day} \cdot {}^\circ C \cdot \text{mg/l})^{-1}$, which is consistent with the laboratory experiments on aerobic decomposition of algae by Jewell and McCarty (1971) and DePinto (1975).

This decay term is not sufficient to explain the rapid declines of the algal blooms observed in Stone Lake and, as might be expected, from high heterotrophic activity. Thus, an additional constraint was placed on the individual algal groups during their domination. The specific growth rate, which is the only positive term in the phytoplankter biomass-determining equation, was arbitrarily set equal to zero when the experimental data indicated that an individual algal group was on the decline. Likewise, when the experimental data indicated that a particular algal group had reached its minimum, the zero growth rate constraint was removed. While a zero growth rate is unlikely, this constraint is about the simplest method of simulating the effect of microbial degradation of algae.

A close inspection of Figure 6.11 reveals that the model incorporating the above elements compares quite favorably with the lake data. Although the exact timing and maximum biomass of each bloom is not reproduced, the important thing to notice is that the relative succession of groups is simulated and their timing and biomass are qualitatively represented. The only drastic difference between the model and the lake data appears to be in the rather low model prediction for the green algal bloom in October. This variation can be easily explained by realizing that the model as applied here considers only the epilimnion of the lake with no external inputs and, therefore, does not contain a mechanism for the addition of nitrogen to the available pool other than recycle from phytoplankton decomposition and zooplankton excretion. As a result, the green bloom, which the model predicts, becomes nitrogen-limited while in reality the October combined inorganic nitrogen levels in upper waters are quite high due to probable input of ammonia from the hypolimnion. This discrepancy is graphically illustrated in Figure 6.12. Note that the model

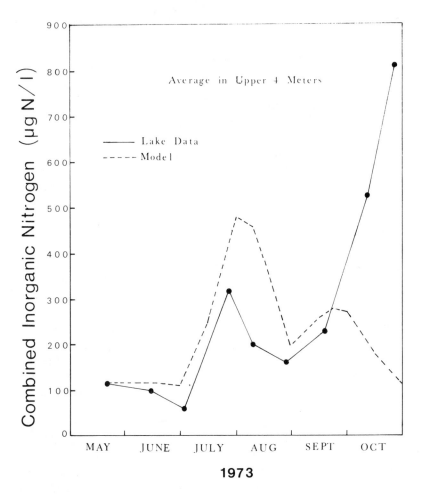

Figure 6.12 Much of the large increase in combined inorganic nitrogen in the surface waters during October is due to factors for which the model cannot account.

predicts the large nitrogen regeneration from the July N_2-fixing bloom. According to the model, however, the green algae rapidly reduce the a available nitrogen level to the point where it becomes limiting.

In order to best see the effect of the nitrogen dynamics and algal death mechanism, it is appropriate to examine model simulations with each of the factors discussed above individually "turned off." Figure 6.13 demonstrates the importance of N_2-fixation in this particular system. It represents the model output of the phytoplankton growth curves when

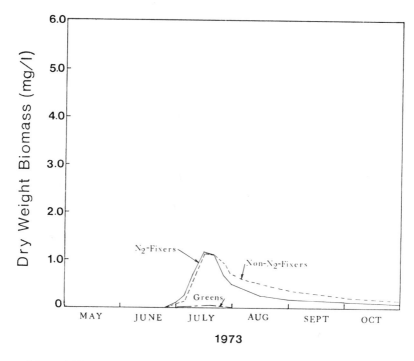

Figure 6.13 The model simulation with N_2-fixation turned off graphically illustrates the importance of this nutrient source in Stone Lake.

the only change in the model is that the N_2-fixing group is *not* allowed to have an unlimited nitrogen source via the atmosphere. From this graph it is obvious that it is nitrogen fixation and the subsequent recycle of this fixed nitrogen that allows the development of the large blooms observed in Stone Lake. Although not shown here, the model predicts that the combined inorganic nitrogen level in the epilimnion of the lake drops to essentially zero by the middle of July. Thus the system becomes severely nitrogen-limited without N_2-fixation as a source.

To see the importance of the regeneration of the fixed nitrogen in this system, the model was run allowing N_2-fixation but with all nitrogen recycle turned off. The only remaining potential source of nitrogen to the available pool was the possible leakage of intracellular nitrogen back to solution under conditions when the external medium concentrations became quite low and the intracellular concentrations of a particular group (most likely the N_2-fixing blue-greens) greatly exceeded the minimum stoichiometric amounts. The model simulation of the algal system under these circumstances is shown in Figure 6.14. Note that the absence of nitrogen recycle has little effect on the N_2-fixing blue-greens. Comparison of Figure 6.14 with Figure 6.11, however, reveals that the key effect of excluding nitrogen recycle is to induce nitrogen limitation on the two algal groups that require NH_3-N or NO_3-N as their nitrogen source. There is a small *Microcystis* bloom in August that develops from the small amount of available nitrogen present in the system at the start of the season. This non-N_2-fixing blue-green peak, however, is only one-third the size of the actual *Microcystis* bloom that occurred in the lake. It nevertheless uses practically the whole available nitrogen pool, thus preventing the development of a green algae group. Some nitrogen does leak back into the medium (~ 50 μg N/l) during the second N_2-fixing bloom, but the amount is too small to aid in the development of a green bloom.

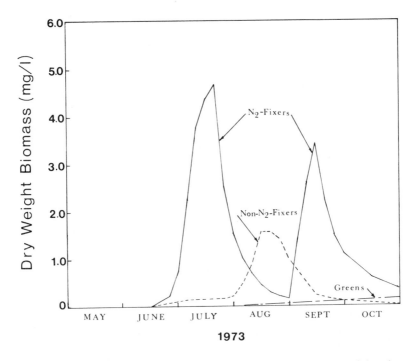

Figure 6.14 Without nitrogen recycle the model predicts that non-N_2-fixing algae never attain significant levels, while the nitrogen recycle has little effect on the algae capable of fixing nitrogen from the atmosphere.

The remainder of the model simulations are intended to demonstrate the importance of a death mechanism in the timing of phytoplankton succession. With this intent the model was run without the specific growth rate constraint previously employed but with the second order decay term. The results of this simulation are shown in Figure 6.15.

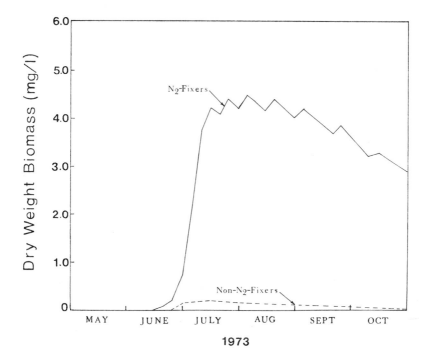

Figure 6.15   This model simulation illustrates that a second order algal decay term alone is not enough to simulate algal bloom declines and the relative timing of succession.

In this run the rise of the N_2-fixing bloom is as before; however, the decay term alone is not enough to induce the rapid decline of this bloom. As a result the N_2-fixers dominate the assemblage throughout the growing season, even though sufficient nitrogen has been introduced into the medium by recycle to stimulate blooms in the other two groups. No such blooms occur, however, primarily because the decay term, the magnitude of which is a function of the total crop, is too large to allow their development. Although interspecies competition and inhibition certainly may exist, they rarely operate to the extent that once a species is dominant it remains so to the exclusion of all others.

Since the above simulation proved to be unrealistic, it was thought that perhaps the decay term should be group specific rather than depending also on the total crop. Therefore, a simulation was made with unconstrained specific growth rates and second order algal decay term proportional to the square of the given algal group biomass to which it is applied ($r_i = - k \cdot T \cdot A_i^2$). In this case (Figure 6.16) all three groups develop a bloom, but the individual decay terms alone are insufficient to simulate the biomass declines. The blooms gradually wane, primarily because of light limitation, but without a specific growth rate constraint these decay mechanisms are not large enough to counteract the phytoplankton growth term.

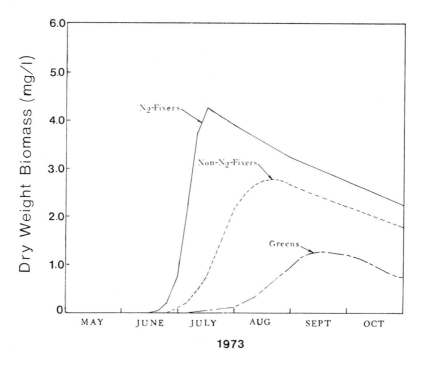

Figure 6.16   An algal group-specific second-order decay term without the growth rate constraint also failed to simulate phytoplankton periodicity.

A better comparison between the two decay terms used is afforded by comparing Figure 6.17 (group-specific decay term) with Figure 6.11. The run that produced Figure 6.17 is identical with the run for Figure 6.11 except that the second-order decay term is a function of the square of the individual group biomass. This simulation is quite reasonable in

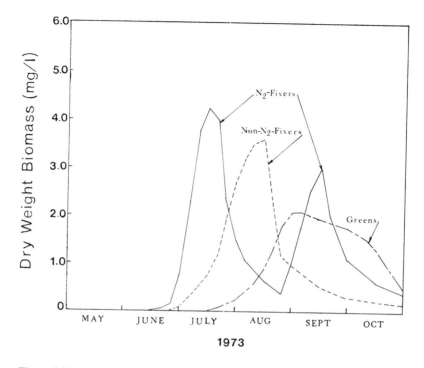

Figure 6.17   This simulation is identical with that shown in Figure 6.11 except that the decay term is group-specific. Note that the green algae develop much sooner and the second N_2-fixer bloom attains roughly three times the biomass.

that it more accurately predicts the development of the first two blooms; however, it also allows the green algae to develop much sooner than in the first run because their decay term is not affected by the total crop at that time. While one might say that Figure 6.11 is a slightly better representation of the lake data, in reality a mixture of the two conditions would probably be best.

The comparison of Figures 6.11 and 6.17 points out a fundamental question that can only be answered by further experimental research: Why can one algal group apparently grow exponentially in the presence of a bacteria community that is destroying another algal group? A definitive answer to this question is not within the scope of this study; however, some possibilities include: different bacterial species degrade different algal species, the bacterial community has become adapted to the original bloom and a lag time is required before the bacteria can affect the new algal group, or other environmental conditions favor the growth of the new algal group to the extent that bacterial decay is a small perturbation on their large growth rate.

DISCUSSION AND CONCLUSIONS

The application of the two-step algal growth model employed in this study brought to light three basic observations about the seasonal phytoplankton succession in Stone Lake (an example of a hypereutrophic lake). The first obvious conclusion, based solely on the data, is that the constant excess of available phosphorus allows all phytoplankton species to pursue their control strategies regardless of any phosphorus stoichiometric differences. It should also be pointed out that the same situation very likely exists for carbon since the average alkalinity of the lake is 120-140 mg/l as $CaCO_3$ and the pH is typically 8 to 9 in the surface waters.

The second observation is that nitrogen fixation and its subsequent recycle is probably the primary reason for the large blooms occurring in Stone Lake. The relatively rapid regeneration of fixed nitrogen appears to play an important role in aquatic systems with low flushing rates. On the other hand, Verhoff, *et al.* (1975) have shown that nutrient recycle does not significantly affect systems with very short retention times (fast moving rivers, for example). Stone Lake, however, has a relatively long hydraulic retention time. In this case the Stone Lake example showed first that the activity of nitrogen-fixing blue-green algae can recharge the upper waters of a lake system with sufficient combined inorganic nitrogen to promote the resurgence of non-N_2-fixer blooms. It also confirmed that the model's output was consistent with the hypothesis that a substantial portion of the nutrients taken up by the plankton in the epilimnion of a stratified lake can be returned to the soluble inorganic state for reuse during the same growing season.

Finally, it should be stressed that our first attempts to describe the Stone Lake data with a relatively sophisticated multiclass model based strictly on phosphorus and nitrogen dynamics [with temperature, light, and zooplankton grazing factors (Biermann 1975)] failed to completely explain the successional pattern. It was obvious that something, unaccounted for by the model, was occurring in the lake and that this interaction was instrumental in the rapid decline of the algal blooms. Controlled laboratory investigations (DePinto 1975) indicated that the magnitude of the heterotrophic bacterial population could be a significant factor in causing the crash of algal blooms and the subsequent nutrient regeneration. In this study it was noted that 53-95% decomposition (as measured by percent reduction in particulate COD) occurred in bacterial-inoculated *Chlorella vulgaris* cultures in less than a two-month, dark incubation period, about one-third of which was a bacterial lag time. The fact that bacteria-free *Chlorella* cultures remained viable in the dark for over 70 days substantiated the contention that the bacteria were the causative agents. Further evidence in support of this hypothesis has recently been reported by Gunnison and Alexander (1975), particularly for cyanophytes. The second-order decay terms and the growth rate constraint were built into the model in an attempt to point out the need for an algal death mechanism. Using the model as a research tool by running comparative simulations without the various death mechanisms emphasized this need.

This study has by no means eliminated alternate possibilities for the decline of algal blooms in Stone Lake, such as trace nutrient limitations, fungi, algal parasites, and toxic materials. It merely points out that the highly unstable, hypereutrophic lakes might better be modeled if the heterotrophic bacterial concentrations in these systems were also modeled. Instead of the two second-order mechanisms used in this study, the decomposition of phytoplankton groups might then be expressed by a function that considers bacteria-phytoplankton interactions explicitly.

REFERENCES

Algal Assay Procedure—Bottle Test. National Eutrophication Research Program, Environmental Protection Agency, Corvallis, Oregon (1971).

Azad, H. S. and J. A. Borchardt. "Variation in Phosphorus Uptake by Algae," *Environ. Sci. Technol.* **4**, 737 (1970).

Barber, J. "The Influx of Potassium into *Chlorella Pyrenoidosa*," *Biochim. Biophys. Acta* **163**, 141 (1968).

Bierman, V. J., Jr. Internal Report, University of Notre Dame (1973).

Bierman, V. J., Jr., F. H. Verhoff, T. L. Poulsen, and M. W. Tenney. "Multi-Nutrient Dynamic Models of Algal Growth and Species Competition in Eutrophic Lakes," *Modeling the Eutrophication Process*, Proceeding of a Workshop, September 5-7, 1973, sponsored by Utah Water Research Laboratory and Environmental Protection Agency, Logan, Utah (1973a), pp. 89-109.

Bierman, V. J., Jr. "Dynamic Mathematical Model of Algae Growth in Eutrophic Freshwater Lakes," Ph.D. thesis, University of Notre Dame, Notre Dame, Indiana (1973b).

Bierman, V. J., Jr. "Mathematical Model of the Selective Enhancement of Blue-Green Algae by Nutrient Enrichment," paper presented before the Division of Environmental Chemistry, American Chemical Society, Philadelphia, Pa., April 6-11, 1975.

Bush, R. M. and E. B. Welch. "Plankton Associations and Related Factors in a Hypereutrophic Lake," *Water Air Soil Pollution* **1**, 257 (1972).

Caperon, J. and J. Meyer. "Nitrogen-Limited Growth of Marine Phytoplankton. I. Changes in Population Characteristics with Steady-State Growth Rate," *Deep Sea Res.* **19**, 601 (1972).

Caperon, J. and J. Meyer. "Nitrogen-Limited Growth of Marine Phytoplankton. II. Uptake Kinetics and Their Role in Nutrient-Limited Growth of Phytoplankton," *Deep Sea Res.* **19**, 619 (1972a).

deNoyelles, F., Jr. "A Stained Organism Filter Technique for Concentration of Phytoplankton," *Limnol. Oceanogr.* **13**, 562 (1968).

DePinto, J. V. "Studies on Phosphorus and Nitrogen Regeneration: The Effect of Aerobic Bacteria on Phytoplankton Decomposition and Succession in Freshwater Lakes," Ph.D. thesis, University of Notre Dame, Notre Dame, Indiana (1975).

Droop, M. R. "Some Thoughts on Nutrient Limitation in Algae," *J. Phycol.* **9**, 264 (1973).

Eppley, R. W. and W. H. Thomas. "Comparison of Half-Saturation Constants for Growth and Nitrate Uptake of Marine Phytoplankton," *J. Phycol.* **5**, 375 (1969).

Fitzgerald, G. P. "Detection of Limiting or Surplus Nitrogen in Algae and Aquatic Weeds," *J. Phycol.* **4**, 121 (1968).

Fitzgerald, G. P. "Some Factors in the Competition or Antagonism Among Bacteria, Algae, and Aquatic Weeds," *J. Phycol.* **5**, 351 (1969).

Fuhs, G. W. "Phosphorus Content and Rate of Growth in the Diatoms *Cyclotella nana* and *Thalassiosira fluviatilis*," *J. Phycol.* **5**, 312 (1969).

Fuhs, G. W., S. D. Demmerle, E. Canelli, and M. Chen. "Characterization of Phosphorus-Limited Planktonic Algae," in *Nutrients and Eutrophication: The Limiting Nutrient Controversy*, Proceedings of a Symposium, American Society of Limnology and Oceanography (1971), pp. 113-132.

Great Lakes Water Quality Board. *Great Lakes Water Quality*, Annual Report to the International Joint Commission (April 1973).

Gunnison, D. and M. Alexander. "Resistance and Susceptibility of Algae to Decomposition by Natural Microbial Communities," *Limnol. Oceanogr.* **20**(1), 64 (1975).

Horne, A. J. and G. E. Fogg. "Nitrogen Fixation in Some English Lakes," *Proc. Royal Society*, London, Series B **175**, 351 (1970).

Hutchinson, G. E. "Limnological Studies in Connecticut. VII. A Critical Examination of the Supposed Relationship Between Phytoplankton Periodicity and Chemical Changes in Lake Waters," *Ecology* **25**, 3 (1944).

Hutchinson, G. E. *A Treatise on Limnology*, Vol. 2 (New York: John Wiley and Sons, Inc., 1967).

Jewell, W. J. and P. L. McCarty. "Aerobic Decomposition of Algae," *Environ. Sci. Technol.* **5**(10), 1023 (1971).

McNabb, A. D. "Enumeration of Freshwater Phytoplankton Concentrated on the Membrane Filter," *Limnol. Oceanogr.* **5**, 57 (1960).

Morton, S. D., P. H. Derse, and R. C. Sernan. "The Carbon Dioxide System and Eutrophication," Office of Research and Monitoring, Environmental Protection Agency (1971).

Ogawa, R. E. and J. F. Carr. "The Influence of Nitrogen on Heterocyst Production in Blue-Green Algae," *Limnol. Oceanogr.* **14**, 342 (1969).

Payne, A. G. "Responses of the Three Test Algae of the Algae Assay Procedure: Bottle Test," paper presented at the Thirty-Sixth Annual Meeting of the American Society of Limnology and Oceanography, Salt Lake City, Utah, June 12, 1973.

Peasall, W. H. "Phytoplankton in the English Lakes. II. The Composition of the Phytoplankton in Relation to Dissolved Substances," *J. Ecol.* **20**, 241 (1932).

Shapiro, J. "Blue-Green Algae: Why They Become Dominant," *Science* **179**, 382 (1973).

Soeder, D. J., H. Muller, H. D. Payer, and N. Schulle. "Mineral Nutrition of Planktonic Algae: Some Considerations, Some Experiments," *International Association of Theoretical and Applied Limnology* **19**, 39 (1971).

Standard Methods for the Examination of Water and Wastewater, 13th ed. (Washington, D. C.: American Public Health Association, 1971).

Stoermer, E. F. "Analysis of Phytoplankton Composition and Abundance During IFYGL," First Annual Reports of the EPA IFYGL Projects, EPA 660/3-73-021 (December, 1973), p. 90.

Tenney, M. W., W. F. Echelberger, and T. C. Griffing. "Effects of Domestic Pollution Abatement on an Eutrophic Lake," Partial report on FWQA grant WPD-126, Department of Civil Eng., University of Notre Dame (1970).

Theis, T. and J. V. DePinto. "Studies on the Reclamation of Stone Lake," Final Report on Grant No. R-801245 from the Environmental Protection Agency, Pacific Northwest Environmental Research Laboratory, Corvallis, Oregon (in press).

Uhlmann, D. "Influence of Dilution, Sinking, and Grazing Rate on Phytoplankton Populations of Hyperfertilized Ponds and Micro-Ecosystems," *Internat. Assoc. Theoret. Appl. Limnol.* **19**, 100 (1971).

Vanderhoef, L. N., C. Huang, R. Musil, and J. Williams. "Nitrogen Fixation (Acetylene Reduction) by Phytoplankton in Green Bay, Lake Michigan, in Relation to Nutrient Concentration," *Limnol. Oceanogr.* **19**(1), 119 (1974).

Verhoff, F. H., K. R. Sundareson, and M. W. Tenney. "A Mechanism of Microbial Growth," *Biotechnol. Bioengin.* **14**, 411 (1972).

Verhoff, F. H., J. B. Carberry, V. J. Bierman, Jr., and M. W. Tenney. "Mass Transport of Metabolites, Especially Phosphorus in Cells," *Amer. Inst. Chem. Engin. Symp. Series 129*, **69**, 227 (1973).

Verhoff, F. H., M. Sandoval, T. H. Cahill. "Mathematical Modeling of Nutrient Cycling in Rivers," Paper presented before the Division of Environmental Chemistry, American Chemical Society, Philadelphia, Pa., April 6-11, 1975.

7

BOISE RIVER ECOLOGICAL MODELING

Carl W. Chen and John T. Wells, Jr.[1]

INTRODUCTION

The diversity of water quality problems occurring on the Boise River
Idaho, mimics those of major river systems in the United States. Upstream
the river is dammed for storage and flow regulation. A large quantity of
water is diverted from the river for irrigation. The reduced flow is left
for degradation by industrial and municipal waste discharges and surface
and subsurface irrigation return flows.

A cooperative study of federal, state and local agencies was initiated
to investigate wastewater management alternatives that would protect and
enhance the water quality of the river. An intensive water sampling program was conducted in August 1973 at selected stations along the Boise
River and its principle tributary, Indian Creek. The primary purpose of
the data collection program was to provide a data base for the calibration
of a water quality ecological model for the Boise River.

The model contemplated would compute not only BOD-DO for waste
load allocation and coliform bacteria for health considerations, but also
nutrients and toxicity and their effects on the higher trophic animals for
ecological evaluations. The model developed provides simulation of 24
water quality and biological parameters including temperature, toxicity,
total suspended solids, coliform bacteria, BOD, dissolved oxygen, NH_3,

[1] Tetra Tech, Inc., Lafayette, California.

NO_2, NO_3, PO_4, alkalinity, pH, floating algae (two types), benthic algae (two types), zooplankton, insects, detritus, organic sediment, benthos and fish (cold water game fish, warm water game fish, and benthic feeders).

This chapter describes the development and calibration of such a model for the Boise River system. The model, which was calibrated with data of the intensive sampling program, served as a framework for the detailed analysis of the ecosystem responses to pollution from both point and nonpoint sources (Chen and Wells 1975). The impacts of reduced river flows and improved irrigation return water were also assessed. The insight gained on the ecosystem behavior has proved invaluable to the planning and formulation of different management alternatives.

THE BOISE RIVER SYSTEM

The Boise River is a 64-mile stretch of stream passing through the urban areas of Boise and Caldwell, Idaho. Originating from the Corps' Lucky Peak Reservoir, located approximately 12 miles upstream of Boise, it flows northwest through Ada and Canyon counties to join the main stem of the Snake River near Parma, Idaho (see Figure 7.1). At Eagle, the river branches into the north and south forks around the island.

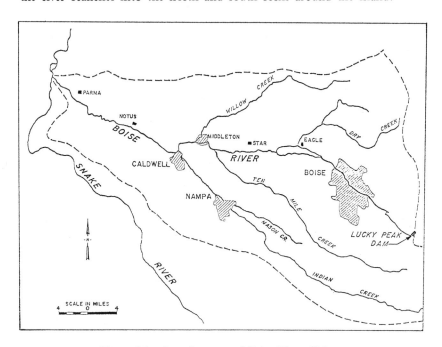

Figure 7.1   Location map of Boise River, Idaho.

Three major dams have been constructed on the headwaters of the Boise River above the study area: Lucky Peak of the Corps and Arrowrock and Anderson Ranch of the U.S. Bureau of Reclamation.

Numerous canals divert irrigation water from the river to the surrounding farm lands, the largest of which is the New York Canal located 2.5 miles below Lucky Peak Dam. During average or below average flow years, the irrigation diversions take nearly all of the Boise River flow. Many drains return agricultural water to the Boise River below Boise. One of the major drains is Indian Creek, which flows in a northwesterly direction, passing through Nampa and Caldwell before joining the Boise River.

The economy of the Boise River area is largely agricultural. The major cities in the study area are Boise (population 90,000), Caldwell (population 14,000), and Nampa (population 21,000). Boise is the capital city of Idaho and the economic, cultural, and governmental center for the region. Nampa and Caldwell are the largest cities in Canyon County, where agriculture and food processing are the principal industries. Other significant population centers include Garden City, Eagle, Star, Middleton, Notus, and Parma. All these cities are situated along the Boise River and Indian Creek.

Hydrology

Table 7.1 summarizes the hydrology of the Boise River for the water year 1972, a year with slightly above average flow. The data show that the Boise River experiences three typical flow regimes throughout an annual cycle:

1. *Low flow in the upper reach and high flow in lower reach.* During the nonirrigation and wet season (from November to February) the flow is low at the headwater, increasing gradually downstream due to storm runoff and ground water accretion.

2. *High flow in upper reach and low flow in lower reach.* April is a transition period from nonirrigation season to the irrigation season. The flow is high below Lucky Peak Dam, decreasing downstream due to diversions and low irrigation returns (water is not drained at the early application of irrigation water).

3. *Gradual change from upstream high to middle section low and then downstream high.* During the irrigation season (April to October), the high water release from Lucky Peak Dam is diverted gradually from the upper reach (above Boise) for irrigation, resulting in a low flow in the middle section (Boise to Star). Downstream of Star, the flow increases again by accretion of irrigation return water.

Table 7.1   Boise River Hydrology for 1972 Water Year, all Flows in CFS

Location	Oct.	Nov.	Dec.	Jan.	Feb.	Mar.	Apr.	May	June	July	Aug.	Sept.
Boise River near Boise	1189	67	108	2105	5634	6338	8367	9436	6965	4866	4415	2955
Boise project—main canal	850	—	—	—	—	—	1362	2698	2598	2753	2568	1721
Penitentiary Canal	3	—	—	—	—	—	2	11	9	10	8	2
Boise River below Div. Dam	336	67	108	2105	5634	6338	7003	6727	4359	2103	1839	1232
Unidentified loss	46	5	15	43	239	132	202	160	143	70	91	30
Boise River at Barber Dam	381	72	122	2063	5396	6206	6801	6567	4215	2033	1748	1202
Ridenbaugh Canal	73	—	—	—	—	—	217	564	545	557	539	446
Boise River below Ridenbaugh	308	72	122	2063	5396	6206	658	6004	3670	1476	1209	756
Misc. diversions to Boise	16	—	—	—	—	—	20	60	66	65	64	43
Unidentified loss or gain	19	5	16	-42	-240	-134	-204	-161	-143	-70	-93	-30
Boise River at Boise	337	77	139	2020	5155	6072	6360	5782	3461	1341	1053	682
Settlers Canal	29	—	—	—	—	—	30	197	165	192	1187	91
Drainage Dist. #3 Mill	11	3	2	—	—	—	9	11	10	13	13	14
Thurman Canal	8	—	—	—	—	—	—	39	37	38	33	19
Farmers Union Canal	29	—	—	—	—	—	61	238	232	249	236	173
Boise Sewer	16	17	16	16	16	15	16	16	17	18	20	20
Unidentified gain	20	42	47	70	169	156	121	95	96	78	90	74

Waste Inputs

Identifiable point sources that contribute to the pollution of the Boise River include five major municipalities—Boise, Caldwell, Garden City, Meridian, and Nampa—and four food processing plants—Armour Meat Packing Company (Nampa), J. R. Simplot Co. (Caldwell), Triangle Dairy (Boise) and Boise Valley Packing Plant (Boise). Only limited data were available to characterize the extent of point source pollution of the Boise River. During the October 18 to November 8, 1971 survey, the U.S. Environmental Protection Agency measured some of the wastewater qualities of the principal point sources (EPA 1973). In the comprehensive sampling program of August 1972, the Corps, in cooperation with the Idaho Department of Environmental and Community Services, provided another set of wastewater quality data (Cooperative Boise Survey 1972). Table 7.2 summarizes the wastewater characteristics as reported by these two sources.

Nonpoint sources waste inputs can be derived from irrigation return waters, ground water seepage, and sludge bank deposits. The wide use of septic tank systems in the area may also contribute to pollution of the ground water that may percolate into the river.

During the Cooperative Boise River Survey in August 1972, the water quality characteristics of surface drains tributary to the Boise River and Indian Creek were measured. Table 7.3 summarizes the results of the survey, which was conducted during a high irrigation period.

The 1971 EPA survey indicated that there were considerable organic deposits downstream of Nampa on Indian Creek and downstream of Caldwell on the Boise River. The organic deposits cause the formation of a sludge bank that serves as an oxygen sink and as a nutrient source to the river water. For the organic deposits on the river bank, the value[1] of 200 to 300 g/m^2 was therefore assumed in the lower reach of Indian Creek where the EPA survey indicated the existence of a sludge blanket. For other areas that are relatively clean, a value of 50 g/m^2 was estimated.

MODEL DEVELOPMENT

The fundamental concepts of ecological modeling have been presented previously by Chen (1970). A detailed description of modeling approaches, mathematical formulation and solution techniques can be found elsewhere (Chen and Orlob 1972). Basically, the river is divided into a series of

---

[1]Oxygen uptake rate for 200 g/m^2 = 200 g/m^2 x 0.01 $\frac{1}{day}$ x $\frac{1}{24}$ $\frac{day}{hr}$

x 2.1 $\frac{gm\ oxygen}{gm\ organic}$ = 0.18 $g/m^2/hr.$

Table 7.2 Point Source Waste Discharges to Boise River, Idaho[a]

Point Source	Date	Discharge cfs	Temp. °C	Coliform MPN/100	BOD mg/l	PO_4P mg/l	ALKA mg/l	pH	NH_3N mg/l	NO_3N mg/l	DO mg/l
Boise STP	8/15/72	19.8	23.0	8.0×10^4	50	6.52	196	7.0	2.2	2.0	2.4
	10/22-25/71	17.5	—	—	27	6.3	—	—	13.0	0.08	—
Garden City STP	8/15/72	7.0	23.0	4.4×10^4	48	2.2	120	6.8	4.0	1.6	4.7
	10/22-25/71	0.7	—	1.0×10^4	37	1.2	—	—	1.7	0.72	—
Boise Valley Packing effluent	8/15/72	—	—	1.8×10^5	98	3.1	152	6.8	23.0	1.2	12.2
Eagle Lagoon effluent	8/15/72	—	22.5	6.5×10^4	17	2.36	216	7.2	4.5	1.3	12.4
Meridian STP	10/22-25/71	0.73	—	480	61	4.3	—	—	7.7	0.45	5.2
	8/15/72	3.4	20.0	5.0×10^4	16	2.93	232	7.1	7.0	3.4	5.2
Star Lagoon effluent	8/15/72	—	21.0	200	18	1.63	124	7.8	1.2	2.7	0.05
Middleton Lagoon effluent	8/15/72	—	24.0	100	12	1.48	176	7.0	3.0	2.2	7.6
Armour Meat Packing effluent	8/15/72	—	23.0	2.6×10^6	158	8.07	544	7.5	45.5	2.5	2.6
Waste Treatment Lagoon effluent	10/18-21/71	0.42	—	1.7×10^4	35	7.4	—	—	13.0	0.08	—
Cooling water effluent	10/18-21/71	0.78	—	—	3.2	0.05	—	—	—	2.84	—
Nampa STP	10/21-25/71	23.4	—	1.2×10^6	141	3.1	—	—	17.0	0.33	—
	8/15/72	—	21.0	80	13	3.02	248	7.5	4.5	23.5	5.2
Caldwell STP	10/18-21/71	6.8	20.0	1.8×10^4	57	3.7	—	—	7.7	0.79	—
	8/14-15/72	13.3	20.0	5000	25	1.71	324	7.3	5.2	4.3	7.2
J. R. Simplot Lagoon effluent	11/4-8/71	6.7	26.0	1.5×10^6	820	10.1	444	6.8	18.0	0.11	0.04
	8/15/72	—	—	8.0×10^7	378	6.36	—	—	21.2	14.0	—
Notus Lagoon effluent	8/15/72	—	22.5	600	36.0	5.79	220	7.3	2.0	8.0	0.4
Parma Lagoon effluent	8/15/72	—	22.0	1540	42.0	6.44	344	7.4	12.8	3.5	0.8
Triangle Dairy	11/23-25/71	0.09	—	3.3×10^6	60	4.8	—	—	0.6	0.13	—

[a]From EPA Report "Water Quality Investigations of Snake River and Principal Tributaries," (February 1973), and Cooperative Boise River Survey, conducted by the Idaho Department of Community Services (August 1972).

Table 7.3 Water Quality Characteristics of Drains in the Study Area[a]

Nonpoint Source	Date	Discharge cfs	Temp °C	TSS mg/l	Coliform MPN/100	BOD mg/l	DO mg/l	PO_4P mg/l	ALKA mg/l	pH	NH_3N mg/l	NO_3N mg/l
Spoil Banks Drain at Eagle, Idaho	8/15	58.0	16.5	26	9000	1.3	5.4	0.21	159	7.4	0.07	0.7
	8/16	58.0	16.0	74	9000	1.1	6.7	0.18	144	7.4	0.02	0.7
Dry Creek at Eagle, Idaho	8/15	3.4	18.5	19	16000	3.6	5.8	0.28	134	7.5	0.15	0.3
	8/16	3.4	18.5	159	11000	0.6	6.8	0.22	101	7.6	0.08	0.3
Thurman Mill Drain near Eagle, Idaho	8/15	37	16.0	5	14000	1.1	4.4	0.22	261	7.7	0.05	1.5
	8/16	37	16.5	12	13000	1.1	9.5	0.17	245	7.7	0.09	1.4
Fifteen Mile Creek at mouth near Middleton	8/15	149	19.0	60	38000	1.7	8.1	0.31	134	7.4	0.08	1.1
	8/16	143	19.0	52	24000	1.7	8.1	0.31	78	7.7	0.04	1.7
Mill Slough at mouth near Middleton	8/15	243	21.0	39	4500	2.2	8.0	0.23	145	7.6	0.08	1.1
	8/16	243	16.0	41	14000	1.2	6.8	0.24	144	7.4	0.10	1.1
Willow Creek at mouth near Middleton	8/15	19	22.5	90	69000	2.6	6.9	0.39	135	7.8	0.14	0.8
	8/16	19	22.5	72	61000	2.4	7.1	0.39	123	7.9	0.14	0.8
Mason Slough at mouth at Caldwell	8/15	32	20.0	63	17000	1.0	6.9	0.34	234	7.8	0.09	1.5
	8/16	24	21.5	34	1000	1.5	6.1	0.32	168	8.2	0.07	1.3
Hartley Drain near Caldwell	8/15	98	22.5	236	17000	1.9	7.5	0.36	198	7.8	0.06	1.3
	8/16	98	17.0	74	12000	1.7	6.7	0.34	198	7.4	0.06	1.3
Mason Creek at mouth at Caldwell	8/15	188	19.0	133	13000	1.9	7.5	0.30	190	7.6	0.08	2.2
	8/16	173	20.0	100	20000	1.3	7.0	–	181	8.2	0.03	2.3
Conway Gulch at mouth near Notus	8/15	44	20.5	140	29000	1.0	7.3	0.36	296	8.1	0.08	2.4
	8/16	44	15.5	116	8500	1.5	8.5	0.37	278	7.9	0.09	2.2
Dixie Slough at mouth near Parma	8/15	225	22.5	54	7.0×10^5	2.1	7.2	0.31	124	7.8	0.06	1.4
	8/16	207	22.0	48	3.0×10^6	2.3	8.5	0.36	154	8.2	0.04	1.2
Sand Hollow Drain at mouth near Parma	8/15	195	24.0	197	1300	1.1	7.2	0.31	124	7.8	0.06	1.4
	8/16	201	24.5	220	17000	1.8	7.1	–	154	7.7	0.2	1.5
South Boise Drain near Parma	8/15	74	25.0	81	7300	2.2	8.5	0.30	210	7.9	0.07	1.5
	8/16	72	18.5	88	2.2×10^5	2.0	7.8	0.31	203	7.8	0.08	1.4
Ross Drain at Mouth near Parma	8/15	2.6	26.7	22	4500	0.9	8.6	0.23	205	8.1	0.09	0.2
	8/16	2.0	17.5	4	13000	0.9	7.3	0.25	290	7.5	0.08	0.1

[a]From the U.S. Army Corps of Engineers Cooperative Boise River Survey, August 1972. Data collected by U.S.G.S.

interconnected segments shown in Figure 7.2. These segments can have variable length. In this application, each segment is approximately two miles long. Connectivity between hydraulic elements can be specified so that the river can have tributary and can branch off and reconnect.

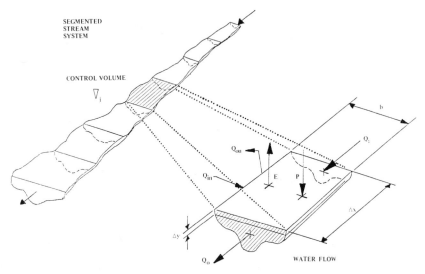

Figure 7.2   Physical representation of a stream.

Mass balance equations are developed around each hydraulic element for each quality constituent in accordance with the following principles:

1. The Law of Conservation of Mass: there is conservation of mass even though a constituent is changed by reactions from one form to another.
2. The Kinetic Principle: the rate of change is equal to the product of a coefficient and one or more constituent concentrations that interact to cause the change.

Ecological Processes

Physical, chemical, and biological processes that may act to change the concentration of water quality parameters were identified. Table 7.4 shows the important processes considered in the Boise River model.

Mass Balance Equations

Consider an idealized hydraulic element where the state of the ecosystem can be defined. The element has a volume $\overline{V}$, a surface area A_s,

Table 7.4   Important Ecological Processes for Modeling

---

1. Physical Processes
   a. Advection between segments
   b. Diffusion between segments
   c. Sedimentation from the segment
   d. External input to the segment
   e. Output to external from the segment
   f. Reaeration
   g. Solar insolation
2. Biochemical transformation, uptake, and release associated with the following:

Bacteria O_2

$NH_3 \rightarrow NO_2 \rightarrow NO_3$

$BOD \rightarrow CO_2$

Detritus $\rightarrow NH_3$, PO_4, CO_2 $\rightarrow$ Algae $\rightarrow$ Zoo $\rightarrow$ Fish

PO_4 Benthos Insect

        Bacteria   Detritus   Benthic Algae

---

and cross-sectional areas Aj's between the adjacent elements. Q_i and Q_o are the advective flows into and out of the element. Q_{in} is the local input from tributary inflows or waste discharges, and Q_{ou} is the local output of pumping exports or natural outflows. Concentrations of constituents C, such as TDS, BOD, DO, algae, zooplankton and fish, are transported by the Q_i and Q_o flows.

In general form, mass balance equations fall into two classes, one for abiotic constituents and another for organic biomass.

Abiotic Constituents

For abiotic substances, mass can be transported where applicable by the processes of advection, dispersion (diffusion), input, output, sedimentation and reaeration. They can be increased or decreased in concentration by decay and transformation. The latter processes include nitrification ($NH_3 \rightarrow NO_2 \rightarrow NO_3$), BOD decay ($BOD \rightarrow CO_2$), and detritus decay (organic$\rightarrow$ NH_3, CO_2, PO_4). Oxygen may be consumed (BOD) or produced (phytosynthesis) in conjunction with biological transformations. Other abiotic substances, such as carbon, nitrogen, and phosphorus are biologically consumed or released.

The above processes act independently and simultaneously. The total change is equal to the sum of the individual processes; thus,

Total Change = ± Advection ± Diffusion + Input − Output
± Sedimentation ± Reaeration − Decay
± Chemical Transformation − Biological Uptake
+ Respiration Release

From the kinetic expressions for each individual term, the corresponding mass balance equation becomes

$$\frac{d(\bar{V}C_1)}{dt} = Q_i C_{li} - Q_o C_1 + \sum_{j=1}^{n} E_j A_j \frac{dC_1}{dx_j} + \Sigma Q_{in} C_{in} - \Sigma Q_{ou} C_1$$

$$- S_1 \frac{\bar{V}}{D} C_1 - K_{r,1} \bar{V} (C_1 - C_1^*) - K_{d,1} \bar{V} C_1$$

$$\pm K_{d,2} \bar{V} C_2 - \Sigma \mu_3 \bar{V} C_3 F_{3,1} + \Sigma R \bar{V} C_3 F_{3,1} \qquad (7.1)$$

The terms that have not been described previously are defined as follows:

$\bar{V}$ = volume of the segment, m^3

Q_i = advective flow from the upstream segment, m^3/sec

Q_o = advective flow to the downstream segment, m^3/sec

C_1 = concentration of the quality constituent, mg/l

C_{li} = concentration of the quality constituent in the upstream segment, mg/l

n = number of adjacent segments

E_j = diffusion coefficients, m^2/day

A_j = cross-sectional area, m^2

dC_1/dx_j = concentration gradient of C_1, mg/l/m

C_{in} = concentration of C_1 in the inflow, mg/l

S_1 = settling rate of C_1, m/day

D = mean depth, m

$K_{r,1}$ = reaeration coefficient for C_1, day^{-1}

C_1^* = saturation concentration of C_1, mg/l

$K_{d,1}$ = decay coefficient of C_1, day^{-1}

C_2 = constituent concentration that may transform to C_1, mg/l

$K_{d,2}$ = decay coefficient of C_2, day^{-1}

C_3 = organism concentration that consumes C_1, mg/l

μ_3 = growth rate of biota C_3, day^{-1}

$F_{3,1}$ = conversion factor between C_1 and C_3

R = respiration rate of biota C_3, day^{-1}

Organic Biomass

Organic biomass is subjected to advection, diffusion, input, output, and sedimentation (algae only). In addition, there are metabolic processes of growth, respiration, and mortality. Depending on the trophic levels involved, there may be a grazing effect. Again, a general mass balance equation can be written by adding individual kinetic expressions (note that subscripts are used only to differentiate species of constituents in the same equation):

$$\frac{d(\bar{V}C_1)}{dt} = Q_i C_{1i} - Q_o C_1 + \sum_{j=1}^{n} E_j A_j \frac{dC_1}{dx_j} + \Sigma Q_{in} C_{in} - \Sigma Q_{ou} C_1 - S_1 \frac{\bar{V}}{D} C_1$$
$$+ (\mu_1 - R_1 - M_1) \bar{V} C_1 - \mu_2 \bar{V} C_2 F_{2,1} \tag{7.2}$$

The two generalized mass balance equations must obviously be adapted to any specific constituents. The adaptation is accomplished by deleting terms that are not applicable. For example, fish biomass will not have the first four terms of Equation (7.2) since fish are assumed not to be advected, diffused or affected by inflow or outflow. In fact, fish biomass must be empirically distributed according to food availability and by consideration of such environmental factors as temperature and dissolved oxygen. TDS on the other hand will only have the first four terms of Equation (7.1).

Heat Budget Equation

For predicting temperature, a heat budget equation analogous to the mass balance equation may be developed as follows:

$$\frac{d(\bar{V}T)}{dt} = Q_i T_i - Q_o T + \sum_{j=1}^{n} E_j A_j \frac{dT}{dx_j}$$
$$+ \Sigma Q_{in} T_{in} - \Sigma Q_{ou} T$$
$$+ (H_s + H_a \pm H_c - H_{br} - H_e) \frac{As}{\bar{V}} \tag{7.3}$$

where

T = water temperature, $^\circ$C
T_i = water temperature of the upstream segment, $^\circ$C
T_{in} = water temperature associated with local inflows (Q_{in}), $^\circ$C
H_s = short wave radiation less the reflection, K Cal/m^2/sec

H_a = long wave atmospheric radiation less the reflection, K Cal/m^2/sec
H_c = heat conductance between water and air, K Cal/m^2/sec
H_{br} = back radiation, K Cal/m^2/sec
H_e = evaporation loss, K Cal/m^2/sec

Heat flux terms are calculated as a function of sun angle and meteorological parameters such as atmospheric pressure, wind speed, dry bulb temperature, wet bulb temperature, and cloud cover. Detailed methodology can be found elsewhere (Edinger and Geyer 1965; Raphael 1962; Water Resources Engineers 1968).

Rate Coefficients

Most of the rate coefficients used in the mass balance equations are prespecified constants (see Table 7.5). Others can be calculated through expressions derived theoretically or empirically. These expressions facilitate the updating of rate coefficients that are known to vary from time to time as the state of the ecosystem changes.

Table 7.5   System Coefficients for Boise River Simulations[a]

Parameter	Value
Decay rate, per day	
BOD	0.20
NH_3-N	0.06
NO_2-N	0.2
Detritus	0.001
Coliform	0.5
Temperature coefficient, Q_{10}	
BOD	1.047
NH_3, NO_2	1.02
Biologic	1.02
Detritus, coliform	1.04
Maximum specific growth rate, per day	
Algae 1	1.5
Algae 2	2.0
Zooplankton	0.25
Fish 1 and Fish 2	0.03
Fish 3	0.025
Bentho	0.05
Insects	0.1

Table 7.5 continued

Parameter		Value	
Stoichiometric equivalence			
O_2/NH_3		3.5	
O_2/NO_2		1.2	
O_2/Detritus		2.0	
O_2/Algae		2.0	
CO_2/BOD		0.2	
Natural mortality, per day per toxicity unit			
Zooplankton		0.005	
Fish		0.001	
Bentho		0.005	
Insects		0.002	
Toxicity mortality, per day per toxicity unit			
Zooplankton		0.03	
Fish		0.025	
Benthos		0.03	
Insects		0.03	
Half saturation constants			
Zoo on algae, mg/l		0.5	
Fish on insects, mg/m^2		500	
Fish on zoo, mg/l		0.05	
Fish on bentho, mg/m^2		500	
Bentho on sediment, mg/m^2		50	
Algae 1			
light, kcal/m^2		0.003	
CO_2, mg/l		0.025	
N, mg/l		0.2	
PO_4, mg/l		0.02	
Algae 2			
light, kcal/m^2		0.006	
CO_2, mg/l		0.03	
N, mg/l		0.1	
PO_4, mg/l		0.05	
Digestive efficiency			
Zooplankton		0.700	
Fish		0.400	
Bentho		0.400	

Chemical composition	C	N	P
Algae	0.500	0.090	0.012
Zooplankton	0.500	0.090	0.012
Fish	0.500	0.090	0.012
Bentho	0.500	0.090	0.012
Detritus	0.200	0.050	0.007

Table 7.5   continued

Parameter		Value
Temperature tolerance limits, °C	Min.	Max.
Zooplankton	5.0	30.0
Algae 1	5.0	25.0
Algae 2	10.0	30.0
Fish 1	5.0	20.0
Fish 2	10.0	30.0
Fish 3	5.0	30.0
Bentho	5.0	30.0
Respiration rates, per day		
Algae		0.2
Zooplankton		0.01
Fish		0.001
Bentho		0.001
Insects		0.01
Algae self-shading factors, per mg/l per meter		0.002
Settling velocity, meter/day		
Detritus		0.05
Algae 1		0.15
Algae 2		0.05
Insect monthly emergence rates[b]		0.4
January		0.4
February		0
March		0.4
April		0
May		0.1
June		0.1
July		0.15
August		0.15
September		0.1
October		0.07
November		0
December		0

[a]Estimated for various sources compiled in "IBP-Desert Biome Aquatic Program (1970); Chen and Orlob (1968); Strickland (1960).

[b]Back-calculated from data collected with artificial substrate.

Temperature Effects

Rate coefficients for algal respiration are known to vary with temperature in accordance with Q_{10} concept

$$R_T = R_{20} \, \Theta^{T-20} \; (\text{day}^{-1}) \tag{7.4}$$

where R_T and R_{20} are respiration rates at T and $20°C$, and Θ is a temperature coefficient ranging from 1.02 to 1.05. Similar equations can be developed for BOD decay and others.

For growth calculations, typical temperature response functions such as the one shown in Figure 7.3 are developed (Bjornn, private communication). The curves state that the growth rate will increase with temperature, reach an optimum and then decline to zero at the temperature tolerance limit. Because of the complexity of the function, the coefficients are read in for every degree centigrade to serve as a look-up table.

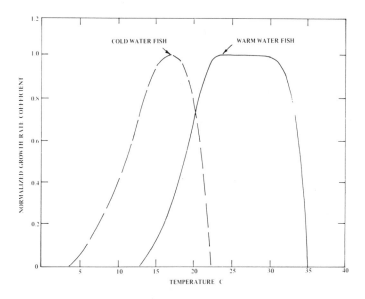

Figure 7.3   Normalized temperature-growth rate relationship.

Specific Growth and Limiting Factors

The specific growth rate of phytoplankton (μ) is determined by the growth limiting factors such as light, carbon, nitrogen, and phosphorus, which serve as independent variables in the following equations:

$$\mu = \overset{\Lambda}{\mu} \; \frac{L}{K_1 + L} \; \frac{C}{K_c + C} \; \frac{N}{K_n + N} \; \frac{P}{K_p + P} \; (\text{day}^{-1}) \qquad (7.5)$$

in which $\overset{\Lambda}{\mu}$ is the maximum possible growth rate of the organisms (day^{-1}); K_1, K_c, K_n, and K_p are half-saturation constants; and the independent variables L, C, N, and P are light intensity, carbon, nitrogen and phosphorus concentrations, respectively.

The growth of zooplankton is determined by an equation similar to Equation (7.5), except that algal density serves as the limiting factor. Fish can graze on either zooplankton or benthic animals, and therefore fish growth rates are a function of the densities of the two food forms. The approach is oriented to consider three types of fish: warm water, cold water, and bottom feeders. The warm and cold water fish feed on zooplankton and insects serving as the independent variables in establishing growth rates. The bottom feeders growth rate depends upon the benthic animals, detritus sediment and benthic algae.

Reaeration

Several physical rate coefficients can be calculated indirectly. For example, the reaeration coefficient of gases (CO_2 and O_2) may be estimated by the expression (Langbein and Durum 1967)

$$K_r = \frac{3.3\bar{U}}{D^{1.33}} \qquad (7.6)$$

where K_r is the reaeration coefficient in per second, $\bar{U}$ is the mean stream velocity in feet per second, and D is the mean stream depth in feet.

Settling

Settling rates of algae and detritus can be estimated by considering the reactor as an ideal settling tank. Under the assumption, the removal rate (S) is equal to

$$S = \frac{S_t}{S_o} \qquad (7.7)$$

where S_t is the settling velocity of the particles in m/sec and S_o is the surface loading rate; *i.e.*, the flow-through rate divided by the surface area, also resulting in an unit of m/sec.

Mortality

For mortality rates, Chen and Selleck (1969) have expressed the toxication rate as a function of wastewater concentration (or toxicity content):

$$M = \alpha + \beta C \qquad (7.8)$$

where M is the mortality rate, α is the natural mortality rate, β is a toxicity coefficient, and C is the toxicity content of the water.

When the dissolved oxygen concentration decreases, fish have to pump more water through their gills to obtain enough oxygen. As a result, their exposure to toxic compounds increases. To account for this effect in the mortality equation, the coefficient β is made inversely proportional to the DO concentration expressed in per cent saturation. The maximum amplification of the coefficient β is 3 when the DO is lower than 50% saturation. The temperature of the water indirectly influences the coefficient β through modification of the DO saturation value.

Hydraulic Properties

At a given river cross section, the width, depth, and velocity change with the flow. Leopold and Maddock (1953) analyzed various river channels throughout the United States and determined the functional relationships among the width, depth, velocity, and discharge. In a natural stream, the relationships were found to take the form of simple power functions:

$$H = h_1 \, Q^{h_2} \qquad (7.9)$$

$$U = u_1 \, Q^{u_2} \qquad (7.10)$$

$$W = w_1 \, Q^{w_2} \qquad (7.11)$$

where H is the mean depth, U the mean velocity, and W the water surface width. The geometric coefficients are evaluated for each specific stream section. Table 7.6 shows the hydraulic geometric coefficients determined for the Boise River.

Self-Shading Effect

The light extinction coefficient of water is calculated as a function of suspended matter as follows:

$$E_1 = E_0 + \Sigma_i \, r_i \, S_i \qquad (7.12)$$

E_1 = light extinction coefficient, m^{-1}

E_0 = light extinction coefficient of pure water, m^{-1}

r_i = shading coefficient, m^{-1} $(mg/l)^{-1}$

s_i = suspended matters including sediment, detritus, algae and
zooplankton.

The same shading coefficient is used for all matter for the lack of data.

Table 7.6   Boise River–Indian Creek Hydraulic Geometry Coefficients

Reach	Segments	Depth		Velocity		Width	
		H_1	H_2	U_1	U_2	W_1	W_2
1	1-2	0.130	0.420	0.200	0.346	42.0	0.230
2	3-9	0.101	0.425	0.240	0.384	41.3	0.191
3	10-12	0.500	0.218	1.00	0.154	2.0	0.628
4	13-16	0.500	0.218	1.00	0.154	2.0	0.628
5	17-26	0.937	0.177	0.016	0.762	66.7	0.061
6	27-33	0.800	0.177	1.04	0.195	1.20	0.628
7	34-42	0.937	0.177	0.016	0.762	66.7	0.061
8	43-44	0.130	0.420	0.200	0.346	42.0	0.230

Computation Sequence

The computer program was coded to perform the simulation in an organized, step-by-step fashion. The computational sequence of the program is outlined in Figure 7.4. The first step in the program is to read control cards, weather information, connectivity and system geometry, and system coefficients. The control cards specify the day to begin simulation, the total number of days of simulation, the number of days in the hydrologic period, the number of weather periods, the time step of computation, the frequency of output, and other control specifications. The weather data provide the necessary information for heat budget computations. Connectivity and system geometry data prescribe the way in which the river is to be segmented and how the water and pollutants are to be routed through the system. System coefficients specify the decay rate of BOD, the growth rate of algae, the response of fish to temperature, and parameters for other important biological activities occurring in the stream.

After the above data are read in, the computational procedure can be initiated. First, a flow balance is performed to determine the discharge throughout the river system. Depth, width, velocity, water volume, and

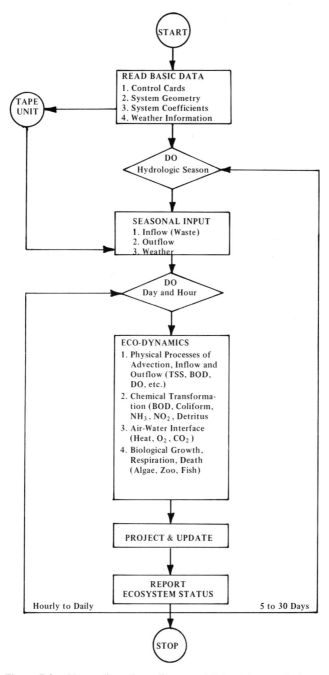

Figure 7.4 Master flow chart diagram of Boise River ecologic model.

reaeration coefficients are calculated for each river segment. The dynamic ecological interactions taking place in the river are then evaluated. The rate of change of each quality parameter resulting from the physical, chemical and biological interactions are computed first. These rates of change are integrated numerically to determine quality concentrations as a function of time and space.

The integration procedure continues until the time period of computation is completed. This can be 5 days, 30 days, or a year depending on the problem set-up. As specified by the control card, the computer is instructed to print a table describing the status of the ecosystem.

Sample Output

Table 7.6 illustrates a typical output describing the status of the ecosystem at a specified time; in this case, the 18th hour of a day in August 1972. The table provides a synoptic view of the whole system.

There is a heading describing the parameter names and the units in which they are expressed. The "Seg" column gives the segment numbers. Water quality and ecological conditions associated with a particular segment can be read along the horizontal line. Twenty-two parameters are tabulated, illustrating the comprehensiveness of the simulation.

By looking at the table, an investigator can determine where BOD rises, where DO drops, and where coliform contamination occurs. All biomass is expressed as dry weight, which is about one-tenth of the wet weight.

Results for fish production require some clarification. The simulation is based strictly on the availability of food (insect, benthic animals, and detritus) and the suitability of environmental conditions (temperature, toxicity, total suspended solids, and dissolved oxygen). It does not consider whether the habitat is good for spawning. In essence, it only reflects the productivity for the "put and take" type of fishery.

Because of the extreme difficulty in estimating fish standing crop, the fish production data are good only for relative comparison. These values can be used to assess which reach of the river is more favorable to fish growth based on the model. For example, during August, the upper reach of the Boise River appears to be more suitable for cold water game fish (type 1). The water is relatively free of pollution, DO is high, and total suspended solids is low. The water is cold, probably too cold for warm water fish (types 2 and 3) to do well. In the lower reach, the water is warm, total suspended solids is high, and DO is low. All those conditions are unfavorable to cold water game fish. Fish production is therefore shifted to the scavenger or fish type 3.

Table 7.6 The Output of August 1972 Simulation, Boise River

Status of Ecosystem			8 month	14 day		18.0 hour														
Seg.	Temp. °C	Toxic μm/l	TSS μg/l	Colifm. mpn/00	BOD mg/l	D O mg/l	D O 0/00	pH	CO₂C mg/l	PO₄P μg/l	NH₃N μg/l	NO₃N μg/l	Algae μg/l	Algae mg/m²	Zoo. μg/l	Detri. mg/l	Benth. mg/m²	Insec. mg/m²	Fish kg/ha/mo	Scour g/m²/d
1	15	0.05	12	4.8+01	1.0	9.1	91	7.6	0.48	70	11	50	205	46	20	2.0	11	409	-0.1	72.3
2	16	0.05	14	4.7+01	1.0	9.1	92	7.6	0.48	70	11	49	210	46	19	2.0	11	408	-0.0	71.2
3	16	0.05	17	4.6+01	1.0	9.2	94	7.6	0.48	70	12	49	214	47	19	2.0	11	408	0.1	82.2
4	16	0.05	20	4.4+01	0.9	9.3	95	7.6	0.47	70	12	49	219	47	18	1.9	11	472	0.1	68.4
5	17	0.05	23	4.3+01	0.9	9.3	97	7.5	0.58	70	13	48	223	48	18	1.9	11	473	0.2	66.7
6	17	0.05	26	6.1+01	0.9	9.2	96	7.5	0.68	71	14	59	222	459	17	1.9	11	473	0.3	61.6
7	18	0.05	30	2.2+02	0.9	9.1	94	7.5	0.82	74	17	75	217	481	16	2.3	11	466	0.4	55.3
8	19	0.09	38	2.8+03	2.6	8.8	94	7.4	1.32	276	118	159	207	590	15	2.2	11	466	0.4	44.3
9	19	0.09	39	2.6+03	2.5	8.7	96	7.4	1.43	270	115	173	205	592	14	2.2	11	574	0.5	44.3
10	19	0.09	42	2.5+03	2.5	8.8	94	7.5	1.14	270	116	172	208	577	13	3.7	11	575	0.6	27.9
11	20	0.11	55	5.1+03	3.0	8.5	96	7.7	1.13	297	283	208	193	610	13	3.7	11	577	0.7	21.7
12	20	0.11	57	5.0+03	2.9	8.6	94	7.7	1.13	296	279	210	196	593	13	2.0	11	578	0.8	18.7
13	17	0.09	39	2.4+03	2.3	8.4	95	7.4	1.77	256	107	208	192	623	12	2.3	11	572	0.6	36.3
14	20	0.08	37	2.7+03	2.0	8.0	91	7.4	2.32	240	99	302	176	647	11	2.3	11	570	0.6	37.6
15	20	0.08	37	2.5+03	1.8	7.9	87	7.5	2.07	230	94	323	167	633	10	2.1	11	568	0.7	36.2
16	20	0.08	40	2.4+03	1.8	8.1	86	7.5	2.04	230	95	322	171	621	9	2.0	11	569	0.7	36.8
17	21	0.09	50	3.6+03	2.3	8.4	89	7.6	1.51	266	194	262	192	566	9	2.9	11	1303	0.9	5.4
18	21	0.11	48	3.2+03	2.2	8.3	93	7.7	1.21	267	196	263	202	551	7	2.8	11	1284	1.0	-7.8
19	21	0.11	47	3.5+03	2.2	8.1	92	7.7	1.23	286	234	273	210	495	6	2.8	11	1270	1.1	-7.4
20	22	0.08	25	2.1+03	1.2	5.8	92	7.3	4.72	205	146	415	125	530	2	1.5	10	2780	1.2	-2.8
21	22	0.07	18	1.6+03	0.8	4.9	63	7.3	5.44	174	112	470	91	526	1	1.0	10	2768	1.1	4.2
22	20	0.06	35	1.4+04	1.2	6.0	55	7.5	3.98	229	78	829	92	590	8	5.0	11	2799	1.0	9.5
23	21	0.06	36	1.1+04	1.3	6.6	67	7.5	4.18	224	80	908	102	1144	10	6.4	11	2575	0.9	24.1
24	21	0.06	39	1.2+04	1.3	6.6	72	7.6	3.35	233	90	906	96	1118	8	6.5	11	2595	1.1	23.8
25	21	0.06	55	1.3+04	1.3	7.0	76	7.6	2.68	265	85	1049	102	1119	11	7.5	11	2650	1.1	30.4
26	19	0.06	55	1.2+04	1.3	7.2	79	7.7	2.14	265	86	1047	115	1101	9	7.5	11	2667	1.1	16.3
27	20	0.20	99	4.4+03	1.4	6.5	81	7.8	1.92	302	54	2298	108	256	19	9.8	109	851	0.8	-5.2
28	20	0.28	82	8.7+04	6.4	6.6	70	7.9	2.00	562	1566	2304	113	255	19	9.8	106	837	1.2	-6.2
29	21	0.28	91	7.4+04	5.6	6.4	72	7.9	2.43	515	1402	2124	109	247	17	8.6	106	816	2.6	0.5
30	21	0.56	94	4.4+04	36.6	5.4	71	7.8	3.12	65	1958	7070	73	175	4	8.9	97	770	-2.0	3.8
31	21	0.54	75	3.4+04	26.1	4.9	60	7.7	3.85	857	1535	5668	59	184	7	6.9	95	754	-5.6	16.4
32	21	0.41	84	2.4+04	20.0	5.3	60	7.6	3.71	740	1098	4629	66	252	8	6.8	101	808	0.8	10.0
33	21	0.19	71	1.6+04	8.2	6.6	74	7.8	2.22	444	468	2369	97	1128	8	7.2	105	791	4.1	34.7
34	21	0.16	65	1.3+04	4.7	6.8	72	7.8	2.21	385	412	1691	110	1251	6	7.5	106	787	4.6	-0.5
35	21	0.15	56	1.0+04	3.9	6.1	68	7.6	3.49	343	358	1523	98	1251	5	6.3	105	909	4.6	8.8
36	21	0.14	47	8.4+03	3.2	5.5	62	7.5	4.38	303	308	1367	84	1230	4	5.2	105	949	4.6	10.5
37	21	0.13	41	6.8+03	2.6	5.1	57	7.5	4.43	271	268	1242	72	1262	4	4.3	104	943	4.6	14.8
38	21	0.13	46	6.9+03	2.2	5.2	59	7.7	4.42	274	228	1282	68	1250	5	4.3	105	1195	4.3	20.1
39	21	0.12	49	5.5+05	2.1	6.4	73	7.7	2.79	300	189	1257	82	467	7	5.9	103	1060	4.3	32.1
40	21	0.12	46	4.7+05	1.9	6.2	68	7.6	3.48	283	179	2247	79	461	6	5.3	103	1060	5.2	30.2
41	21	0.12	44	4.2+05	1.7	6.1	68	7.6	3.50	266	170	1197	76	461	6	4.9	104	1096	5.2	35.2
42	21	0.12	42	3.7+05	1.6	6.0	67	7.6	3.51	256	162	1105	73	457	6	4.5	104	1123	5.2	39.9
43	21	0.11	41	3.3+05	1.4	6.0	68	7.6	3.45	246	161	1069	71	463	5	4.1	104	1138	5.3	25.2
44	22	0.11	40	2.9+05	1.3	6.0	68	7.6	3.45	237	159	1038	68	447	4	3.8	104	1141	5.5	26.8

The last column indicates whether suspended sediments are scoured from or deposited on the river bed. Positive values indicate scouring. The scouring rate is assumed to be proportional to the flow velocity, reaching a peak at a velocity of 8 ft/sec. The maximum scouring rate in $mg/m^2/sec$ is input to the model. It does not consider increased erosion as a result of higher river stages. The sedimentation rate is determined according to an estimated settling velocity of $2x10^{-5}$ ft/sec under turbulent conditions.

MODEL CALIBRATION

August Case

The hydrologic regime of August 1972 represents the irrigation season. During that period, Lucky Peak reservoir released 4415 cfs to the Boise River. Most of the water was diverted for irrigation. The drainage water returned to Boise River downstream of Star. Field data were available for comparing the observed and calculated diurnal variations of temperature and dissolved oxygen. Figures 7.5 and 7.6 show such comparisons for six locations selected along the Boise River.

In the lower reach, the model was found to predict smaller diurnal fluctuations of DO than those observed in the field. Several factors, the most important being benthic algae, may contribute to the discrepancy. The input data might have underestimated the standing crop of benthic algae and thus their oxygen production, particularly in the vicinity of Notus where the periphyton density was high. It is also possible that the DO of agriculture return water entering the downstream reach of the river varies diurnally due to weeds in the drains. Such a variation was not considered in the model input, but a field program has been initiated to investigate the problem.

Figures 7.7 and 7.8 show comparisons of calculated and observed concentration profiles of PO_4-P, NH_3-N, NO_3-N, coliform, temperature, and benthic algae. Diurnal fluctuations of these parameters are negligible under steady waste load input conditions. With the exception of NH_3-N, the model results followed the observed data reasonably well.

A large amount of the ammonia nitrogen that was discharged to Indian Creek and the Boise River disappeared from the system. Ammonia can decrease due to nitrification by bacteria or photosynthetic uptake of algae. Increasing the nitrification rate could not account for the rapid disappearance of ammonia. Bacterial assimilation and ammonia volatilization, not considered in the model, might have occurred.

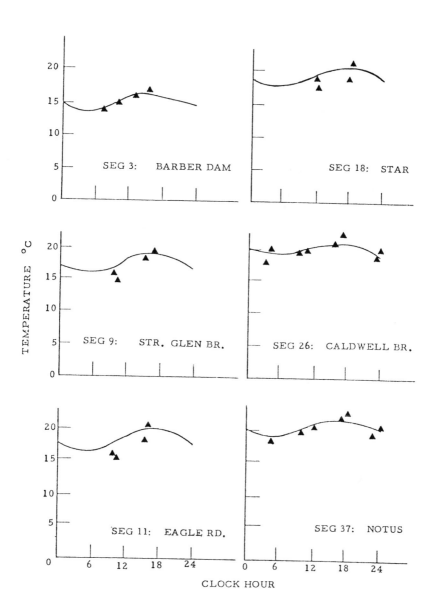

Figure 7.5 Observed and calculated diurnal variation of temperature in Boise River, August 1972.

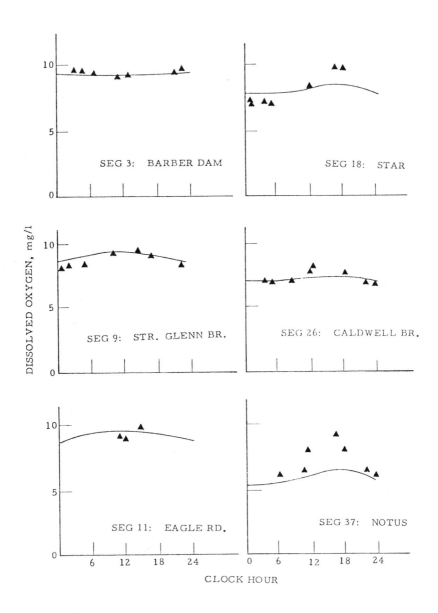

Figure 7.6 Observed and calculated dirunal variation of DO in Boise River,
August 1972.

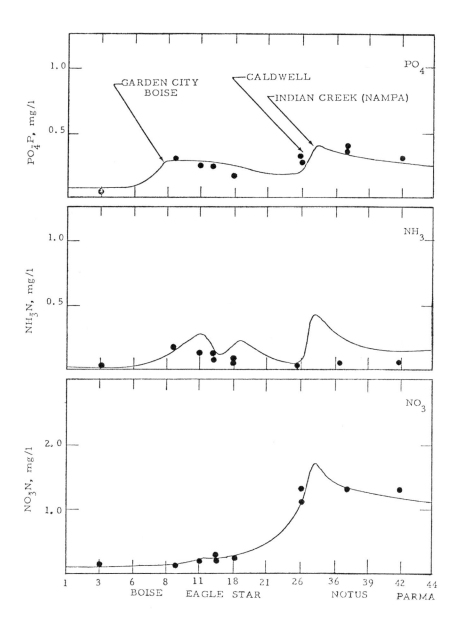

Figure 7.7 Observed and calculated concentration profiles of nutrients
in Boise River, August 1972.

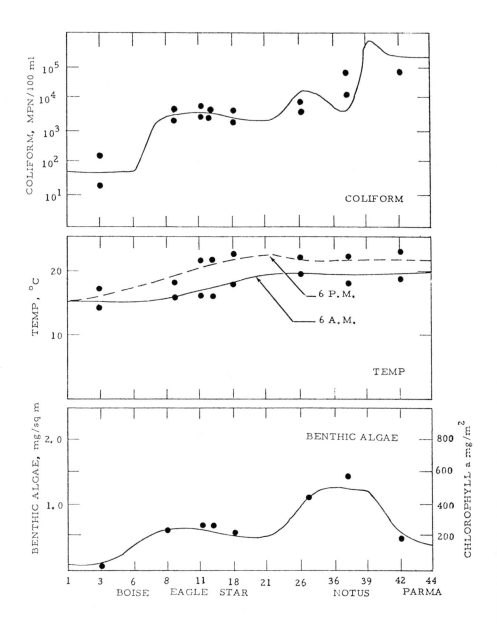

Figure 7.8 Observed and calculated concentration profiles of coliform, temperature, and benthic algae in Boise River, August 1972.

During the early phase of calibration work, the model tended to over-exaggerate the pollutional state of the Boise River downstream of Caldwell. Since mass had to be conserved, this situation could happen only if waste loads were overestimated or if pollutant-removing mechanisms in the real system were not accounted for in the model. Detailed analyses revealed that J. R. Simplot Co. discharged their waste intermittently. During the sampling period, it did not release wastewater to the river. Due to the short residence time of the river water, the pollutional effect of J. R. Simplot Co. did not persist long enough to be monitored by the sampling program.

November Case

The hydrology of November 1971 was typically low flow in the upper reach and high flow in the lower reach of the river. During that period, however, there were two hydrologic regimes. In early November, Lucky Peak Dam released 120 cfs of water to the Boise River. Local drainage added to this discharge so that the flow was 1000 cfs near the river mouth. During the latter half of November releases from Lucky Peak Dam were held near zero due to inspection of the outlet works.

The water quality conditions during both hydrologic regimes were available. The EPA survey collected samples during the period when the Lucky Peak discharge was 120 cfs. A cooperative survey conducted jointly by the U.S. Geological Survey and state and local agencies in the Boise region sampled the river two weeks later (Ada Council of Governments 1973). The latter set of data corresponded to the period when there was no reservoir release with the exception of minor seepage.

Figures 7.9 and 7.10 show the comparisons between the observed and calculated concentration profiles for various water quality parameters in the Boise River, November 4-10, 1971, during normal operations (*i.e.,* 120 cfs). The EPA data were used in this comparison. The predicted curves for BOD-DO generally follow the observed data, with the exception of a few points. At those points, however, the EPA data showed a wide range of variation for many water quality parameters. The model responded with less diurnal variation in DO than the prototype data.

Phosphorus and nitrate concentrations followed the observed data reasonably well. The predicted NH_3-N was higher than the observed values. Analysis indicated that the high values were caused by the ammonia concentration of the Boise treatment plant effluent. According to the EPA, the ammonia content of the Boise treatment plant effluent was 11 mg/l as nitrogen. Sampling data for the period indicates that the ammonia concentration of the Boise River near Boise was about 1 mg/l

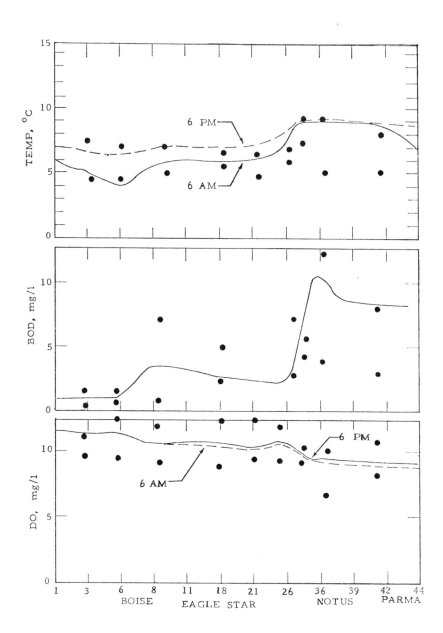

Figure 7.9   Observed and calculated concentration profiles of temperature, BOD and DO in Boise River, November 4-10, 1971, with 120 cfs of reservoir release.

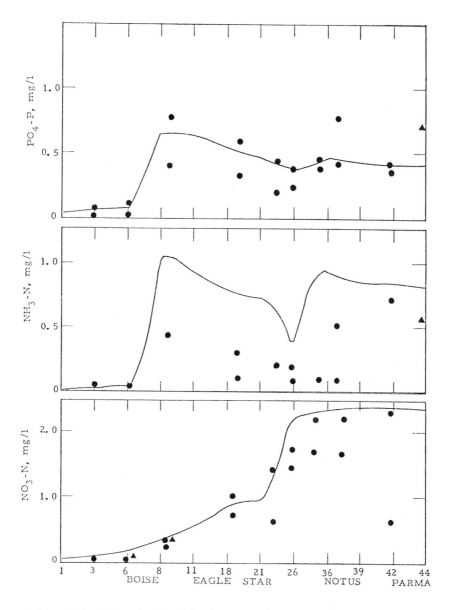

Figure 7.10 Observed and calculated concentration profiles of plant nutrients in Boise River, November 4-10, 1971, with 120 cfs of reservoir release.

as nitrogen. With the Boise treatment plant discharging 17 cfs of effluent and with available dilution of 10 to 1, the resulting ammonia concentration of the Boise River near Boise was calculated to be greater than 1 mg/l. If the ammonia concentration of Boise effluent was 5 mg/l, a value typically observed in the nearby Caldwell plant, the comparison would be better.

MODEL APPLICATIONS

Modeling exercises were undertaken to reach some understanding of the Boise River ecosystem and its response to pollution. Model results were analyzed to determine (1) the algal bloom potential in Boise River upon removal of suspended solid from the agriculture return water, (2) impact of diversion and agriculture return waters, (3) desired toxicity removal from wastewater, and (4) the relative fish productivity in different reaches of the river.

A detailed discussion of the results would be too lengthy for this chapter. Major findings, however, will be summarized.

1. In all cases tested both in winter and summer months, Indian Creek was affected by the industrial waste discharge of a meat packing company and the waste discharge of the Nampa Treatment Plant. The dissolved oxygen in certain reaches of the creek dropped below 6 mg/l.

2. Boise River water quality was affected below Caldwell, primarily as a result of the waste loads contributed by Caldwell and Indian Creek, under both winter and summer conditions.

3. Diversion of Boise River water from the upstream reaches had the following effects on the river ecology: (a) The amount of water available for diluting the waste discharges from Boise, Garden City, Eagle, and Star was reduced; (b) The water depth from the city of Eagle to a point two or three miles upstream of Caldwell was reduced. To an extent, the reduced flow created a riffle and pool habitat that might be suitable for benthic algae and insects and, therefore, for fish production. Over-reduction of water depth may deprive fish of living space and therefore may have an adverse effect on fish production.

4. Agricultural wastewater returns to the river in both surface and subsurface flows: (a) Surface drainage that contained high suspended solids increased the turbidity of the water from an upstream low of 3 JTU to a downstream high of 20 JTU during the irrigation season. Such a high turbidity level could make the water in a downstream area esthetically unappealing. Excessive suspended solids can inhibit the

production of cold water game fish; (b) Subsurface drainage, which contain low DO, BOD, and suspended solids, did exert considerable influence on the DO resources of the river water, as indicated by the model results.

5. Floating algae could not thrive in the river due to the short hydraulic residence time of the water. This was the case for all of the hydrologic conditions analyzed. Algal blooms would not occur if suspended solids were removed from irrigation return waters.

6. Boise River water was too cold for fish growth in the winter months. In the summer, the reach from Boise to 2 or 3 miles upstream of Caldwell would be suitable for cold water game fish production. Downstream, conditions of low DO, high TSS, toxicity and temperature, make the habitat unsuitable for cold water game fish. For fish that are benthic feeders, maximum production would occur downstream of Caldwell where the food density is higher; i.e., where there is a high population of benthic worms even though other environmental factors such as DO, TSS, and toxicity are not as favorable. Benthic feeders would not populate the upper reach of the river because of their intolerance of low temperatures.

7. The lower reach of Indian Creek was too polluted to support any type of fish production, according to the model results.

8. The Lucky Peak shutdown in November reduced the water available for diluting waste discharges. According to the model, from Boise to Star the water quality indicators such as BOD, nutrients, and coliform bacteria showed an increase in concentration of more than 200% because of the shutdown. In the same reach, the toxicity level increased anywhere from 50 to 120%, depending on the specific location within the reach. Unfavorable environmental factors induced a fish mortality rate of 30 to 100% above the natural rate. The water quality effects due to the reservoir shutdown became nearly insignificant in the reach of the Boise River below Caldwell because of the dilutional effects of irrigation return flows during the November period.

ACKNOWLEDGMENT

The basic model and computation scheme used in this study were adapted from the work sponsored by the Office of Water Resources Research, U.S. Department of the Interior, authorized under Title II of the Water Resources Act of 1964, as amended.

The current project was sponsored jointly by the Corps of Engineers, Walla Walla District, the Idaho Water Resources Board, and the Idaho

Department of Environmental and Community Services. Mr. Vic Armacost of the Corps and Mr. Wayne Haas of the Idaho Department of Environmental and Community Services were project officers, representing their respective agencies and coordinating problems of their mutual interest. Mr. Tom Davis of the Ada Council of Governments and Mr. Arlo Nelson of the Canyon Development Council provided coordination for local inputs to the study.

Mr. Bob Gifford of the Corps of Engineers oversaw the project on a day-to-day basis. The hydrology of the river was prepared by Mr. Dave Reese of the Corps and Mr. Bob Sutter of the Idaho Water Resources Board. Mr. Bob Minter of the Ada Council of Governments furnished insect data and other information derived from his substrate study. Dr. Wayne Minshall of the Idaho State University served as a special consultant, advising in the area of food web relationships.

Many individuals from the Environmental Protection Agency, the U.S. Bureau of Reclamation, the U.S. Bureau of Sport Fisheries and Wildlife, and the U.S. Geological Survey provided assistance directly or indirectly. Their contributions to the project are gratefully acknowledged.

REFERENCES

Ada Council of Governments. Canyon Development Council and U.S. Army Corps of Engineers, Walla Walla District, "Boise Valley, Idaho Regional Water Management Study," (June 1973).

Chen, C. W. "Concepts and Utilities of Ecological Model," *J. San. Eng. Div. ASCE* **96** (SA5) (October 1970).

Chen, C. W. and G. T. Orlob. "A Proposed Ecologic Model for a Eutrophying Environment," Report to the FWQA, Water Resources Engineers, Inc. (1968).

Chen, C. W. and G. T. Orlob. "Ecologic Simulations of Aquatic Environments," Final Report to the Office of Water Resources Research, Water Resources Engineers, Inc., Walnut Creek, California (November 1972).

Chen, C. W. and R. G. Selleck. "A Kinetic Model of Fish Toxicity Threshold," *J. Water Poll. Control Fed., Res. Suppl.* (August 1969).

Chen, C. W. and J. T. Wells, Jr. "Boise River Water Quality-Ecologic Model for Urban Planning Study," Report to the Corps of Engineers, Tetra Tech, Inc., Lafayette, Ca. (February 1975).

Cooperative Boise River Survey. (August 1972), data sheets.

Edinger, J. E. and J. C. Geyer. "Heat Exchange in the Environment," Report to the Edison Electric Institute, The Johns Hopkins University, (June 1965).

Environmental Protection Agency. "Water Quality Investigation of Snake River and Principal Tributaries from Walters Ferry to Weiser, Idaho," (February 1973).

"IBP-Desert Biome Aquatic Program," G. Wayne Minshall, Aquatic Coordinator, Idaho State University (August 1970).

Langbein, W. B. and W. H. Durum. "The Aeration Capacity of Streams," Geological Survey Circular 542, Washington, D.C. (1967).

Leopold, L. B. and T. Maddock. "The Hydraulic Geometry of Stream Channels and Some Physiographic Implications," Geological Survey Professional Paper 252, Washington, D.C. (1953).

Raphael, J. M. "Prediction of Temperature in Rivers and Reservoirs," *Proceedings ASCE, J. Power Div.* **88(PO2)** (July 1962).

Strickland, J. D. H. "Measuring the Production of Marine Phytoplankton," Bulletin No. 122, The Fisheries Research Board of Canada, Ottawa (1960).

"Temperature Respose Curves for Cold and Warm Water Fish." A private communication with Professor Bjornn, University of Idaho, Moscow, Idaho.

Water Resources Engineers. "Prediction of Thermal Energy Distribution in Streams and Reservoirs," Report to the California Department of Fish and Game, Water Resources Engineers, Inc., Walnut Creek, California (August 1968), revised.

8

MATHEMATICAL MODELING
OF NUTRIENT CYCLING IN RIVERS

Moises Sandoval, F. H. Verhoff and T. H. Cahill[1]

INTRODUCTION

In the past ten years considerable effort has been expended in the development and application of mathematical models for the prediction of water quality as a function of the point and nonpoint pollutant inputs to lakes or streams. This effort hopefully was to produce a universal model that could be used on any river or lake (sometimes with "calibration") to predict quantitatively the consequences of increased pollutional load on water quality or preferably to indicate the desirable improvement of water quality with the implementation of abatement technology. Unfortunately, no such universal model has been developed. Generally, *a priori* quantitative predictions of changing water quality resulting from input alterations have been unattainable except for a few restricted instances. In particular, predictions concerning a conservative substance such as chloride, which does not interact with the biota significantly, have been successful. Also "calibrated" models have been adjusted so that they (quantitatively) predicted changing water quality after these changes have been monitored.

The predictive inability of present models results from inadequate knowledge of the fundamental processes that determine water quality. For example, it is known that algal growth is important for both oxygen concentration calculation and for algal mass generation, yet the fundamental dependency of the dynamic algal growth rate on the nutrient

---

[1] Universidad Del Valle, Cali, Columbia; West Virginia University, Morgantown, West Virginia; Resource Management Associates, West Chester, Pennsylvania, respectively.

concentration or the nutrient concentration history is not well understood. Further complications are involved in the microbial degradation of this algal mass. In general, the knowledge of kinetics and mass transfer in natural water systems can be classified as poorly understood. As a consequence, this modeling effort emphasizes the only law in which confidence can be placed, *i.e.,* the conservation of mass.

The four most important elements involved in excessive nutrient loading of rivers are phosphorus, nitrogen, oxygen, and carbon. The model presented here contains complete mass balances on these four elements. The compartments or variables defined in the model, such as categorizing total algal concentration as one quantifiable variable, are conceptual as well as physical segregations of the mass in the system. Since there are an extremely large number of possible variables, *i.e.,* each microbial cell or, more broadly, each microbial species could be a compartment, it is necessary for the modeler to select a few variables into which many separate physical individuals can be classified. Certain other physical entities, such as weeds and fish, that contain the elements of interest, will be totally excluded from the model. Further, approximate kinetics and stoichiometry will be employed. In summary the model could be described as exact mass balances on carbon, oxygen, phosphorus, and nitrogen in the microbial organisms and the nutrients associated with these organisms. The kinetics and stoichiometry quantifying the transfer of mass between these organisms is approximate.

PREVIOUS INVESTIGATIONS

The literature on mathematical models of rivers dates back to the original oxygen sag equation of Streeter and Phelps. Subsequently, there followed many variations of this equation, *e.g.,* the unsteady equation of Li (1962). The use of mathematical oxygen mass balance descriptions for the quantitative analysis, prediction, and planning related to water bodies had increased substantially during the last decade principally because of the development of rapid digital computation (O'Connor 1970). A reasonably comprehensive list of authors who have investigated mathematical modeling of water bodies and a discussion of distinctions between each of these investigators would require a monograph. To give some indication of the variety of modeling research, the following papers are cited: O'Connor, *et al.* (1970); Parker (1968); Tenney, *et al.* (1972); Verhoff, *et al.* (1971); Grantham, *et al.* (1971); Davidson, *et al.* (1966); Goodman, *et al.* (1971); and Chen (1970). All of these authors are concerned only with the conservation of oxygen, except Chen who considers mass balances of other elements. However, Chen does not complete the cycle for carbon and nitrogen.

QUANTITATIVE CONSIDERATIONS
IN THE MODEL

All mathematical models involve assumptions and simplifications. Nutrient cycling in rivers is a complex phenomenon; hence significant simplifications and assumptions are required. The analysis and predictions resulting from the model depend greatly upon these considerations. Often there is no objective procedure for model development; rather the process depends upon the intuition of the modeler and his views about the important physical processes. In the case of nutrient cycling in rivers it was thought that the principle biota affecting the water quality were the algae, the aerobic microorganisms, and the anaerobic heterotrophic organisms. These classes of organisms were divided into categories depending upon the particular form of combined or free nitrogen that they used. The nutrients of concern for this model are those that cycle between the classes of microorganisms—phosphate, nitrogen, ammonia, nitrate (nitrite), oxygen, carbon dioxide, and carbonaceous compounds associated with the organisms, including organic acids and methane of anaerobic digestion. The concentrations of these substances then will constitute the state variables or components of the system.

The physical aspects of the river must be described mathematically. In addition to the velocity variation in time and all three dimensions, there are spatial and temporal variations of temperature and light conditions. To simplify the velocity, only the component in the direction of the river will be considered. Further the river will be segmented in the direction of flow and divided into a top and a bottom layer. The river velocity will be assigned in both the top and bottom layer for each segment of the river because the velocity of the bottom layer is significantly slower than the top. Both daily and yearly temperature variations were fit with a sine curve. Sine curves or partial sine curves also were used to represent the changes in average intensity of light in both daily and yearly considerations; no attempt was made to mimic the sunlight variation due to clouds. Because of the computer time limitations, it was necessary to consider nutrient cycling on two time scales, one that mimicked diurnal occurrences and one that simulated seasonal effects. For the diurnal model the simulation was performed for a period of thirty days, whereas for the yearly model the simulation ran for five years, with average values of sunlight and temperature employed for each day. Results were considered after transients had subsided.

After the conceptual picture of nutrient cycling has been determined, the time-dependent mass balances are formed on each component employing the associated kinetics and stoichiometry. The stoichiometry, particularly for the biotic reactions, is formulated so that oxygen, carbon, nitrogen, and phosphorus are conserved. For simplified stoichiometric representation, the following symbolism is used: CPN is the compound $C_aH_bO_cN_dP$, which is the composition of all microbial organisms as well as the immediate product of lysing. Degradation will yield CN, the above compound minus a phosphate ion, or CP, the above compound minus an ammonia molecule, and CC, the above compound minus both a phosphate ion and an ammonia molecule. Kinetics for the processes can be categorized as chemical or microbial. All chemical reactions are assumed to be first order in the compound decomposing. The microbial kinetics are based upon a separation of the transport of nutrients into the cell from the chemical reactions occurring there. These factors will be discussed quantitatively.

MODEL DERIVATION

Mechanism of Microbial Cell Growth

To describe cell growth kinetics a model proposed by Verhoff et al. (1971), describing dependence on limiting substrate, has been used. It is assumed that cell growth occurs in two steps. The first step involves absorption (accumulation) of nutrients by the microbe; the second step involves the ingestion (conversion of nutrients) of the absorbed nutrients by the organism and possibly a subsequent cell division. This process can be illustrated by the following stoichiometric relationship:

$$A + nS_s \underset{r_2}{\overset{r_1}{\rightleftharpoons}} B \overset{r_3}{\rightarrow} (1+e)A + P_2(n-e)S_s \tag{8.1}$$

where

$$
\begin{aligned}
A &= \text{organism (1 unit mass)} \\
S_s &= \text{limiting substrate (1 unit mass)} \\
B &= \text{ingesting cell (1 unit mass)} \\
n &= \text{number of mass units of substrate absorbed per mass unit of A} \\
e &= \text{mass units of substrate ingested per initial mass unit of A} \\
P_2(n-e) &= \text{represents the mass units of substrate returned to the medium as substrate per initial mass units of A.}
\end{aligned}
$$

The absorption process is expected to be mass transfer limited, either in the cell membrane or in a liquid film around the cell. Then, the rate of absorption per unit volume can be written

$$r_1 = k_1 A(S_s - S_c) \tag{8.2}$$

where

k_1 = rate constant (independent of all substances whose concentrations change during growth)

S_c = substrate concentration maintained by the cell; because it is small, $S_s - S_c = S_s$.

It is assumed that the ingestion rate is independent of the concentration of limiting substrate in the solution.

$$r_3 = k_3 B \qquad (8.3)$$

where

r_3 = rate of ingestion and cell division per unit volume

k_3 = rate constant independent of all substances whose concentrations change during growth.

Also the lysing rates of the organisms is specified, as a first order dependence in the respective cell concentration.

Chemical Composition of Bacteria and Algae

The elemental composition of bacteria and algae is not well known, and different compositions are to be found in the literature. Stumm (1963) suggests an average stoichiometric relation between carbon, nitrogen, and phosphorus of about 106 to 16 to 1 atoms, in other words $C_{106}H_{108}O_{45}N_{16}P$. McCarty (1965) suggests that an empirical formulation for bacteria cells is $C_5H_7O_2N$; extrapolating this to the nitrogen-to-phosphorus ratio (16:1) gives the formula $C_{80}H_{112}O_{32}N_{16}P$ for the bacteria and algae cells. Extrapolating this formulation to the carbon-to-phosphorus ratio (105:1) gives the formula $C_{105}H_{147}O_{42}N_{21}P$. The study of Verhoff *et al.* (1971) presents the results of a sensitivity analysis of the stoichiometric constants by the simulation of a simplified model of the surface water of a lake. It was observed in that study that the composition of algae and bacteria causes changes in concentration of algae, bacteria and nutrients and also causes changes in the shapes of the concentration-time curves.

Mass Transfer Rates Used in the Model

It is possible to consider three types of mass transfers of nutrients in a river system.

Convection

For the upper and bottom water the convective transfer term comes out in the derivation of the equation of conservation of mass as

$$\frac{U \partial C_i}{\partial x} \tag{8.4}$$

where

U = stream velocity; it depends upon time and position (due to precipitation, wastewater discharges, runoff, tributaries, slope, and transverse area)

C_i = concentration of component i

x = position in the stream

Diffusion

In the derivation of the equation of conservation of mass, the diffusion in the direction of flow will be neglected in each subsystem, since the convection term dominates.

There are three nutrients diffusively transferred between the air and upper water, and it will be assumed that the main resistance to mass transfer is in the water. These nutrients are O_2, N_2, and CO_2 or HCO_3^-. Carbon dioxide can be considered as CO_2, HCO_3^-, or other, since all are proportional for a given pH. No solubility is assumed for CH_4 and upon production it goes directly to the atmosphere. The rates of gas transfer per unit volume between air and the upper river waters are quantified by the equation

$$r_i = K_i(Y_{si} - Y_i) \tag{8.5}$$

where

r_i = rate of mass accumulations of N_2, HCO_3^- or O_2 per unit volume of upper waters

K_i = mass transfer coefficients for N_2, HCO_3, and O_2

Y_{si} = saturation mass concentration of N_2, CO_2 or O_2 in equilibrium, mg/e with the air (temperature dependent)

Y_i = mass concentration of N_2, CO_2 or O_2 in the upper waters.

Diffusional transfer mechanism of nutrients between upper and bottom waters is also under consideration. The nutrient flux between these two subsystems is determined by equations like the following:

$$r_j = K_j(X_j - Y_j) \tag{8.6}$$

where

r_j = net mass transfer rate of NH_4^+, NO_3^-, $H_2PO_4^-$, O_2, or CH_4 between upper and bottom waters per unit volume of upper waters

K_j = mass transfer coefficients; vary with time and position due to temperature, velocity and mixing variations

Y_j = upper waters mass concentration
X_j = bottom water mass concentration

$$r'_j = \emptyset\, K_j(X_j\text{-}Y_j) \tag{8.7}$$

where

r'_j = net mass transfer rate of NH_4^+, NO_3^-, HCO_3^-, $H_2PO_4^-$, O_2 or CH_4 between upper and bottom waters per unit volume of bottom waters

$\emptyset$ = ratio of lower water volume to upper water volume

Sedimentation

The settling of CPN, CP, CN, and CC are taken into account, and are assumed as proportional to the concentration of the respective nutrient in the upper waters. These rates are typified by

$$r_k = K_k Y \tag{8.8}$$

where

r_k = mass rate at which settling occurs
K_k = mass settling coefficient, which may vary with velocity and temperature; assumed the same for CPN, CP, CN, CC.

It is important to notice that no interaction between living organisms of upper and bottom waters is included. Figure 8.1 gives an overall picture of the interrelationship of nutrient cycling between air, upper waters and bottom waters.

Chemical Kinetics Used in the Model

It is assumed that there are four homogeneous chemical reactions occurring in the upper and bottom water. The carbon phosphorus material (CPN) decomposes into carbonaceous material with nitrogen (CN) and phosphate ($H_2PO_4^-$). The CPN also can decompose into carbonaceous material with phosphorus and without nitrogen (CP), plus NH_4^+. The CP decomposes into only carbonaceous material (CC) and $H_2PO_4^-$. The CN decomposes into CC and NH_4^+. All these decomposition of organics are assumed to be first order, and the rates of the reaction are given by the following formula:

$$r_1 = K_1 Y \tag{8.9}$$

where

r_1 = mass of CPN, CPN, CP, or CN per unit volume decomposing to CP, CN, CC, or CC, respectively

K_1 = first-order reaction constants (temperature dependent)

Y = mass concentration of particular carbonaceous material.

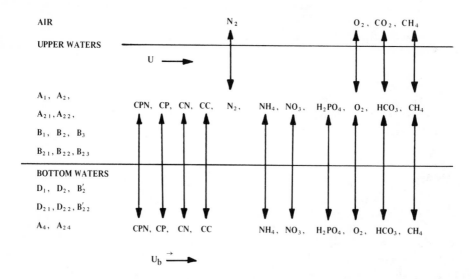

Figure 8.1 Components and associated transport included in the river model.

Biological Reactions

In order to keep the model from becoming difficult to handle, the algae and bacteria populations will be lumped in different groups based upon the type of nutrient transformations accomplished by the microbe. The mechanism of biological reactions will divide the algae and bacterial masses into two parts, the absorbing and the ingesting part. The rate functions describing the absorption of nutrients from the water by algae and bacteria are assumed to be first order in mass concentration of organism and respective nutrients. The disappearance of ingested algae or bacteria is assumed to be first order in mass concentration of ingested cells.

The lysing rates of the algae and bacteria also will be assumed to be first order in algae or bacteria concentration. In this term is included the effect of natural death and perhaps predation.

Temperature Variation

The temperature of the river varies with time and position. Since the variation of temperature with position is negligible, only the variation with time is taken into account; through this variation the parameters affected by temperature are correlated to time. Data found in the literature (Thomann *et al.* 1970), indicate that temperature exhibits diurnal and annual periodicity, and it is advantageous to fit the data to a sine function with either a one-day cycle or a yearly cycle for the daily or yearly model, respectively.

The parameters affected by temperature variation are: the first-order rate constants for the ingestion process of algae, the first-order rate constant for the ingestion process of bacteria, the mass transfer coefficient for transfer of O_2, N_2, and CO_2 between air and upper water (Dobbins 1964; Fair *et al.* 1968), the saturation mass concentration of O_2, N_2 and HCO_3^- in equilibrium with the air phase, the mass transfer coefficients for the diffusion of nutrients between the upper and bottom waters, and the rate constant for the lysing process of algae and bacteria.

Sunlight Variation

Another of the environmental variables is incident solar radiation, which varies with time and position. Solar radiation is the energy source for the photosynthetic growth of the phytoplankton. In a natural environment the light intensity to which the phytoplankton are exposed is not uniformly at the optimum value, but it varies as a function of depth due to turbidity present (self-shading) and as a function of time of day. Thus the phytoplankton in the lower layers are exposed to intensities below the optimum and those at the surface may be exposed to intensities above the optimum so that their growth rate would be inhibited.

According to Beer's law, the solar energy reaching the water surface is attenuated exponentially with depth influenced by the suspended phytoplankton. Therefore a sinusoidal model as the function of photoperiod and intensity during the photoperiod was also used in this study (Chen 1970).

The hourly distribution of solar intensity was derived from a plot of mean daily solar radiation (Ly/day) *vs.* time (days) presented by DiToro and O'Connor (1970) for the San Joaquin River, Mossdale, 1966-67. The photoperiod distribution was derived from an equation presented by Parker (1968).

The parameters that take into account the light intensity variation are the first-order rate constants for the growth process of algae A_{21}, A_{22} and A_{24}. These parameters are made functions of the light intensity in the same manner as DiToro and O'Connor (1970) quantify the algal growth rate.

Nutrients

Algal growth depends on the availability of suitable nutrient elements. The three elements that are present to the greatest proportion by weight in plants are oxygen, carbon, and hydrogen, with nitrogen and phosphorus as the ones most likely to be in limited supply in natural waters (Committee Report 1970). In this study HCO_3^- was chosen as the source of inorganic carbon because a plot of the distribution of dissolved carbonate forms as a function of pH shows that the predominant specie is the HCO_3^- over the pH range of 6.4 to 10.4.

Soluble orthophosphate is readily assimilated by plants and other aquatic organisms, forming particulate organic phosphorus. Either during growth or death of biological life, soluble compounds that contain organic phosphorus may be excreted into solution. These compounds are either reassimilated to form particulate organic phosphorus or, through degradation, are converted back to inorganic orthophosphates. From a plot of the distribution of dissolved orthophosphate forms as a function of pH, it is shown that the predominant dissolved orthophosphate species over the pH range 5 to 9 are $H_2PO_4^-$ and HOP_4^{2-} (Committee Report 1970).

Molecular nitrogen derived from the atmosphere may be reduced and converted to organic nitrogen by certain nitrogen-fixing bacteria and algae. Natural waters contain dissolved N_2, ammonia, and salts of the nitrate and nitrite ions. In addition, waters contain organic nitrogen compounds, primarily attributable to the presence of aquatic life.

The nitrification process may be considered in two steps. In the first step ammonia may be oxidized under aerobic conditions to nitrite nitrogen by species of autotrophic bacteria of the genus *Nitrosomonas*. In the second step nitrite is oxidized to nitrate under aerobic conditions by bacteria of the genus *Nitrobacter*. Kinetic data for the second step in the nitrification process indicate that this step is very fast, and for this reason nitrite nitrogen is the low concentration nitrogen species that is readily oxidized to nitrate nitrogen. In this study the nitrification process is considered as one process, the ammonia being oxidized to nitrate nitrogen and nitrite nitrogen not considered because it seldom exists in a considerable concentration in surface waters due to its transitory nature. Thus the model derived here includes only molecular nitrogen, ammonia, and nitrate.

MATHEMATICAL FORMULATION
OF THE SYSTEM

The conservation of mass, which describes the relationship between the transport of a substance through a water volume and the sources and sink of mass within it, is fundamental to the analysis of any water quality problem. This equation is derived by writing a mass balance over a differential volume. It is stated by the following equation.

| time rate of change of mass in volume element | = | rate of mass in | − | rate of mass out | + | the difference between that produced by the sources and that removed by the sinks |

It is found after mathematical manipulation that an equation valid at a point in the water stream has the following form:

$$\frac{\partial Y}{\partial t} + U(x,t)\,\frac{\partial Y}{\partial x} = S(Y,X,x,t) \qquad \text{upper waters} \qquad (8.10)$$

and

$$\frac{\partial X}{\partial t} + U_b(x,t)\,\frac{\partial X}{\partial x} = S(Y,X,x,t) \qquad \text{bottom waters} \qquad (8.11)$$

where

Y = concentration of upper water quality variables (mg/l)
X = concentration of bottom water quality variables (mg/l)
S = sources and sinks of the substance being considered
U = upper velocity of the stream (miles/day)
U_b = bottom velocity of the stream (miles/day)
x = distance downstream (miles)
t = time (day)

The systems of equations will be solved numerically with the boundary conditions at $x = 0$ and initial conditions at $t = 0$.

Applying the equation of conservation of mass to the upper water for each of the 20 different components results in 20 partial, first-order, nonlinear differential equations derived for the following substances: two kinds of absorbing algae and their respective two groups of ingesting algae, three groups of absorbing heterotropes and three associated groups of ingesting heterotropes, NO_3^-, NH_4^+, N_2, CPN, CP, CN, CC, HCO_3^-, $H_2PO_4^-$, and O_2.

Applying the equation of conservation of mass to the bottom waters for each of the 19 components results in 19 partial, first-order, nonlinear differential equations developed for the following substances: one kind of algae, three types of bacteria, and NO_3^-, NH_4^+, CPN, CP, CN, CC, OC, CH_4, HCO_3^-, $H_2PO_4^-$, and O_2. Again note that each microorganism type returned two differential equations.

The postulation of microorganism groups for this simulation is based upon the transformation of nutrients, particularly the various forms of nitrogen. In the upper waters, two types of algae are presumed to exist; one grows utilizing ammonia or nitrates and the other utilizes atmospheric nitrogen. Three types of heterotrophic organisms are assumed for the upper waters; one simply grows on either ammonia or nitrate, the second oxidizes ammonia to nitrate, and the third converts nitrate to atmospheric nitrogen. The lower waters contain an algal group that grows on ammonia or nitrate, an aerobic heterotrophic organism group that utilizes ammonia or nitrate for growth, and two groups of anaerobic organisms that convert organic materials to organic acids and thence to methane, respectively. It should be noted that each of these groups does not conform to specific species of microorganisms. For example, facultative bacteria could be considered in two groups depending upon whether their immediate environment is aerobic or anaerobic. Additionally, a group defined above might be comprised of quite diverse species, such as fungi, yeasts, or bacteria.

EQUATION SOLUTION TECHNIQUES

The general differential equations are hyperbolic, first order, nonlinear partial differential equations. Since analytical solutions of equations are impossible, numerical techniques must be employed. The total stream is discretized by dividing it into reaches that enable a numerical method to be applied to the mass balance equations. In each of these length intervals, in the numerical scheme for the solution of the mass balance equations, all properties are assumed constant. For any given time interval, each length interval behaves as a completely mixed reactor, allowing the mass of water quality constituents to be transported in and out by convection, diffusion, and settling.

Each of these length intervals is considered divided in two parts, in this way representing the upper and bottom waters in which each subsegment has constant properties. The state of the entire system is described at any desired time by the aggregation of behavior in the discrete elements. Reach nodes are established at every major or significant change in the system. Significant changes are the confluence of a high order tributary or the discharge point of a waste load.

Finite difference representations for partial derivatives are the most frequent approach to the numerical integration of partial differential equations. This technique consists of replacing the partial derivatives by difference quotients in the independent variables. In the solution of Equation (8.10) and (8.11), the x, t plane is divided into a rectangular grid and the derivatives with respect to x and t are expressed as differences as shown in Figure 8.2. The explicit technique for the method of finite differences is applied for the solution of the dynamic model since it is the most convenient to use for this type of problem.

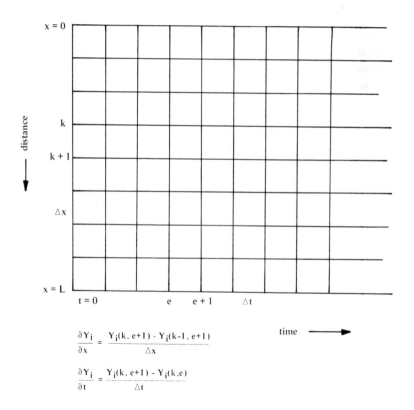

$$\frac{\partial Y_i}{\partial x} = \frac{Y_i(k, e+1) - Y_i(k-1, e+1)}{\Delta x}$$

$$\frac{\partial Y_i}{\partial t} = \frac{Y_i(k, e+1) - Y_i(k, e)}{\Delta t}$$

Figure 8.2 Numerical grid and approximating finite differences.

Computer Implementation

A formal stability analysis of numerical techniques is currently only possible for linear partial differential equations. The results of these investigations can be used on nonlinear problems, but with no assurance that they will be applicable. With the 39 equations, it is impossible to establish an exact stability criterion due to the magnitude and the nonlinearity characteristics of the system. However, it was shown by Birkhoff (1965) that a stability ratio $\alpha = U(\Delta t/\Delta x)$ satisfying $|\alpha| < 1$ gives good success in computing approximate solution of the equation

$$\frac{\partial v}{\partial t} + U(x,t)\,\frac{\partial v}{\partial x} = S(x,t)$$

with an explicit four-point central difference approximation. In this study the stability of the system is studied by a variation of Δt (time increment) until instability is shown by the computer output.

This set of finite difference equations is solved using the IBM 370 computer. Initial conditions are required for solving the mathematical model and to simulate the various subsequent behaviors in the system. Initial river water quality, climatological data, flow quality and quantity, rate constants of biological, chemical, and physical activities, and physical dimensions of the river are necessary. Other input data to the program include DELT (the time-step), N (the number of increments DELX), and TMAX (the desired total simulated time).

Excessive output is avoided by printing the results only periodically after a certain number of time-steps have elapsed. It also uses certain plotting subroutines available for displaying graphically the computed concentration *vs.* time for a chosen position in the river.

APPLICATION TO BRANDYWINE CREEK

The Brandywine Creek begins in southeastern Pennsylvania and flows in a general southeastern direction from the Piedmont Plateau, across the fall line and into the Atlantic coastal plain. The Creek has a drainage area of 317 square miles and an overall length of about 45 miles from its headwaters to the mouth below Wilmington, Delaware. It is a tributary to the Christina River, which is a tributary to the Delaware River (Figure 8.3). At its highest elevation, approximately 900 feet above sea level on the southeast slope of the Welsh Mountains, the Brandywine Creek is basically a spring-fed stream. The Brandywine Creek near its mouth serves as the principal source of water supply for the city of Wilmington, Delaware. The average withdrawal by the city for domestic supply is approximately 30 million gallons per day.

Figure 8.3 Map of the Brandywine River.

During a 25-year-period of record, several thousand chemical samples have been collected within the Brandywine basin. This extensive inventory of historical data is now available, having been compiled by the Tri-County Conservancy during 1971. The network of previous grab sampling stations is now integrated in the EPA "STORET" Data System.

In addition to the grab sampling records, there are three continuously operated electronic monitoring stations in operation in the basin. The locations of these stations are shown in Figure 8.3. Station No. 1 on the East Branch, below the urbanized area of Downingtown, has been in operation for almost two years.

Since the full drainage basin of the Brandywine is over 300 square miles, a portion of the basin was selected to develop procedures and simplify sampling and measurement techniques. The East Branch of the Brandywine Creek from above the urbanized area of Downingtown to the forks of the East Branch with the main stem near Wawaset (an 11-mile reach of stream with a drainage area of approximately 63 square miles) was chosen as the pilot area. This reach of stream and subbasin encompasses five major and a number of minor point discharges and also provides the greatest quantity of historical water quality records and effluent samples within the basin during the past two and a half years.

Beginning on March 1, 1972, chemical samples for some 40 parameters were collected at three key points on the main stem of the East Branch. These locations represented the top, middle, and lower end of the reach under study. As the program continued, additional sampling stations were set up on the four major tributaries flowing into this 11-mile reach of stream. These four subbasins account for some 73% of the drainage area. In addition, the five major point waste discharges were also sampled. All stations were sampled on a weekly basis, and corresponding discharge measurements were made at each location simultaneously. In addition, the three major stream stations were sampled on two-hour intervals over a 24-hour period. Since it is necessary to quantify all chemical sampling measurements, accurate measurements of discharge had been made at each sampling location within the stream network for every sampling period.

RESULTS OF THE MODEL

After having developed the mathematical formulation and having established the solution scheme by the finite difference method, computer runs were made of the programs to assure their validity. This assurance rested on comparison between values predicted by the model and the observed field data for the Lower East Branch of Brandywine Creek in southeastern Pennsylvania.

The daily and yearly models simulated water variables for the year 1972 since records from the laboratory and field experiments were available. The part of the river under study received discharges from several water treatment plants (Brandywine Paper Co., Downingtown Paper Co., and Downingtown Municipal Plant), and several tributaries (Beaver Creek, Parker Run, Valley Creek and Taylor Run). These measured inputs were supplied as data to the program.

Two different types of computer out-puts were obtained: one showing concentration *vs.* position (miles) for the river reach under study, and the other showing concentration *vs.* time for the position mile 6 where the monitoring station No. 1 is situated.

Comparison of Predictions with Experimental Data

A spate of parameters in excess of 220 must be assigned values before the 39 differential equations can be solved. These parameters ranged from kinetic values to tributary inputs and although too numerous to list here they are listed and discussed in Sandoval (1974). In most cases the choice of parameter values is confined to a small range of reasonable values; however, a few constants could range over several orders of magnitude. This large number of parameters does not permit the simulation to vary in any manner; rather the conservation of mass and many reasonably known kinetic factors force the system of equations to behave in certain ways. Thus the problem of finding parameters that were reasonable and yet fit the experimental data was not extremely difficult.

Since there are 39 differential equations, comparisons with experiment would be possible on 39 variables. Most of these variables are unmeasurable and hence comparisons were made on those few that are experimentally available. For daily simulations, comparisons were made on nitrate (Figure 8.4), phosphate (Figure 8.5), and several other variables. Comparisons of oxygen (Figure 8.6), ammonia (Figure 8.7), and carbonate (Figure 8.8), and others were performed for the yearly model. As can be seen from the graphs the comparison between predictions and measurement is reasonable, i.e., the order of magnitude is correct. The randomness exhibited by the experimental data probably results from the inherent stoichastic nature of the weather, certain waste inputs, and some experimental error. A periodic nature is observed in some of the data; this

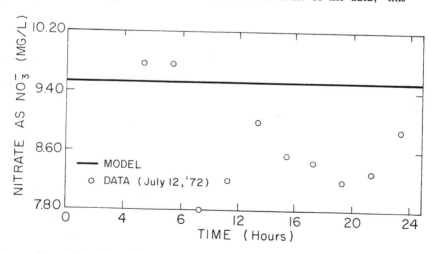

Figure 8.4 Comparison of model predictions and measured nitrate values on July 12, 1972.

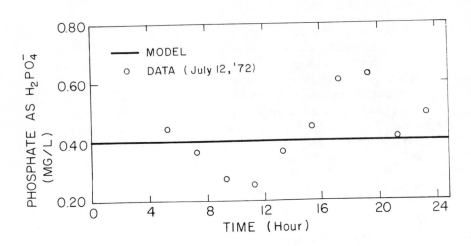

Figure 8.5 Phosphate concentration on July 12, 1972.

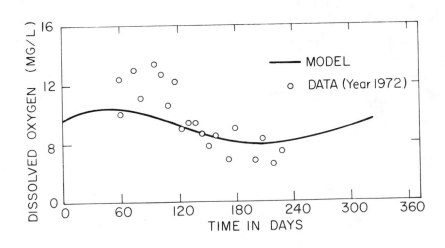

Figure 8.6 Yearly variation of the oxygen concentration in the Brandywine.

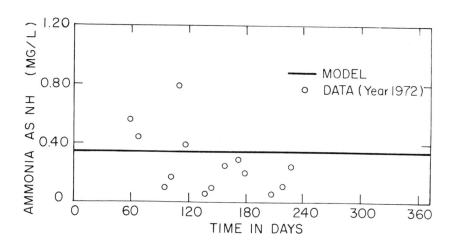

Figure 8.7   Ammonia concentration for the Brandywine in 1972.

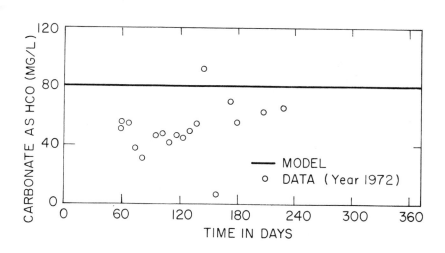

Figure 8.8   Carbonate concentration during 1972 in the Brandywine.

probably can be traced to either daily or yearly temperature and light variations or to the periodic heavy loadings received by wastewater treatment plants. The model does not include these phenomena and hence cannot simulate them.

In general, it appears that the variability of measured water quality parameters in flowing streams results not from the metabolic activity of living organisms but rather from the inputs to the receiving water. Two notable exceptions are the oxygen and ammonia concentrations in the lower waters during low flow; in this case, the aerobic organisms reduce the oxygen concentration to one-third its saturation value, as will be discussed later.

To illustrate the predictions of unmeasured water quality variables, Figures 8.9 and 8.10 depict heterotrophic activity and the carbonaceous material concentration as a function of distance downstream. The rapid

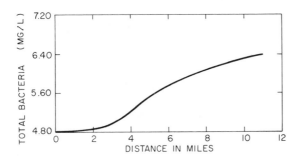

Figure 8.9   Predicted heterotroph concentration at noon during July in the Brandywine.

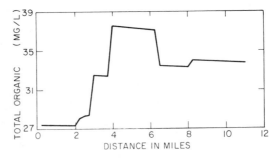

Figure 8.10   Total organic concentration as a function of river mile predicted in the Brandywine during July.

rise at river mile 3 is caused by the paper company, and the rise at river mile 4 results from sewage treatment plant effluent. At river mile 6 a stream of relatively good water quality dilutes the instream water yielding lower carbonaceous concentration. However, at river mile 8, a small stream with a sewage treatment plant again raises the carbonaceous concentration.

As stated earlier the structure of the model, or the mass balance and approximate kinetics, determines the general types of responses obtained in simulation and these general properties cannot be altered by varying the parameters. However, it is instructive to perform a sensitivity analysis to ascertain which of the parameters significantly affects the predicted water quality. The nutrient absorption rate of algae, $K(l)$, was varied by an order of magnitude, and the predicted response of the algae is presented in Figure 8.11. As can be seen, the effect of this parameter in the response of the system is insignificant. Similar small changes were observed for changes in the absorption rate constant for the heterotrophs and for the algae growth rate.

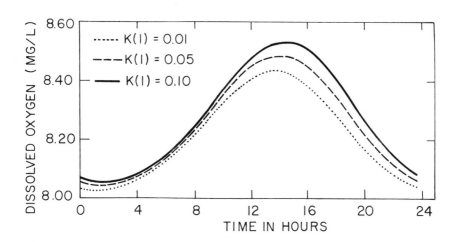

Figure 8.11 Sensitivity of algal concentration to algal absorption rate constant.

Cloud cover and river shading by trees can alter the daily variation of light intensity on a stream. Similarly, differing weather conditions can generate shifts in the diurnal temperature fluctuation. The amplitude factor of both the diurnal temperature function and the diurnal light intensity function were investigated separately. The output from the simulation indicated small shifts in the levels of most variables; however, these shifts were unimportant compared with the random fluctuations of these variables.

Figure 8.12 contains a sensitivity analysis of the algal lysing constant, $K°(3)$, which shows an intermediate influence upon the algal concentration. It has less effect upon other water quality variables. Similar moderate changes in various concentrations were observed to occur with alterations of the air-water transport coefficient.

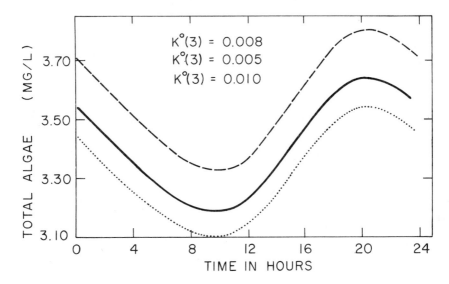

Figure 8.12   Sensitivity of algal concentration to algal lysing rate constant.

The three most significant parameters were the transport coefficient between upper and lower waters, the sedimentation rate, and the velocity of the stream.   Figure 8.13 contains a plot of the ammonia and oxygen concentrations in the bottom waters as a function of the mass transport coefficient between upper and bottom waters.   This coefficient can range over several orders of magnitude depending upon the turbulence of the stream.   For a two-order-of-magnitude drop in the coefficient value, the bottom oxygen concentration can decrease from near saturation to 3 ppm. The intake for the continuous monitoring station was located approximately one foot from the bottom, and occasionally it would measure extremely low oxygen concentrations in the dry summer.

The settling rate sensitivity depends upon the mass transfer coefficient between upper and bottom waters.   As long as the transfer coefficient was large enough, the sedimentation had little effect upon concentrations

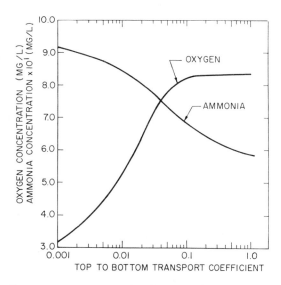

Figure 8.13   Bottom water concentration for ammonia and oxygen as a function
of mass transfer coefficient for 12:00 noon July 12, 1975
at Monitoring Station 1.

in the stream. However, with low transfer coefficients between upper and
bottom waters, small changes in sedimentation rates caused significant
concentration alterations in the bottom waters. One could say that a
synergistic relationship existed between sedimentation and mass transfer
between top and bottom waters. Under conditions of low transfer coefficient and high sedimentation, the water quality of the stream was
seriously impaired.

Increasing stream velocity causes increased upper water oxygen concentration (also bottom water oxygen concentration), decreased phosphate
concentration, and decreased ammonia concentration (Figure 8.14). The
phosphate and ammonia concentration dependency upon stream velocity
is called the dilutional effect. Most of these nutrients come from the
point sources and with increased flow the concentrations are diluted.
This is only true during approximately steady flow conditions, which are
assumed in the model. The slow velocity itself causes decreased oxygen
concentration and when coupled with decreased mass transfer coefficient
between upper and bottom waters significant water quality problems can
result. This is precisely what is found in the Brandywine and other
streams; with low flows, dissolved oxygen deficits are observed.

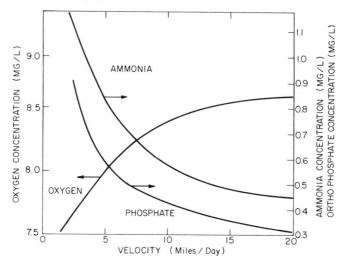

Figure 8.14   Oxygen, ammonia, and phosphate concentration (upper waters) as a function of stream velocity for 12:00 noon July 12, 1975 at Monitoring Station 1.

The system was simulated with no inputs from the point sources to determine what possible water quality improvements could be expected with the implementation of such a program. The major improvements in water quality occurred in the inorganic nutrient concentrations, particularly the ammonia and the orthophosphate, which exhibited decreases of about 80%. The difference between the case of river inputs to that of no inputs is shown in Table 8.1. The most improvement was seen in the bottom water oxygen and ammonia concentrations, where the reduced inputs would make the dissolved oxygen deficit negligible in the bottom waters.

CONCLUSIONS

Essentially three conclusions result from this study. The first relates to expected water quality with the complete elimination of the point source inputs to the river, and the other two concern the modeling technique utilized in this study. First, a complete halt to the point source inputs to the Brandywine River in the upper East Branch will cause significant reductions in the inorganic nutrient levels found in the stream, but most significantly, this point course reduction will yield a stream that does not have intolerable dissolved oxygen deficits in the bottom waters during steady flow conditions in the summer.

Table 8.1 Comparison of Daily Average Nutrient Concentrations in the Upper Waters at Monitoring Station 1 for July 12, 1975.

	Concentrations mg/l	
Parameter	Inputs of Nutrients	No Inputs of Nutrients
Ammonia	0.572	0.121
Oxygen	8.3	8.9
Nitrate	9.26	8.76
Phosphate	0.40	0.074
Carbonate	83.1	86.8
Total algae	3.8	3.2
Total heterotroph	5.7	5.0
Total organic	35.7	25.6

Second, this modeling effort indicates that it is not necessary to consider the complete recycle of most nutrients between algae and heterotrophs for a river model to yield reasonable results. This is in contrast to lake models that must contain such recycle to be realistic. Thus the inorganic nutrient levels are primarily determined by upstream and tributary influxes and not by the activity of organisms. This is shown in Figures 8.9 and 8.10 wherein the organisms do not follow the organic concentrations. Also Table 8.1 indicates that large changes in nutrient inputs do not extensively alter the organism concentration in the upper waters. The notable exception is the oxygen consumed by heterotrophs and generated by algae. Thus a model containing only oxygen recycle/and input mass balances on other nutrients would have been satisfactory. This conclusion is true for the Brandywine and many streams of similar character but may not hold true for other streams (e.g., the Mississippi).

The last conclusion is that the sedimentation rate, mass transfer between top and bottom, and the velocity of the stream are the most important parameters in affecting the water quality of the stream. It is ironic that the worst of all three parameters occur during summer low flow; i.e., the velocity is the lowest, the mass transfer between upper and bottom waters is lowest and the sedimentation rate is the highest. Thus one would expect the significant water quality problems during this time period, as is found in many streams including the Brandywine.

ACKNOWLEDGMENTS

The authors wish to acknowledge the financial assistance of The Tri-County Conservancy of the Brandywine in the form of a predoctoral fellowship (to M.S.).

REFERENCES

Birkhoff, G. "Partial Difference Method in Procedures, IBM," Scientific Computer Symposium Large-Scale Problems in Physics, IBM Corp. (1965), pp. 20-21.

Chen, C. W. "Concepts and Utilities of Ecologic Model," *J. San. Eng. Div., Amer. Soc. Civil Eng.* (SA5) 1083 (1970).

Committee Report. "Chemistry of Nitrogen and Phosphorous in Water," *Scien. Amer. J. Amer. Water Works Assoc;*, 127 (1970).

Davidson, R. S. and A. B. Clymer. "The Desirability and Applicability of Simulating Ecosystems," New York Acad. Sci. **128**, 790 (1966).

DiToro, D. M., D. J. O'Connor, and R. V. Thomann. "A Dynamic Model of Phytoplankton Populations in Natural Waters," Environmental Engineering and Science Program, Manhattan College, Bronx, New York (June 1970).

Dobbins, W. E. "BOD and Oxygen Relationships in Streams," *J. San. Eng. Div. Amer. Soc. Civil Eng.* (SA3), 53 (1964).

Fair, B. M., J. C. Geyer, and D. A. Okum. *Water and Wastewater Treatment and Disposal.* (New York: John Wiley and Sons, Inc., 1968).

Goodman, A. S. and R. J. Tucker. "Time Varying Mathematical Model for Water Quality," *Water Res.* **5**, 227 (1971).

Grantham, G. S. and J. C. Schaake, Jr. "Water Quality Simulation Model," *Proc. Amer. Soc. Civil Eng.* (SA5), 569 (1971).

McCarty, P. L. *Second International Conference on Water Pollution Research.* (New York: Pergamon Press, 1965), p. 169.

O'Connor, D. J. "Mathematical Modeling of Natural Systems," Environmental Engineering and Science Program, Manhattan College, Bronx, New York (1970).

O'Connor, D. J. and D. M. DiToro. "Photosynthesis and Oxygen Balance in Streams," *J. San. Eng. Div. Amer. Soc. Civil Eng.* **96**(SA2), 547 (1970).

Parker, R. A. "Simulation of an Aquatic Ecosystem," *Biometrics* **24**, 803 (1968).

Sandoval, M. "Mathematical Modeling of Nutrient Effects in River Systems," Ph.D. Thesis, University of Notre Dame (1974).

Stumm, W. and M. W. Tenney. "Waste Treatment for the Control of Heterotrophic and Autotrophic Activity in Receiving Waters," Twelfth South Municipal and Industrial Waste Conference, Raleigh, N.C. (1963).

Tenney, M. W., W. F. Echelberger, P. C. Singer, F. H. Verhoff, W. A. Garvey. "Biogeochemical Modeling of Eutrophic Lakes for Water Quality Improvement," *Fourth International Conference on Water Pollution Research.* (New York: Pergamon Press, 1972).

Thomann, R. V., D. J. O'Connor, and D. M. DiToro. "Effect of Nitrification on the Dissolved Oxygen of Streams and Estuaries," Environmental Engineering and Science Program, Manhattan College, Bronx, New York (1970).

Verhoff, F. H., W. F. Echelberger, M. W. Tenney, P. C. Singer, and C. F. Cordeiro. "Lake Water Quality. Prediction Through Systems Modelings," *Proceedings 1971 Summer Computer Simulation Conference*, Boston Mass. (1971), p. 1014.

Wen-Hsiung, Li. "Unsteady Dissolved Oxygen Sag in a Stream," *J. San. Eng. Div. Amer. Soc. Civil Eng.* (SA3), 75 (1962).

9

COMBINING CHEMICAL EQUILIBRIUM
AND PHYTOPLANKTON MODELS—
A GENERAL METHODOLOGY

Dominic M. Di Toro[1]

INTRODUCTION

The seasonal development of phytoplankton biomass can have a marked effect on the inorganic chemical regime of a body of water. The uptake of dissolved inorganic phosphorus, silica, carbon and either hydrogen or hydroxyl ions are the major interactions. The subsequent effects on the pH and the changes in the equilibria can be substantial and interact with the phytoplankton growth kinetics as well as provide sources or sinks of nutrients.

The case of the carbon dioxide—bicarbonate—carbonate equilibria has been incorporated previously into phytoplankton and other water quality models (Chen and Orlob 1972, Yeasted and Shane 1974). The total inorganic carbon and alkalinity is calculated using conventional mass balance equations, and the algebraic equations of the chemical equilibria are then solved. This chapter will present both the theoretical justification for this procedure using the carbonate equilibria as an example as well as some interesting results for this system. A generalized methodology that is applicable to any chemical equilibria model is then elaborated.

The method depends on the existence of a transformation matrix that is orthogonal to the linear vector space spanned by the rows of the reaction matrix of the chemical model. It is shown that such a transformation

---

[1] Environmental Engineering and Science Program, Manhattan College, Bronx, New York.

is easily available and can be made to correspond to well-known chemical properties of the system such as alkalinity. The theory can be easily applied to situations for which the transport mechanisms are dependent on the chemical species concentrations. For example, precipitated minerals in open waters have a sinking velocity whereas dissolved species do not, or dissolved species in pore waters diffuse whereas the precipitates do not. The equations that apply in these cases are presented and discussed, as are the computational advantages of this method, and an estimate is made of the computational burden imposed by the calculations.

DISSOLVED CARBON DIOXIDE

The inclusion of chemical reactions into models of water quality, by contrast to the biochemical and biological reactions usually considered, poses a new set of problems by virtue of the rapid rates and the highly nonlinear forms of the governing reactions. Nevertheless, it appears that these problems can be handled nicely within the context of conservation of mass equations. The investigation of purely chemical equilibrium models has been underway for quite some time. An altogether satisfactory presentation of these results is available (Stumm and Morgan 1970), and application of certain of these models to Great Lakes water chemistry has been undertaken (Kramer 1964, Kramer 1967). From this work and others (Verduin 1956) it is clear that biological reactions affect the chemical systems primarily via the assimilation of dissolved carbon dioxide during phytoplankton growth.

From the point of view of the eutrophication problem, at issue are the purely chemical reactions that phosphorus undergoes, for example precipitation or dissolution of phosphate minerals. These reactions may be of practical importance in terms of the overall dynamics of the phosphorus in the lakes. The other chemically active phytoplankton nutrient, silica, may also be affected by purely chemical reactions involving the alumino-silicate minerals. Thus there are possible practical consequences of these purely chemical reactions. The presentation given below is based primarily on the carbon dioxide equilibria because of its importance and because it can serve as a prototype for all other such reactions.

EQUILIBRIUM ANALYSIS

The carbonate equilibria is treated in the conventional way (Stumm and Morgan 1970, Trussell and Thomas 1971). The major species considered are aqueous or dissolved CO_2 (the concentration of hydrated CO_2 being small), bicarbonate (HCO_3^-) and carbonate ($CO_3^=$) ion, together with the hydrogen (H^+) and hydroxyl (OH^-) ions. At first glance one is tempted to write mass balance equations for each of these species. However, this is complicated by the fact that they undergo reversible ionization reactions, *e.g.,* $CO_2 + H_2O \rightleftharpoons H^+ + HCO_3^-$, which occur very rapidly relative to the time scales of mixing and gas transfer (Kern 1960). In fact it is clear that the individual species are each extremely reactive so that a direct mass balance formulation results in equations that are nonlinear and numerically quite badly behaved.

The very fact that the ionization reactions are rapid leads to a much more tractable formulation in terms of quantities that are conservative relative to the ionization reactions. A formal mathematical discussion is given subsequently. However, it is easy to see the principle involved. Consider the total inorganic carbon concentration, C_T:

$$C_T = [CO_2] + [HCO_3^-] + [CO_3^=] \tag{9.1}$$

It is clear that any reversible ionization reaction will not affect the concentration of C_T (we assume no carbonate precipitates are forming or dissolving). Thus a mass balance equation for C_T can ignore the ionization reactions since they are neither sources nor sinks of C_T.

The electroneutrality requirement can be used to obtain a second conservative quantity. Let M and L be the concentration in equivalents of the metal ions and ligands, *e.g.,* N_a^+, K^+, C_a^{++}, ..., and Cl^-, $SO_4^=$, ..., that are present and assumed not to react appreciably with the carbonate or hydrogen-hydroxide species. Then it is clear that a charge balance equation:

$$M - L = [HCO_3^-] + 2[CO_3^=] + [OH^-] - [H^+] \tag{9.2}$$

is unaffected by ionization reactions so that the quantity called alkalinity

$$Alk = M - L \tag{9.3}$$

is also conservative in the sense that a mass balance equation for *Alk* need not consider ionization sources or sinks. It can be shown there are no other independent conservative quantities for this system; the other common conservative quantities (Stumm 1970) are linear combinations of C_T and *Alk.*

If C_T and *Alk* are known, the concentration of all the species being considered is easily obtained. In fact, using the equilibrium equations (Stumm 1970, Trussell and Thomas 1971)

$$[H^+] \ [OH^-] \quad = K_W \tag{9.4}$$
$$[H^+] \ [HCO_3^-] = K_1 [CO_2]$$
$$[H^+] \ [CO_3^=] \ = K_2 [HCO_3^-]$$

where K_W, K_1, and K_2 are equilibrium constants. It is true that

$$\text{Alk} = \alpha(H) \ C_T + \gamma(H) \tag{9.5}$$

where

$$\alpha(H) = \frac{1 + 2K_2/[H^+]}{1 + [H^+]/K_1 + K_2/[H^+]} \tag{9.6}$$

and

$$\gamma(H) = K_W/[H^+] - [H^+] \tag{9.7}$$

If the alkalinity concentration is large relative to $\gamma(H) = [OH^-] - [H^+]$, which is the common situation, then Equation (9.5) becomes

$$\frac{\text{Alk}}{C_T} = \alpha(H) \tag{9.8}$$

so that only the ratio of alkalinity to total inorganic carbon is important in determining the resulting pH whenever $\text{Alk} \simeq \text{Alk} + [H^+] - [OH^-]$. Using tabulated values of K_1 and K_2 for pure water Figure 9.1 illustrates the pH, Alk/C_T relationship as a function of temperature. These figures are most useful since they represent the carbonate equilibria completely as a function of the ratio of the critical conservative quantities. Taken together with the conventional log concentration plot of the species *vs.* pH, they provide a complete picture of the behavior of the equilibria.

The equations governing the distribution of Alk and C_T are available as mass balances, independent of the ionization reactions. For example, in a one-dimensional vertical lake model the equations would be

$$\frac{\partial [\text{Alk}]}{\partial t} - \frac{1}{A} \frac{\partial}{\partial z} \ (EA \ \frac{\partial [\text{Alk}]}{\partial z}) = S_{\text{Alk}} \tag{9.9}$$

$$\frac{\partial [C_T]}{\partial t} - \frac{1}{A} \frac{\partial}{\partial z} \ (EA \ \frac{\partial [C_T]}{\partial z}) = S_{C_T} \tag{9.10}$$

where $A(z)$ is the area of the lake normal to z, $E(z)$ is the vertical dispersion coefficient, S_{Alk} and S_{C_T} are the internal sources and sinks of

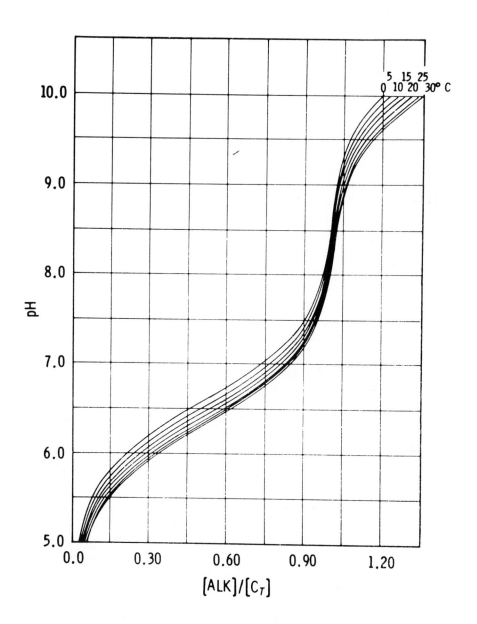

Figure 9.1   Relationship of pH and the ratio, Alk/C_T.

alkalinity and total inorganic carbon. A zero mass flux boundary condition at z=0 is required for alkalinity. The air-water exchange condition for C_T, representing the exchange of carbon dioxide is

$$-E \frac{\partial [C_T]}{\partial z} = K_L \, (CO_2(g) - CO_2(aq)) \qquad (9.11)$$

where K_L is a mass transfer coefficient, $CO_2(g)$ and $CO_2(aq)$ are the concentrations of gaseous and dissolved carbon dioxide. This is a non-linear boundary condition since $CO_2(aq)$ is a nonlinear function of the dependent variables of the problem [Alk] and [C_T], so that a numerical procedure is required.

The internal sources and sinks of alkalinity and inorganic carbon are due to biological reactions. For example, the nitrification reaction (if bacterial synthesis is ignored) can be represented as

$$NH_4^+ + 2 \, O_2 \rightarrow NO_3 + H_2O + 2H^+ \qquad (9.12)$$

so that as ammonia is oxidized to nitrate, alkalinity decreases (*i.e.,* H^+ is produced) and a term representing this sink of alkalinity would be included in the expression of S_{Alk}.

Similarly as phytoplankton grow they utilize inorganic carbon, and this sink would be included in S_{C_T}. These mass balance equations would then be solved in conjunction with the equations for the other species of interest.

LINEAR APPROXIMATION

It is clear that mass balance equations that ignore the ionization reactions are only correct if the dependent variable is conservative relative to these reactions. Consider the conservative quantity $[C_T]$-[Alk] = [CO_2 - Acidity]. Clearly it is conservative since it is the difference of conservative quantities. From Equations (9.1), (9.2) and (9.3), it is the case that

$$[CO_2\text{-Acy}] = [CO_2] + [H^+] - [CO_3^=] - [OH^-] \qquad (9.13)$$

However, this equation implies that over a certain range of pH and C_T, the major component of CO_2-acidity is [CO_2] itself. Figure 9.2 presents the percentage of error in the approximation $[CO_2\text{-Acy}] \simeq [CO_2]$. The chemical reasoning that leads to the same conclusion begins with the observation that in the above pH range, the alkalinity is primarily bicarbonate ion. If a source introduces only carbon dioxide, this does not change the alkalinity; therefore the concentration of bicarbonate remains constant.

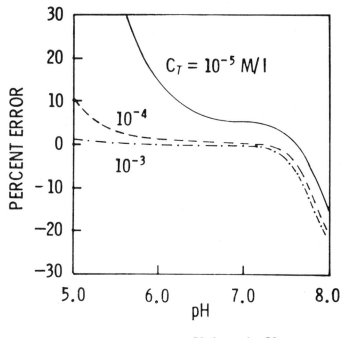

Figure 9.2   Percentage error $\dfrac{CO_2(approx) - CO_2}{CO_2}$ 100%

Hence the introduced CO_2 does not ionize to HCO_3^- and H^+ but rather remains as CO_2. Hence ionization reactions can be ignored. Thus from Equations (9.9) and (9.10) the mass balance equation for CO_2-acidity is

$$\frac{\partial[CO_2\text{-Acy}]}{\partial t} - \frac{1}{A}\frac{\partial}{\partial z}\left(EA\frac{\partial[CO_2\text{-Acy}]}{\partial z}\right) = S_{C_T} - S_{Alk} \qquad (9.14)$$

and the reaeration source of $[CO_2\text{-Acy}]$ can be approximated as K_L $([CO_2(g)] - [CO_2 \cdot Acy])$. Thus within the errors in Figure 9.2 the equation for $[CO_2\text{-acidity}]$ is linear. And since the alkalinity equation is also linear, simple analytical solutions are available and superposition can be applied to buildup a solution when many sources and sinks are active. This is in contrast to the nonlinear C_T equation, which must be solved with all sources and sinks at once, an inconvenient situation.

Unfortunately this linear region does not include the pH range typically encountered so that this simplification is not available. However, for more acidic situations this simplification provides a useful and direct avenue for the solution of the inorganic carbon–alkalinity mass balance equation.

THEORETICAL CONSIDERATIONS

A formal derivation of Equations (9.9) and (9.10) requires that ionization reactions be considered explicitly as sources and sinks of the components. Assume that the following three reactions are the mechanisms by which hydration and ionization take place:

$$CO_2 + H_2O \rightleftharpoons H^+ + HCO_3^- \quad ; \quad r_1 = k_{v1}(K_1[CO_2] - [H^+][HCO_3^-])$$

$$HCO_3^- \rightleftharpoons H^+ + CO_3^= ; \quad r_2 = k_{v2}(K_2[HCO_3^-] - [H^+][CO_3^=])$$

$$H_2O \rightleftharpoons H^+ + OH^- \quad ; \quad r_3 = k_{v3}(K_W - [H^+][OH^-]) \tag{9.15}$$

where the net rates, r, are the rates of the reactions proceeding to the right, with velocity coefficients k_v and that the law of mass action correctly describes the rates of reaction as indicated. For notational convenience let D denote the mass transport differential operation, i.e., the lefthand side of Equations (9.9) and (9.10) are denoted by $D[Alk]$ and $D[C_T]$. The conservation of mass equations for each species is obtained by equating mass transport to sources and sinks:

$$D[CO_2] = -r_1 + S_{CO_2} \tag{9.16}$$

$$D[H^+] = r_1 + r_2 + r_3 + S_H \tag{9.17}$$

$$D[HCO_3^-] = r_1 - r_2 + S_{HCO_3} \tag{9.18}$$

$$D[CO_3^=] = r_2 + S_{CO_3} \tag{9.19}$$

$$D[OH^-] = r_3 + S_{OH} \tag{9.20}$$

where S equals whatever direct sources of the species exist, and r is the rate associated with the assumed reactions. It is, of course, possible in principle to solve these simultaneous nonlinear differential equations numerically. The values chosen for k_v would be large enough so that the solution is unaffected by their magnitude.

This procedure has been tried and found to be quite unstable numerically, the cause being the "stiffness" of the equations due to the large reaction rates that make convergence difficult to achieve. Also the k_v's are, in a sense, artifacts of the formulation since they are present only to introduce the ionizations correctly in a formulation requiring such reactions to be included as kinetic expressions. Therefore it would be convenient to obtain equations devoid of the reaction rates entirely. This can be done by defining new variables that are appropriate sums of the species in such a way that the reaction terms cancel out. Thus if $C_T = [CO_2] + [HCO_3^-] + [CO_3^=]$, then by adding Equations (9.16), (9.18) and (9.19) the result is

$$DC_T = S_{CO_2} + S_{HCO_3} + S_{CO_3} \qquad (9.21)$$

The other variable is, of course,

$$Alk = [HCO_3^-] + 2 [CO_3^=] + [OH^-] - [H^+] \qquad (9.22)$$

and since D is a linear operator,

$$D[Alk] = D[HCO_3^-] + 2D [CO_3^=] + D[OH^-] - D[H^+] \qquad (9.23)$$

and using Equations (9.17), (9.18), (9.19), and (9.20), the result is Equation (9.9) as before, with the r's cancelling out in a satisfying way.

So far no approximations have been made; the equations for Alk and C_T are exact, regardless of the magnitude of the k_v's. They are the invariants of the reactions (Shapiro 1962) in the sense that their values are unchanged by the kinetics. The approximation, and it is an excellent one for this application, is to compute the concentrations of the species from the equilibrium expressions, $e.g.$, (9.4), rather than the differential Equations (9.16-9.20).

There is a slight theoretical flaw in the above derivation since the explicit forms are assumed for the reactions considered and a reaction of the form

$$CO_2 + OH^- = HCO_3^-$$

was not explicitly considered. However, it is a simple matter to verify that the inclusion of this reaction does not affect the validity of the resulting equations for Alk and C_T, but no guarantee has been presented that such a reaction does not exist (Shapiro 1962).

DEFINITIONS

The definitions used below conform to common usage for chemical equilibrium formulations (Van Zeggeren and Storey 1970, DeHaven 1968). Let A_i, i = 1, ..., N_s be the names of the N_s species of interest, be they molecules, ions, or complexes. Examples are CO_2, H^+, and $CaHCO_3^+$. Their concentrations are denoted by $[A_i]$. Let B_j, $j = 1, ..., N_c$ be the names of the N_c components used to specify the composition of the N_s species. For gas reactions the common choices are the elements that make up the species, $e.g.$, C, H, O. For aqueous reactions simple ions or molecules are usually chosen, $e.g.$, CO_2 or H^+, H_2O. It is shown subsequently that the specific set of components chosen is immaterial. The concentrations of the components are denoted by $[B_j]$.

The relationships between the species A_i and the components B_j are specified by the formula matrix a_{ij}, with dimensions N_s by N_c.

$$A_i = \sum_{j=1}^{N_c} a_{ij} B_j \qquad i = 1, ..., N_s \qquad (9.24)$$

Typically a_{ij} are integers. For example, if $A_1 = CaHCO_3^+$, then the equation corresponding to Equation (9.24) is

$$CaHCO_3^+ = 1.0 \ Ca^{++} + 1.0 \ H^+ + 1.0 \ CO_3^= \qquad (9.25)$$

assuming the components are as indicated.

Suppose that there exist N_r chemical reactions that involve the species A_i and that they are represented by the equations

$$\sum_{i=1}^{N_s} \nu_{ji} A_i = 0 \qquad j = 1, ..., N_r \qquad (9.26)$$

where ν_{ji} is the stoichiometric coefficient of species A_i in reaction j. The matrix ν_{ji} is the reaction matrix with dimensions N_r by N_s. An example of such an equation is $HCO_3^- - H^+ - CO_3^= = 0$ which represents the reaction

$$HCO_3^- = H^+ + CO_3^= \qquad (9.27)$$

These fast chemical reactions are assumed to be at equilibrium on the time and space scales of interest. Typically these reactions satisfy equilibrium mass action laws of the form

$$\prod_{i=1}^{N_s} [A_i]^{\nu_{ji}} = K_j \qquad j = 1, ..., N_r \qquad (9.28)$$

where $[A_i]$ is the molar concentration of A_i and K_j is the equilibrium constant of reaction j.

For the purposes of the subsequent derivation, assume that the rates of these reactions, while fast, are finite, and the rate at which A_i is produced in reaction j is $\nu_{ji}R_j$. Using the above bicarbonate ionization as an example and assuming it follows a mass action law, the reaction rate is

$$R = k_2[H^+][CO_3^=] - k_1[HCO_3^-] \tag{9.29}$$

where k_1 and k_2 are the rates of the forward and backward reaction respectively. The sign convention is chosen so that if $R > 0$, the reaction is producing HCO_3^- since the reaction is going to the left and the rate of HCO_3^- production is 1.0 R; it is "producing" H^+ and $CO_3^=$ at rates - 1.0R. This corresponds to the signs of the ν for this reaction and the definition of R.

Let S_i be the net source of A_i due to all the other reactions occurring at slow rates and not explicitly included in the chemical reactions; *e.g.*, if $A_i = CO_2$, then a part of S_i would be the production of CO_2 by the decay of organic carbon via bacterial oxidation.

CONTINUITY OF MASS EQUATIONS

If the concept of fast reactions at equilibrium is not employed, it is possible to write directly the mass balance equations for each A_i. Using the general mass conservation equation, with dispersion matrix E and velocity vector v, the equations are

$$\frac{\partial[A_i]}{\partial t} + \nabla \circ (-E\nabla[A_i] + v[A_i]) = S_i + \sum_{j=1}^{N_r} \nu_{ji} R_j \tag{9.30}$$

The source term includes each reaction j that is producing A_i at the rate $\nu_{ji}R_j$. The problem with solving these equations as they stand is that the forward and backward reaction rates for each reaction correspond to a very short time scale and the numerical techniques to handle such equations are unwieldy even for ordinary differential equations (Fowler and Wanten 1967). Further, all that is usually known for these reactions is the ratio of forward to backward reaction rate, *i.e.*, the equilibrium constant. Finally there are N_s such equations to be solved simultaneously.

CONSERVATION EQUATIONS FOR THE COMPONENTS

It has been suggested (Shapiro 1962; Prober, *et al.* 1971; Galant and Appleton 1973) that various transformations be employed to make

Equation (9.30) more tractable. The crux of the idea is to eliminate the terms $\sum_j \nu_{ji} R_j$ which cause all the trouble and replace Equation (9.30) with an alternate set of equations. The following fact leads to a convenient choice for the transformation.

Theorem (1): If the component concentrations are conserved by the reactions then the formula matrix a_{ij} is orthogonal* to the reaction matrix ν_{ji}, *i.e.*

$$\sum_{k=1}^{N_S} \nu_{jk}\, a_{ki} = 0 \quad i=1, ..., N_c; \; j=1, ..., N_r \tag{9.31}$$

Proof: Consider a short interval of time Δt and assume each fast reaction advances an amount $\Delta \xi_j$ so that the change in $[A_k]$ during Δt is

$$\Delta\,[A_k] = \sum_{j=1}^{N_r} \nu_{jk} \Delta\xi_j \tag{9.32}$$

During this interval the change in each component B_i is

$$\Delta[B_i] = \sum_{k=1}^{N_S} a_{ki}\,\Delta[A_k] = \sum_{k=1}^{N_S} a_{ki} \sum_{j=1}^{N_r} \nu_{jk}\Delta\xi_j \tag{9.33}$$

but $\Delta[B_i]$ must be zero for each component B_i, $i=1, ..., N_c$ because the concentration $[B_i]$ is fixed by the number of gram atoms that exist for each component. For example the total amount of carbon atoms is fixed (excluding nuclear reactions), and it is assumed that during Δt there are no slow reactions taking place to supply or remove any carbon. Hence

$$\sum_{j=1}^{N_r} \Delta\xi_j \sum_{k=1}^{N_S} a_{ki}\, \nu_{jk} = 0 \tag{9.34}$$

But the choice of $\Delta\xi_j$ is arbitrary, and the above equation must be satisfied for all possible choices, which implies that Equation (9.31) must be the case.

---

*The term "orthogonal" is used as a shorthand description of the property that each of the rows of one matrix, as vectors, are orthogonal to each of the columns of the other matrix. The terminology is adopted from the definition of orthogonal subspaces. The theorem states that the subspace spanned by the rows of the reaction matrix is orthogonal to the subspace spanned by the columns of the formula matrix.

This fact suggests that the conservation equation should be transformed by the formula matrix. Thus multiplying the N_s conservation equation (9.30) by the transpose of the matrix a_{ij} yields

$$D \sum_{i=1}^{N_s} a_{ik}[A_i] = \sum_{i=1}^{N_s} a_{ik} S_i + \sum_{j=1}^{N_r} R_j \sum_{i=1}^{N_s} a_{ik} \nu_{ji} \quad k=1, ..., N_s \quad (9.35)$$

where $D = \partial/\partial t + \nabla \circ (-E\nabla + v)$. But the above theorem implies that the terms involving R_j are zero. Further, the terms

$$\sum_{i=1}^{N_s} a_{ik} [A_k] = [B_k] \qquad k=1, ..., N_n \qquad (9.36)$$

are the concentration of the kth component B_k, so that the conservation of mass equations for the components B_k

$$\frac{\partial [B_k]}{\partial t} + \nabla \circ (-E\nabla[B_k]+v[B_k]) = \sum_{i=1}^{N_s} a_{ik} S_i \qquad (9.37)$$

are independent of the fast reactions. It is the case that once the N_c concentrations of the components B_k are known, the N_s- N_c equilibrium equations* for A_i, Equation (9.28) can be used in conjunction with the N_c component mass balance Equation (9.36) to solve for $[A_i]$. Thus the time history of the species A_i is computed by solving simultaneously N_c differential Equation (9.37) which gives the component mass balances, the N_s species equilibrium Equation (9.28), and the N_c component Equation (9.36) which are algebraic. This is a numerical task that is much more tractable than the N_s differential equations for the species themselves.

It is important to realize that the equations for the components contain the sources and sinks of the species due to the slow reactions so that the production and utilization of the chemical species by biological reactions, for example, is correctly accounted for. And since the concentrations of the chemical species are being calculated from the equilibrium equations, these concentrations can be used as dependent variables in the formulation of the slow reaction kinetics.

The results of this analysis are quite in keeping with the structure of chemical equilibrium models. The component concentrations and the equilibrium constants completely specify an equilibrium solution, so it is

---

*This assumes that $N_s = N_r + N_c$ which is true (Van Zeggeren and Storey 1970) for nonredundant component specifications. A redundant component specifications. A redundant specification occurs if, for example, two components are always in a fixed proportion in every species.

quite natural to expect the component concentrations to play an important role. What has been shown is that the components can be treated in exactly the same way as any other water quality variable, which is a pleasing and fortunate result.

EXAMPLE—THE CO_2–H_2O SYSTEM

In order to make the general notations and results specific, consider the carbon dioxide system as an example. For the components choose CO_2, H^+, and H_2O. The formula matrix for the species of interest is then

A_i	B_j	CO_2	H^+	H_2O
CO_2		1	0	0
HCO_3^-		1	-1	1
$CO_3^=$		1	-2	1
H^+		0	1	0
H_2O		0	0	1
OH^-		0	-1	1

The reaction matrix corresponding to the three reactions listed in Equation (9.15) is

Reaction	A_i	CO_2	HCO_3^-	$CO_3^=$	H^+	H_2O	OH^-
1		1	-1	0	-1	1	0
2		0	1	-1	-1	0	0
3		0	0	0	-1	1	-1

To show that the formula matrix ν_{ji} is orthogonal to the reaction matrix, simply perform the multiplication

$$
\begin{bmatrix}
1 & -1 & 0 & -1 & 1 & 0 \\
0 & 1 & -1 & -1 & 0 & 0 \\
0 & 0 & 0 & -1 & 0 & -1
\end{bmatrix}
\begin{bmatrix}
1 & 0 & 0 \\
1 & -1 & 1 \\
1 & -2 & 1 \\
0 & 1 & 0 \\
0 & 0 & 1 \\
0 & -1 & 1
\end{bmatrix}
=
\begin{bmatrix}
0 & 0 & 0 \\
0 & 0 & 0 \\
0 & 0 & 0
\end{bmatrix}
\tag{9.38}
$$

In fact it is clear that using CO_2 and H^+ as the components of the species leads to equations for total inorganic carbon for the CO_2 equation and negative alkalinity for the H^+ equation. The H_2O equation is, of course, not solved in practice since the amounts of H_2O produced or utilized are small relative to the amount present.

GENERAL REACTIONS

The above argument is incomplete since it appears to depend on the reaction matrix ν_{ji}; yet in the final equation it is not to be found. However, nothing specific was required of the reaction matrix, and, in fact, it can be shown that any reaction matrix with certain properties will do.

To make this specific, consider the rank of the formula matrix a_{ij}. Recalling that the rank of the matrix is the number of linearly independent rows and columns, it follows that a_{ij} should be chosen so that its rank is N_c. That is, the components are not redundant in some way, for example if two components always appear in a fixed proportion in each species. If the rank of $a_{ij} < N_c$, then the redundant components can be removed until the rank of $a_{ij} = N_c$.

The number of species involved is N_S. The components provide N_C linearly independent equations; thus there must be at least $N_s - N_c = N_r$ linearly independent reactions to specify the equilibrium problem correctly. Let ν_{ji} be such a matrix with rank N_r. Theorem (1) guarantees that this matrix is orthogonal to the formula matrix. Now suppose that other possible reactions exist. For example, HCO_3^- can be formed by two reactions (Miller, et al. 1971):

$$HCO_3^- + H^+ = CO_2 + H_2O \tag{9.39a}$$

$$HCO_3^- = CO_2 + OH^- \tag{9.39b}$$

depending on the pH of the solution. Let μ_{ji} be another reaction matrix, with dimensions N_{r*} by N_s representing N_{r*} additional reactions. Theorem (1) indicates that μ_{ji} must also be orthogonal to a_{ij} since the theorem applies to any set of reactions that conserve components. Since a_{ij} is orthogonal to ν_{ji} and rank (a_{ij}) + rank $(\nu_{ji}) = N_c + N_r = N_s$, the N_c columns of a_{ij} and the N_r rows of ν_{ji} span an N_s dimensional Euclidian linear vector space. Thus the N_{r*} rows of μ_{ji} can be written as linear combinations of these vectors. However, since ν_{ji} is orthogonal to a_{ij}, only the rows of ν_{ji} are required. Thus there exists a matrix β_{ij} so that

$$\mu_{ji} = \sum_{k=1}^{N_r} \beta_{jk}\nu_{ki} \tag{9.40}$$

Letting R'_j be the rates of the μ_{ji} fast reactions the conservation Equation (9.30) becomes

$$D[A_i] = S_i = \sum_{j=1}^{N_r} \nu_{ji} R_j + \sum_{j=1}^{N_r*} \mu'_{ji} R'_{j'} \tag{9.41}$$

And using Equation (9.40) yields

$$D[A_i] = S_i + \sum_{j=1}^{N_r} \nu_{ji} (R_j + \sum_{j'=1}^{N_r*} \beta_{j'j} R'_{j'}) \tag{9.42}$$

or

$$D[A_i] = S_i + \sum_{j=1}^{N_r} \nu_{ji} R''_j \tag{9.43}$$

where

$$R''_j = R_j + \sum_{j'=1}^{N_r*} \beta_{j'j} R'_{j'} \tag{9.44}$$

and the preceding arguments apply.

Thus the details of the reaction schemes chosen do not affect the choice of the transformation and the resulting use of the components in the mass balance equations. Also it can be shown (Shapiro 1962) that changing the reaction scheme modifies the equilibrium constants in a simple way and makes no fundamental changes.

The choice of the specific components to be used is arbitrary also, so long as they correctly represent the species. A new choice of components implies a new formula matrix, α_{ij}. But α_{ij} must be orthogonal to the reaction matrix by Theorem (1). Hence α_{ij} is related to a_{ij} by a linear transformation by reasoning analogous to that which established the existence of β_{ij}. This transformation can then be applied to the conservation law, Equation (9.30), yielding an analogous equation for the new components.

EXAMPLE–CONTINUED

Consider the reaction

$$HCO_3^- = CO_2 + OH^- \tag{9.45}$$

which has a reaction matrix μ.

$$[\ -1 \quad 1 \quad 0 \quad 0 \quad 0 \quad -1 \ ] \tag{9.46}$$

It is then a simple matter to verify that the β matrix for this case is

$$\beta = [\ -1 \quad 0 \quad 1\ ] \qquad (9.47)$$

which indicates that this new reaction can be expressed as the difference between reaction (3) and reaction (1). Thus it is redundant. In fact it can be shown by direct calculation that the rank of the reaction matrix ν is three and the rank of the formula matrix a is also three; since they are orthogonal they span the six dimensional vector space corresponding to the six species of interest.

To see that the choice of components has no effect on the formulation consider the same system but now use the atoms, C, H, O as the components. The formula matrix is

A_i	B_j	C	H	O
CO_2		1	0	2
HCO_3^-		1	1	3
$CO_3^=$		1	0	3
H^+		0	1	0
H_2O		0	2	1
OH^-		0	1	1

The reaction matrix is as previously and the orthogonality follows by direct calculation:

$$\begin{bmatrix} 1 & -1 & 0 & -1 & 1 & 0 \\ 0 & 1 & -1 & -1 & 0 & 0 \\ 0 & 0 & 0 & -1 & 1 & -1 \end{bmatrix} \begin{bmatrix} 1 & 0 & 2 \\ 1 & 1 & 3 \\ 1 & 0 & 3 \\ 0 & 1 & 0 \\ 0 & 2 & 1 \\ 0 & 1 & 1 \end{bmatrix} = \begin{bmatrix} 0 & 0 & 0 \\ 0 & 0 & 0 \\ 0 & 0 & 0 \end{bmatrix} \qquad (9.48)$$

However this choice is less desirable from a practical point of view since it is not clear how to ignore the "H_2O" component mass balance equation. For this reason it is advisable to pick the components that correspond to important chemical facts, such as the large concentration of H_2O relative to all other components or the utility of the concept of alkalinity and the desirability of having it explicitly in the formulation.

SPECIES-DEPENDENT SLOW REACTIONS

It is often the case that important slow reactions exist that depend on the concentration $[A_i]$ of the species themselves rather than the component concentration $[B_j]$. For example the exchange of CO_2 across the air-water interface is a function of CO_2 (aq) and not the total inorganic carbon, a usual choice for a component in the carbonate equilibrium. In principle, species-dependent slow reactions are not a problem since as the component concentrations $[B_j]$ are being calculated in time and space, the chemical equilibrium can be obtained for each spatial location at each time step by solving the chemical equilibrium equations with the appropriate $[B_j]$. However, this can be a massive amount of calculation.

An approximate method is available that is based on a set of linear equations relating the changes in species composition to component changes. Thus for an equilibrium solution $[A_j]$ based on the component concentrations $[B_j]$ it is possible to compute the Jacobian matrix: $J_{ij} = \partial[A_i]/\partial[B_j]$ about the solution $[A_i]$. The formulas for the Jacobian are (Clasen 1965)

$$J_{ij} = [A_i] \sum_k a_{ik} \, r_{kj}^{-1} \qquad (9.49)$$

where r_{kj}^{-1} are the elements of the inverse of the matrix with elements

$$r_{kj} = \sum_i a_{ik} \, a_{ij} [A_i] \qquad (9.50)$$

Consider the small changes δB_j that occur at a location over a time interval, as computed during the solution of Equation (9.37). In order to evaluate the new species concentration $[A'_i]$, a Taylor series approximation is employed

$$[A'_i] = [A_i] + \sum_j J_{ij} \, \delta B_j \qquad (9.51)$$

The errors in this approximation are due to the higher order terms in Equation (9.51), which have been neglected, and the Jacobian itself, since it is computed around the solution $[A_i]$. If Equation (9.51) is used to propagate $[A'_i]$ to $[A''_i]$ due to $\delta B'_i$

$$[A''_i] = [A'_i] + \sum_j J_{ij} \, \delta B'_j \qquad (9.52)$$

the Jacobian ought to be computed around A'_i. It is possible to compute this new Jacobian using $[A'_i]$ by solving Equation (9.49). However to do even this every time step Δt may be a burden. Therefore it seems reasonable to use J_{ij} as long as it appears to be accurate and then either update the Jacobian, in which case all species concentrations must be propagated using Equation (9.51), or resolve the chemical equilibrium and compute the Jacobian. Which one is practical depends on the particular application.

SPECIES-DEPENDENT TRANSPORT

It is also possible that the transport mechanisms are affected by species concentration. An example is the vertical settling of a precipitated mineral. In this case a settling velocity exists only if the species is insoluble, e.g., $CaCO_3(s)$. This case is slightly more complex if only velocities are involved. Consider a velocity v_i in the z direction, which is a function of A_i, e.g., $v_i = \bar{v}$ for soluble species and $v_i = \bar{v} + v_s$ for precipitates. Let D be all other species-independent transport mechanisms. Then the mass transport Equation (9.30) becomes, after transformation,

$$\frac{\partial [B_k]}{\partial t} + \sum_{i=1}^{N_s} v_i a_{ik} \frac{\partial [A_i]}{\partial z} + D[B_k] = \sum_{i=1}^{N_s} a_{ik} S_i \tag{9.53}$$

In order to find the gradients of $[A_i]$, the chain rule is employed:

$$\frac{\partial [A_i]}{\partial z} = \sum_{j=1}^{N_c} \frac{\partial [A_i]}{\partial [B_j]} \frac{\partial [B_j]}{\partial z} = \sum_{j=1}^{N_c} J_{ij} \frac{\partial [B_j]}{\partial z} \tag{9.54}$$

Thus Equation (9.53) becomes

$$\frac{\partial [B_k]}{\partial t} + \sum_{j=1}^{N_c} (\sum_{i=1}^{N_c} v_i a_{ik} J_{ik}) \frac{\partial [B_j]}{\partial z} + D[B_k] = \sum_{i=1}^{N_s} a_{ik} S_i \tag{9.55}$$

Therefore the conservation of mass equations for $[B_k]$ are coupled through the velocity term. If B is the vector of component $[B_k]$ and S is the vector of sources, S_i, then the set of Equations (9.55) can be written

$$\frac{\partial B}{\partial t} + \tau \frac{\partial B}{\partial z} + DB = AS \tag{9.56}$$

where τ is a matrix with k, j^{th} element $[\Sigma_i v_i a_{ik} J_{ij}]$. Since τ is a function of B through J_{ij} (see Equations 9.50, 9.49, 9.36, and 9.28), this equation is nonlinear, although as mentioned above τ is approximately constant for small changes in B. The solution technique is similar to that discussed above.

The situation for a species-dependent dispersion coefficient, E_i, is quite complex and is included for completeness. Transforming Equation (9.30) leads to a term of the form

$$\sum_{i=1}^{N_s} E_i \, a_{ik} \, \frac{\partial^2 [A_i]}{\partial z^2} \tag{9.57}$$

and by differentiating Equation (9.54) it is true that

$$\frac{\partial^2 [A_i]}{\partial z^2} = \sum_j J_{ij} \, \frac{\partial^2 [B_j]}{\partial z^2} + \sum_j \frac{\partial J_{ij}}{\partial z} \, \frac{\partial [B_j]}{\partial z} \tag{9.58}$$

Thus the gradient of the Jacobian is required. Using Equation (9.49) the gradient of the Jacobian is

$$\frac{\partial J_{ij}}{\partial z} = (\sum_k a_{ik} r_{kj}^{-1}) \, (\sum_k J_{ik} \, \frac{\partial [B_k]}{\partial z}) + [A_i] \, \sum_k a_{ik} \, \frac{\partial}{\partial z} \, r_{kj}^{-1} \tag{9.59}$$

which involves the gradient of the elements of an inverse matrix. If $R = \{r_{kj}\}$ in the matrix then it is true (Bellman 1960) that

$$\frac{\partial}{\partial z} R^{-1} = -R^{-1} \, \frac{\partial R}{\partial z} \, R^{-1} \tag{9.60}$$

The gradient of R itself is found from its definition:

$$\frac{\partial r_{kj}}{\partial z} = \sum_i a_{ik} a_{ij} \, \sum_k J_{ik} \, \frac{\partial [B_k]}{\partial z} \tag{9.61}$$

Thus all terms in the conservation equations are in terms of the derivatives of $[B_k]$ and the Jacobian. Although this is a complex set of relationships, in principle it correctly accounts for the differing dispersion coefficients for each species.

AN ALTERNATE METHOD

For applications involving only a few species, an alternate method of solution is available (Clasen 1967). Although it appears we can generalize to cases involving transport, it is included here in the form derived for kinetic equations. Without transport terms, Equation (9.30) becomes

$$\frac{d[A_i]}{dt} = S_i + \sum_{j=1}^{N_r} \nu_{ji} \, R_j \qquad i=1, ..., N_s \tag{9.62}$$

Since the fast reactions are assumed to be at equilibrium it is true that

$$\prod_{i=1}^{N_s} [A_i]^{\nu_{ji}} = K_j \qquad j=1, ..., N_r \tag{9.63}$$

The unknowns in the problem are the reaction rates R_j. Consider the result of taking the time derivative of the above Equation (9.63)

$$\sum_{i=1}^{N_s} \nu_{ji} [A_i]^{\nu_{ji}-1} \frac{d[A_i]}{dt} \prod_{\substack{k=1 \\ k \neq i}}^{N_s} [A_k]^{\nu_{jk}} = 0 \qquad (9.64)$$

and combining terms and using Equation (9.63) yields

$$K_j \sum_{i=1}^{N_s} \nu_{ji} [A_i]^{-1} \frac{d[A_i]}{dt} = 0 \qquad j=1, ..., N_r \qquad (9.65)$$

If the kinetic Equations (9.62) are substituted into this equation the result is

$$\sum_{i=1}^{N_s} \nu_{ji} [A_i]^{-1} (S_i + \sum_{k=1}^{N_r} \nu_{ki} R_k) = 0 \qquad j=1, ..., N_r \qquad (9.66)$$

or, in standard form

$$\sum_{k=1}^{N_r} R_k \sum_{i=1}^{N_s} \nu_{ji}\nu_{ki} [A_i]^{-1} = -\sum_{i=1}^{N_s} \nu_{ji} [A_i]^{-1} S_i \qquad j=1, ..., N_r \qquad (9.67)$$

The above equation is a set of linear algebraic equations (with a symmetric coefficient matrix) which can be solved for the unknown R_k. It is interesting to note that the equilibrium constants, K_j, do not appear in either this equation or the kinetic Equation (9.62). The solution is started with initial conditions that are in equilibrium, however, and the Equations (9.67) keep the solution at equilibrium throughout the integration of the kinetic equations. Of course Equations (9.67) are solved at each integration step of the kinetic Equations (9.62).

A great deal of computation is needed for applications involving more than a few species since a differential equation is required for each species in contrast to the preceding methods, which require conservation equations for the components only.

The comparison of the species method to the component method is made in Table 9.1. The former requires the solution of N_s differential equations and $N_r = N_s - N_c$ linear equations; the component method requires the solution of N_c differential equations and N_c nonlinear algebraic equations. However for typical applications a very good starting solution is available for the nonlinear equations (the solution at the previous time step) and it is estimated that no more than five interactions of a Newton-Raphson method would be required for solution. Table 9.1b presents this comparison in terms of the ratio $\alpha = N_s/N_c$. The

Table 9.1   Comparison of Computational Effort

a

Method	Number of Differential Equations	Number of Linear Equations	Number of Iterations	Number of Operations[a]
Species	N_S	$N_S\text{-}N_C$	1	$(N_S\text{-}N_C)^3$
Components	N_C	N_C	~ 5	$5N^3{}_C$

b

$\alpha = N_S/N_C$

Method				
Species	αN_C	$(\alpha\text{-}1)N_C$	1	$(\alpha\text{-}1)^3 N_C^3$
Components	N_C	N_C	~ 5	$5N_C^3$

[a]For the linear equations only, assuming N^3 operations for N equations.

computational efforts for the linear equation portion of the method are equal for $\alpha = 2.71$ with the species method becoming inferior as α increases. The number of differential equations is always greater for the species method. Thus for even moderately complex models, $\alpha > 3$, the component method is favored.

ACKNOWLEDGMENTS

The assistance and cooperation of the members of our research group—Donald J. O'Connor, Robert V. Thomann and Richard P. Winfield—as well as the members of the Grosse Ile Laboratory of EPA—Nelson Thomas, Tudor Davies, and William Richardson—is gratefully appreciated and acknowledged. This work was performed under EPA Research Grant No. R 800610 as part of the International Field Year on the Great Lakes (IFYGL) investigations.

REFERENCES

Bellman, R. *Introduction to Matrix Analysis*. (New York: McGraw-Hill, 1960).

Chen, C. W., G. T. Orlob. "Ecological Simulation for Aquatic Environments," Report to Office of Water Resources Research OWRR C-2044, Water Resources Engineers Inc., Walnut Creek, California (December 1972).

Clasen, R. J. "The Numerical Solution of the Chemical Equilibrium Problem," Rand Corp., Santa Monica, California RM-4345-PR, January 1965.

Clasen, R. J. "The Numerical Integration of Kinetic Equations for Chemical Systems Having Both Slow and Fast Reactions," Rand Corp., Santa Monica, California, P-3547, September 1967.

DeHaven, J. C. "Prerequisites for Chemical Thermodynamic Models of Living Systems," Rand Corp., Santa Monica, California, RM-5691-PR, November 1968.

Fowler, M. E. and R. M. Warten. "A Numerical Technique for Ordinary Differential Equations with Widely Separated Eigen Values," *IBM, J. Res. Develop.* (September 1967).

Galant, S. and J. P. Appleton. "The Rate-Controlled Method of Constrained Equilibrium Applied to Chemical Reactive Open Systems," Fluid Methanics Laboratory, MIT, No. 73-6, July 1973.

Kern, D. M. "The Hydration of Carbon Dioxide," *J. Chem. Ed.* **37**(1), 14 (1960).

Kramer, J. R. "Theoretical Model for the Chemical Composition of Fresh Water with Application to the Great Lakes," Great Lakes Research Division, **11**, 147 (1964).

Kramer, J. R. "Equilibrium Models and Composition of the Great Lakes," in *Equilibrium Concepts in Natural Water Systems.* Advances in Chemistry Series No. 67 (Washington, D.C.: American Chemical Society, 1967).

Miller, R. F., D. C. Berkshire, J. J. Kelley, and D. W. Hood. "Method for Determination of Reaction Rates of Carbon Dioxide with Water and Hydroxyl Ion in Sea Water," *Environ. Sci. Technol.* **5**, 2 (1971).

Prober, R., Y. Y. Haimes, M. Teraquichi, and W. Moss. *An Ecosystem Model of Lake Algae Blooms.* (Cleveland, Ohio: Case Western Reserve University, 1971).

Shapiro, N. "Analysis by Migration in the Presence of Chemical Reaction," Rand Corp., Santa Monica, California, P-2596, June 1962.

Stumm, W. and J. J. Morgan. *Aquatic Chemistry.* (New York: Wiley-Interscience, 1970).

Trussell, R. R. and J. F. Thomas. "A Discussion of the Chemical Character of Water Mixtures," *J. Amer. Water Works Assoc.* **63**(1), 49 (1971).

VanZeggeren, F. and S. H. Storey. *The Computation of Chemical Equilibria* (Cambridge: Cambridge University Press, 1970).

Verduin, J. "Energy Fixation and Utilization by Natural Communities in Western Lake Erie," *Ecol.* **37**, 40 (1956).

Yeasted, J. G. and R. M. Shane. "pH Profiles in a Free-Flowing River System with Multiple Acid Loads," Dept. of Civil Engineering, Report, Carnegie-Mellon University, Pittsburgh, Pennsylvania (1974).

10

DYNAMICS OF AN ALGAL-PROTOZOAN GRAZING INTERACTION

Fredric G. Bader, A. G. Fredrickson, and H. M. Tsuchiya[1]

INTRODUCTION

One of the apparent goals of the study of microbial population dynamics is to uncover basic generalized principles regarding microbial growth, microbial interactions with their environments, and interactions between microorganisms. This appears to apply to those who engage in pure mathematical modeling, pure, mixed, or impure culture laboratory studies, studies of natural ecosystems, or combinations of the above. However, proponents of generalized principles seem to be frequently coming into conflict. As a result, it is becoming more and more necessary to recognize that each organism, each environment, and each interaction may have its own peculiarities, some of which we are not yet capable of handling experimentally or mathematically. One might state that although phosphorus may be the element that limits algal growth in many lakes, it is unlikely that it is in all lakes; although the Monod model may approximate some substance-limited growth rate relationships, it does not apply to all; although some predator-prey relationships may be inherently unstable, others may be quite stable. This does not say that the search for generalized principles is wrong. Quite the contrary, such a search is a fundamental part of science. It does say that one must be both cautious and open-minded in the application of generalized principles to a given system.

---

[1] Department of Chemical Engineering and Materials Science, Department of Microbiology, University of Minnesota, Minneapolis, Minnesota 55455.

In the following discussion, a number of different mathematical models for the interaction between predator and prey will be compared. Many of these have already been published separately, while others are a direct result of existing theory. This comparison will show that (1) the stability of the predator-prey interaction is highly dependent upon the growth kinetics of the predator, and (2) increased effectiveness of the predator at low prey concentrations tends to destabilize the predator-prey interaction.

An experimental laboratory study of the predator-prey relationship between the blue-green algae *Anacystis nidulans* and the ciliated protozoan *Colpoda steinii* will then be briefly discussed. The details of the experimental study may be found elsewhere (Bader, 1974). The key feature of the experimental study that will be noted here is that the predator growth rate was a function of both the algal "substrate" concentration and the predator population density, itself. This laboratory study is included to demonstrate that the unique properties of each organism, each environment, and each interaction create difficulties in mathematical and experimental treatment. Hence, caution is required in drawing general conclusions regarding the stability of such systems.

PREDATOR-PREY MODELS

General Form of Predator-Prey Model

The simplest mathematical model that may be used to describe a predator-prey interaction in continuous culture must account for growth of both species, hydrodynamic washout of both species, and the predator growth-related removal of the prey. A general form of this model may be written as

$$\frac{da}{dt} = G_a - a/\theta - b_2\, G_p \tag{10.1}$$

$$\frac{dp}{dt} = G_p - p/\theta \tag{10.2}$$

where

G_a = growth rate of the prey (a)
G_p = growth rate of its predator (p)
b_2 = stoichiometric coefficient (= 1/yield coefficient)
θ = holding time of the culture vessel (= 1/dilution rate)

Substrate-Limited Growth of Prey

When dealing with substrate-limited prey growth, where the substrate is fed to the chemostat (continuous culture vessel) as part of the feed stream, an additional equation is required that accounts for the input of substrate with the feed, the hydrodynamic washout of substrate, and the prey growth-related removal of substrate from the vessel. This equation is

$$\frac{dS}{dt} = \frac{(S_f - S)}{\theta} - b_1 \, G_a \qquad (10.3)$$

In this equation, S_f is the feed substrate concentration, S is the substrate concentration in the chemostat, and b_1 is another stoichiometric coefficient (= 1/yield coefficient).

The above equations (10.1, 10.2, 10.3) in general ignore death, maintenance, and complex cell life cycles and assume a homogeneous distribution of cells and substrates and a single limiting substrate.

For prey that are capable of exponential growth, the growth rate (G_a) may be written in general form as

$$G_a = \nu' a \qquad (10.4)$$

where ν' is the specific growth rate that is at least a function of the maximum specific growth rate ν and the substrate concentration S. If we assume that the substrate dependence is of hyperbolic form, then we have the Monod (1942, 1949) model for substrate limited growth, or

$$G_a = \nu \, \frac{Sa}{K_1 + S} \qquad (10.5)$$

where K_1 is the so-called Michaelis constant. Other models for substrate-limited prey growth have been discussed by Blackman (1905), Dabes, Finn and Wilke (1973), Jost (1973) and others. Only the Monod model will be considered here.

Predator Growth Models

For a predator that is capable of exponential growth, the rate G_p may be written as

$$G_p = \mu' p \qquad (10.6)$$

where μ' is the specific growth rate. For the simple models that will be considered here, it may be assumed that μ' is a function of only the maximum specific growth rate of the predator μ and the prey population density a.

Three different models that relate the predator specific growth rate to the prey population density will be considered. The relationship between μ' and a for these three models is shown in Figure 10.1.

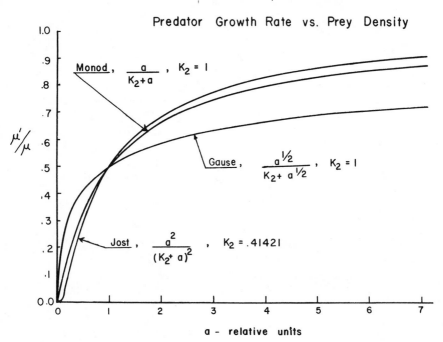

Figure 10.1 The functional response curves for three predator models. The fraction of the maximum predator growth rate is plotted *vs.* the prey population density for three models for predator growth. The Michaelis constants (K_2) have been chosen so that when $a=1$, $\mu'/\mu=\frac{1}{2}$.

The first model is the Monod model, which may be written as

$$\mu' = \mu \frac{a}{K_2 + a}; \qquad G_p = \mu \frac{a \, p}{K_2 + a} \qquad (10.7)$$

where K_2 is the Michaelis constant. As shown in Figure 10.1, the Monod model predicts that at low prey concentrations, the predation rate is linearly dependent upon the prey concentration. (This is essentially equivalent to the Lotka-Volterra model with the exception that the

Monod model recognizes the well-known biological fact that growth rate reaches a maximum level at some saturating substrate concentration). This is what one would expect if predation occurs by random encounters.

The Monod model has been used by Proper and Garver (1966) to describe the growth of *Colpoda steinii* on *E. coli*, and by Tsuchiya *et al.* (1972) to describe the growth of the cellular slime mold *Dictyostelium discoideum* on *E. coli*. In the work of Tsuchiya *et al.*, the Monod model was used to describe the predator-prey interaction in continuous culture. It was found that this model agreed well with the experimental data in that continuous oscillations of the predator and prey populations were predicted. These oscillations were independent of initial conditions, and predicted amplitudes and periods of the oscillations agreed well with those obtained experimentally.

Jost *et al.* (1973a) found objections to the Monod model in studying predator prey interactions between *Tetrahymena pyriformis* and the bacteria *E. coli* and *Azotobacter vinelandii*. These objections were based upon the following experimental observations. First, the minimum prey concentrations predicted by the Monod model during oscillations of the predator and prey were much lower than was experimentally observed. Second, the Monod model for predator growth predicted continuous oscillations of the predator and prey populations at long holding times (slow dilution rates). Experimentally, Jost *et al.* (1973a) observed that, at holding times greater than 20 hours, damped oscillations were obtained.

Due to these observations, Jost *et al.* (1973a) developed their so-called multiple saturation kinetic model which is

$$\mu' = \frac{\mu\ a^2}{(K_2 + a)\ (K_3 + a)}; \quad G_p = \frac{\mu\ a^2\ p}{(K_2 + a)\ (K_3 + a)} \tag{10.8}$$

where K_2 and K_3 are Michaelis constants. If K_3 is set equal to zero, then this model reduces down to the Monod model. This model differs most from the Monod model when $K_3 = K_2$; it is this limiting case of the model that will be used in the analyses that follow. As can be seen in Figure 10.1, the predator specific growth rate drops off as the square of the prey concentration for low prey concentrations. This essentially states that the predator is less effective in capturing prey at low prey population densities. In essence, a psuedo-threshold value of prey is required before any significant predator growth is observed. This phenomenon may be partially explained in biological terms by assuming that, at low prey population densities, the predator has to spend an increasing amount of time and energy in hunting, between captures of prey.

The third case that we will consider was used by Gause (1934) to describe his predation experiments between *Didinium* and *Paramecium.* Gause described the predator growth rate as

$$G_p = \mu\, p\sqrt{a} \qquad (10.9)$$

By recognizing the fact that predator growth rate reaches a maximal value at high prey concentrations, Gause's model may be generalized as

$$\mu' = \frac{\mu\, a^{\frac{1}{2}}}{(K_2 + a^{\frac{1}{2}})}\; ; \qquad G_p = \frac{\mu\, a^{\frac{1}{2}}\, p}{(K_2 + a^{\frac{1}{2}})} \qquad (10.10)$$

As shown in Figure 10.1, this model predicts that the predator becomes more effective in capturing prey at low prey concentrations. Biologically, this might occur if chemotaxis plays a significant role in the predator's search for food or if the predator became more active in its search for food.

In order to determine which model for predator growth should be used for a particular experimental system, accurate data for predator growth as a function of prey concentration must be obtained. Such data are characteristically very difficult to obtain directly. Hence, the use of a particular model is frequently determined by making various assumptions that may or may not be valid. In addition, certain aspects of the experimental system such as (1) wall growth of the predator and/or prey, (2) morphological changes of predator and/or prey, (3) changes in yield coefficients, (4) changes in the abiotic phase of the experimental environment, (5) changes in maintenance requirements, (6) changes in the physiological state of the predator and/or prey, and (7) the mixing rate of the culture, may affect the relationship between the predator growth rate and the prey population density. It is interesting to note that the experiments of Gause (1934) were not stirred while those of Tsuchiya *et al.* (1972) and Jost *et al.* (1973a) were well stirred. It is also interesting to note that Gause provided his *Paramecium* (the prey) with pulses of bacterial food. This substrate for his prey was not one of the experimentally measured parameters.

Stability Analysis

If we substitute the Monod model (Equation 10.5) for substrate-limited growth of the prey into Equations (10.1) and (10.3), and substitute one of the three prey-limited predator growth models (Equations 10.7, 10.8, or 10.10) into Equations (10.1) and (10.2), then we can apply stability analysis to the set of equations. From the stability analysis, we obtain operating diagrams that show the predicted nature of the predator-prey interaction at various values of holding time θ and feed substrate concentration S_f. The three types of steady states that may occur are (1) washout of predator and prey, (2) washout of predator but not prey, and (3) the normal steady state. The stability analysis tells us (see Lotka 1925) whether a particular steady state is:

1. a saddle point—steady state is unstable and will never be reached.
2. a stable node—steady state is approached directly.
3. an unstable node—steady state cannot be reached.
4. a stable focus—steady state is approached in a damped oscillatory fashion.
5. an unstable focus—steady state cannot be reached. In general, the system will oscillate with the amplitude of the oscillations increasing until a limit cycle is reached or one of the species is driven to extinction.

The stability analysis is essentially used to determine how the system of differential equations will behave if a small perturbation from the steady state is introduced. One linearizes the perturbations about a singular point and determines, by solving for the eigenvalues of the system of equations, whether the perturbations will increase or decrease with time. Because of the linearization, the stability analysis only holds in a strict sense for small perturbations from the steady state. One must check, by numerical integration, whether the stability analysis also holds for large perturbations from the steady state. In general, large perturbations show the same stability as small perturbations, but each case must be checked and verified. The operating diagrams are generated by performing a computer scan of the eigenvalues at various conditions of holding time and feed substrate concentration.

The detailed mathematics of the stability analyses will not be presented here. These may be found in Bader (1974). Some of the analyses may also be found elsewhere (such as Jost *et al.* 1973b). The values for the various constants that are used in the models may also be found in Bader (1974). Jost (1972) has stated that the general shape of an operating diagram does not change significantly with reasonable variations in the values of the constants in the equations. This has also been observed in Bader (1974) for the cases presented below.

Figure 10.2 shows the operating diagram for the case where we have substrate-limited prey growth and the Gause model for predator growth (Equation 10.10). For the bulk of the operating region, the normal steady state is an unstable node. Numerical integration of the predator-prey equations in this region shows that the predator drives the prey to extinction, which agrees with the observations of Gause (1934).

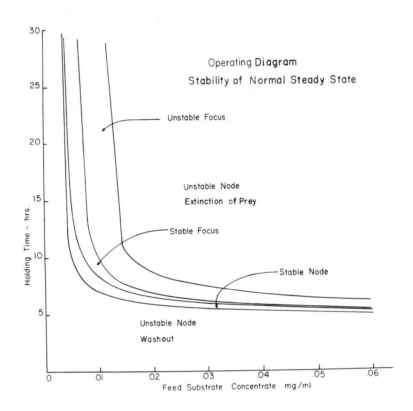

Figure 10.2 The normal steady-state operating diagram for substrate-limited growth of the prey (Monod model) and prey-limited growth of the predator (Gause model). For the bulk of the operating region, the prey is driven to extinction.

Figure 10.3 shows the operating diagram for the same predator-prey interaction but with the Monod model used for predator growth (Equation 10.7). For the bulk of the operating region, the normal steady state is an unstable focus where numerical integration of the equations shows that continuous oscillations of the predator and prey populations occur. This agrees with similar analyses by Canale (1970) and with the experimental observations of Tsuchiya *et al.* (1972).

Figure 10.4 shows the operating diagram for the same predator-prey interaction again except with the multiple saturation (Jost) model for predator growth (Equation 10.8). Here, the bulk of the operating region

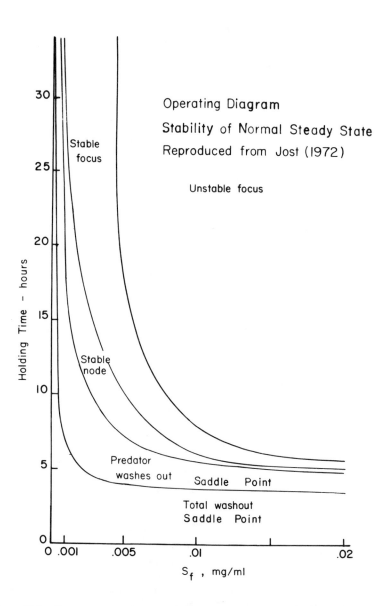

Figure 10.3 The normal steady-state operating diagram for substrate-limited growth of the prey (Monod model) and prey-limited growth of the predator (Monod model). For the bulk of the operating region, the two populations coexist in a continuous oscillatory fashion. Reproduced from Jost (1972).

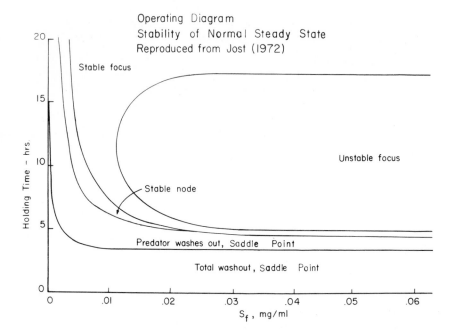

Figure 10.4 The normal steady-state operating diagram for substrate-limited growth
of the prey (Monod model) and prey-limited growth of the predator (Jost model).
At low holding times, continuous oscillations occur. At higher holding times,
damped oscillations occur that converge on the steady state.
Reproduced from Jost (1972).

is divided into two parts. At short holding times, the normal steady
state is an unstable focus, with continuous oscillation of the predator
and prey populations occurring. For longer holding times, the oscillations
become damped (a stable focus) so that the steady state should actually
be reached. This agrees with the experimental results of Jost *et al.*
(1973a).

Figures 10.2, 10.3, and 10.4 demonstrate the following two main
points:

1. The stability of the predator-prey interaction is highly dependent
   upon the growth kinetics of the predator.
2. Increased effectiveness of the predator at low prey concentrations
   tends to destabilize the predator-prey interaction.

Light-Limited Growth of Prey

Certain types of nutrients may be fed to a chemostat independent of the main liquid feed stream. Two common examples of this would be gases (CO_2, O_2) and light. When one deals with light-limited algal growth, the light is supplied continuously from outside the chemostat. With low algal population densities, the algal specific growth rate is a function of the external light intensity I_O only. When the algal population density becomes large, the algal culture growth rate becomes a function of the incident light intensity and the algal population density. Mutual shading occurs and the algal culture shifts from the exponential growth that occurs in dilute cultures to linear growth. Linear growth occurs when the algal population absorbs all of the light incident upon the culture. Both modes of growth may be accounted for in models for light-limited growth such as the exponential model (Bracket 1935, van Oorschot 1955), the Tamiya model (Tamiya *et al.* 1953) and the Steele model (Steele 1965). The Steele model appears to be the model of preference since it accounts for the decrease in algal growth rate at high light intensities. Since light is independent of the feed stream, and its utilization can be directly accounted for in the algal growth model, Equation (10.3) of the predator-prey interaction equations has no meaning and is disregarded.

Figure 10.5 shows a plot of specific algal growth rate as a function of light intensity for dilute algal populations as predicted by the Steele model. This model may be written as

$$G_a = a \nu \frac{I}{I_m} e^{(1 - \frac{I}{I_m})} \tag{10.11}$$

where ν is the maximum specific growth rate of the algal prey a, I is the light intensity observed by the algae, and I_m is the light intensity at which the maximum exponential growth rate is obtained.

For cultures of finite thickness, the absorption of light by the algal culture follows the Beer-Lambert law:

$$I = I_o e^{-\epsilon ax} \tag{10.12}$$

where I_o is the light intensity at the surface of the culture, I is the light intensity at some distance x from the surface, and ϵ is the extinction coefficient of the algal suspension. By substituting Equation (10.12) into Equation (10.11), and integrating with respect to x, the average rate of growth of the algal population in a vessel of finite surface area A and thickness W may be obtained as

$$G_a = \frac{\nu}{\epsilon W} \; [e^{1 - \frac{I_o}{I_m} e^{-\epsilon aw}} - e^{1 - \frac{I_o}{I_m}} \;]$$ (10.13)

This expression assumes that the algal culture is well mixed. Note that in Equation (10.13), G_a is proportional to a for small values of a and is independent of a for large values of a.

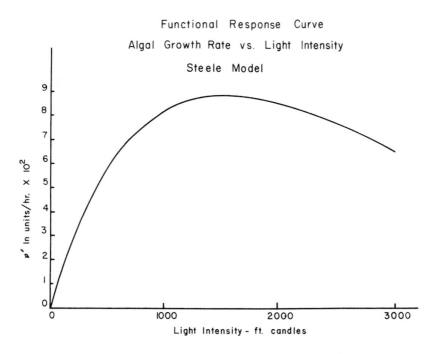

Figure 10.5   The functional response curve for the light-limited growth of the prey as predicted by the Steele model.

Stability Analysis

We can now substitute the Steele model (Equation 10.13) for light-limited growth of the prey into Equation (10.1), and substitute one of the three prey-limited predator growth models (Equations 10.7, 10.8 or 10.10) into Equations (10.1) and (10.2) and apply stability analysis to the equations as before. The stability analyses will yield operating diagrams that show the predicted nature of the predator-prey interaction at various values of holding time θ and external light intensity I_o. The results are shown in Figures 10.6, 10.7, and 10.8.

Figure 10.6 shows the operating diagram for the case of light-limited prey and the Gause model for predator growth (Equation 10.10). For the bulk of the operating region, the normal steady state is an unstable focus. Numerical integration shows that for this case, the oscillations of the predator and prey populations would normally increase in amplitude until the prey is driven to extinction. The unstable node region that occurs at long holding times also represents a region where the prey is

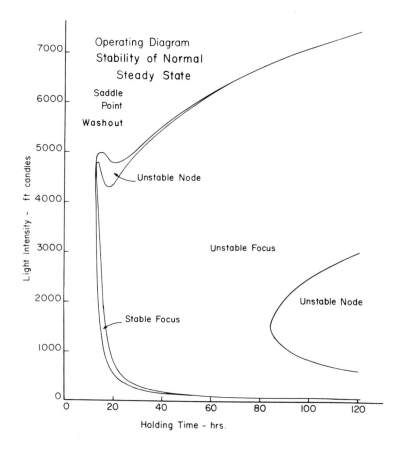

Figure 10.6 The normal steady-state operating diagram for light-limited growth of the prey (Steele model) and prey-limited growth of the predator (Gause model). For the bulk of the operating region, the predator drives the prey to extinction.

driven to extinction. The total washout steady state (saddle point) that occurs at high light intensities is due to the inhibiting effect of high light intensity on algal growth.

The operating diagram for the same predator-prey interaction only with the Monod model for predator growth (Equation 10.7) is shown in Figure 10.7. Here, the bulk of the operating region shows that the normal steady state is an unstable focus. Numerical integration of the equations shows that continuous oscillations (limit cycle behavior) occurs in this case. The unstable node region that occurred in Figure 10.6 does not appear.

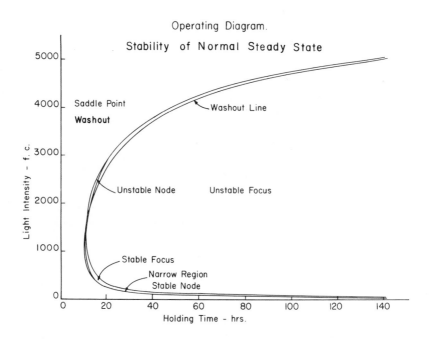

Figure 10.7 The normal steady-state operating diagram for light-limited growth of the prey (Steele model) and prey-limited growth of the predator (Monod model). For the bulk of the operating region, the two populations coexist in a continuous oscillatory fashion.

Figure 10.8 shows the operating diagram for the predator-prey interaction with the Jost model for predator growth (Equation 10.8). Here the bulk of the operating region for the normal steady-state is a stable focus (damped oscillations) with a small region of unstable focus (limit cycle) behavior.

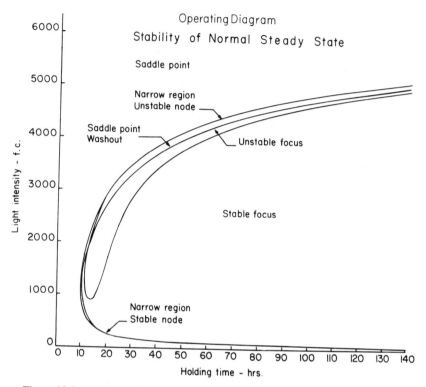

Figure 10.8 The normal steady-state operating diagram for light-limited growth of the prey (Steele model) and prey-limited growth of the predator (Jost model). For the bulk of the operating region, damped oscillations occur that converge on the steady state.

The two observations that were made with respect to the substrate-limited predator-prey interactions (Figure 10.2, 10.3, 10.4) also hold true for the light-limited predator-prey interactions (Figures 10.6, 10.7, 10.8). These were that the stability of the predator-prey interaction is highly dependent upon the growth kinetics of the predator; and increased effectiveness of the predator at low prey concentrations tends to destabilize the predator-prey interaction. A comparison of the substrate-limited and the light-limited predator-prey interactions is shown in Table 10.1. It is interesting to note that the nature of the prey kinetic-growth model has little effect upon the stability of the predator-prey interaction.

Table 10.1 Stability of Theoretical Predator-Prey Interactions

Prey Growth Rate Model	Predator Growth Rate Model	Figure	Response
I. Substrate-limited growth of prey			
Monod $G_a = \nu as/(K_1 + s)$	Gause $\dfrac{\mu p a^{1/2}}{(K_2 + a^{1/2})}$	10.2	Unstable node, extinction of prey
	Monod $\dfrac{\mu p a}{(K_2 + a)}$	10.3	Unstable focus, continuous oscillatory coexistence
	Jost $\dfrac{\mu p a^2}{(K_2 + a)^2}$	10.4	Low holding time—unstable focus; long holding time—stable focus, continuous or damped oscillations
II. Light-limited growth of prey			
Steele $G_a = \dfrac{\nu}{\epsilon W}\left[e^{1 - \frac{I_o}{I_m} e^{-\epsilon a W}} - e^{1 - \frac{I_o}{I_m}} \right]$	Gause	10.6	Unstable focus, extinction of prey
	Monod	10.7	Unstable focus, continuous oscillatory existence
	Jost	10.8	Stable focus, damped oscillatory coexistence

EXPERIMENTAL OBSERVATIONS

The models for the light-limited predator-prey interaction, which are presented here, were initially studied in preparation for a laboratory study of a light-limited algal-protozoan interaction. The above models predicted that the predator may drive the prey to extinction, or that the predator and prey may coexist with damped or continuous oscillations occurring, depending upon the growth kinetics of the predator.

In the experimental study, the predator was the ciliated protozoan *Colpoda steinii* and the prey was the blue-green alga *Anacystis nidulans*. This protozoan-algal interaction was studied in chemostat type vessels under conditions of batch and continuous operation with light limitation of algal growth. The details of the experimental study have been presented elsewhere (Bader 1974). Only two experiments will be presented here.

Figure 10.9 shows the results of a batch experiment where the external light intensity was 1100 footcandles. This is near the experimentally observed optimum of 1400 footcandles. At time zero, the *Colpoda* cells were inoculated into the batch culture vessel, which already contained a dense population of the alga, *Anacystic nidulans* ($\sim 2 \times 10^{-3}$ g/ml).

The *Colpoda* initially grew exponentially to a high population density where crowding or some other form of self-inhibition occurred. The *Colpoda* population then began to encyst, with approximately 99% of the population becoming encysted. The cysts stick to the wall of the culture vessel and are, therefore, removed from the homogeneous liquid phase. This accounts for the decline in the *Colpoda* population. Following encystment, the *Colpoda* population began to increase due to growth and/or excystment of cysts from the wall. This occurred until the population, again, became self-inhibitory. The *Colpoda* population oscillates in batch growth in this fashion. However, the degree of encystment lessened and the minimum in the *Colpoda* population became progressively higher following each peak in the *Colpoda* population.

During the oscillations of the *Colpoda* population, the *Anacystis* population density remains very high and is not significantly affected by the growth of the predator. Enough algae are always present to support the growth of the predator, so encystment is not due to starvation for lack of food. Eventually, both algal and protozoan populations reach some steady-state level where the algal population remains very dense and the protozoan population remains very close to the encystment level.

When the *Colpoda* are inoculated into *Anacystis* cultures that are initially dilute and growing exponentially, the same basic results occur.

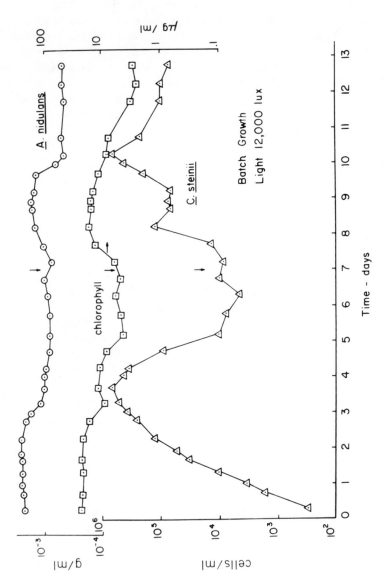

Figure 10.9 A batch growth curve for the light-limited predator-prey interaction between the blue-green alga *Anacystis nidulans* and the ciliated protozoan *Colpoda steinii*. The algal population is so dense that it is not significantly affected by the large protozoan population. The *Colpoda* encyst as a response to their own population density.

However, the number and amplitude of the protozoan oscillations are decreased and the populations approach a steady-state more directly.

It is interesting that a steady-state appears to eventually occur in the batch system. Why the oscillations in the *Colpoda* population tend to dampen out is not clearly known. However, the most probable explanation for this is that a balance between encystment and excystment eventually develops so that the net flux of cysts from the liquid phase to the wall is counter-balanced by excystment of cysts on the wall.

Figure 10.10 shows the results of a continuous culture experiment where the incident light intensity is 1100 footcandles and the holding time is 36 hours (dilution rate = 0.0278). Again, the encystment of the *Colpoda* population is observed in the presence of sufficient algal population to support growth. The interesting feature of this and similar continuous culture data that was obtained is that the system appears to be highly damped. The two populations converge upon their steady-state levels much more directly than predicted by the earlier models.

DISCUSSION

As summarized in Table 10.1, the above theoretical analysis demonstrates that a reasonably wide variety of predator-prey interactions are potentially possible for even simple systems that would fall within the constraints of the models discussed. The only models that were presented above that predict that predator and prey cannot coexist are the cases where we have the Gause model for predator growth (Figures 10.2 and 10.6). The other four models presented predict that predator and prey can coexist over a wide range of operating conditions. Their coexistence, however, may be oscillatory.

It is then apparent that, at least theoretically, coexistence of predator and prey can occur. This is contrary to the initial observations of Gause, which have been generalized by others. The concept that heterogeneity is necessary for predator-prey coexistence, as proposed by van den Ende (1973), is theoretically not necessary. In addition to the theory, the fact that predator and prey can coexist has been verified experimentally by Jost *et al.* (1973a), Tsuchiya *et al.* (1972), Canale *et al.* (1973) and others.

Generally, it is assumed that nonhomogeneities in a system such as a surface provide the prey with a place to hide from the predator. However, the experimental observations presented above show a case where the opposite is true. The wall, in this case, provides a refuge for the predator during conditions that are unfavorable to its growth. In addition, the population of cysts on the wall of the vessel prevents the predator

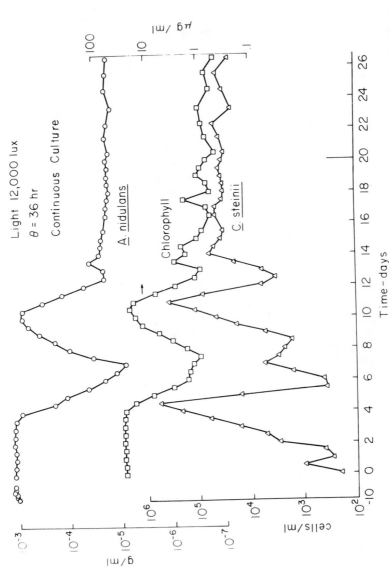

Figure 10.10 A continuous culture study of the light-limited predator-prey interaction between *Colpoda* and *Anacystis*. The interaction is highly damped and converges to a steady state much more rapidly than predicted by any of the models. Encystment, attachment of cysts to the walls of the vessel, and probable excystment of the *Colpoda* is observed.

from washing out of the system and allows the predator to respond more rapidly to favorable growth conditions, such as an increase in the algal population density. It is interesting to note that this type of encystment-excystment response was also detected by Salt (1967) for his *Woodruffia-Paramecium* system. It would then appear that nonhomogeneities can benefit either or both populations in a given system. In the experimental work of van den Ende (1973) as well as that presented briefly in Figures 10.9 and 10.10, the interaction of cultures with the wall appears to have a stabilizing effect on the system.

Certainly, one should not assume that all predator-prey interactions lead to coexistence. When either predator or prey populations become very low, stochastic processes may lead to the extinction of either species. Bartlett (1957), for instance, has shown that the stochastic version of the Lotka-Volterra equation will lead to extinction of the predators. Other predator-prey interactions may, by their nature, result in the extinction of either species. In general, the above models would indicate that the prey would be driven to extinction in cases where the predator is extremely effective in hunting prey at low prey population densities, as demonstrated by the Gause model for predator growth kinetics. Higher organisms, with more complex life cycles, may also behave differently than the more simple organisms to which the above models apply. Addition of higher trophic levels, additional predator or prey species, the production of cysts, spores, and eggs will certainly affect the nature of a specific predator-prey interaction. The response of a predator to its own population density, as mentioned above, provides a potentially interesting case.

The response of predators to their own population density has been discussed by Salt (1967). In general, there appears to be an upper limit to most predator population densities. In social predators, certain number densities may be beneficial. However, as predator density increases, competition for food and territory tends to decrease the efficiency of the predators. For aquatic predators, such as *Colpoda*, the upper limit may occur due to metabolite accumulation, competition for food and physical crowding.

If an upper limit to the population density of a predator exists as a general phenomenon, then the results shown in Figures 10.9 and 10.10 may have general limnological implication. These experiments demonstrate that dense algal populations may not be controllable by predation due to the self-limiting response of the predator. If the algal population density in Figure 10.9 was limited by a nutrient such as phosphate to a maximum algal density of 10^{-4} g/ml, then clearly the protozoa could remove most of the algae from the ecosystem. However, if additional phosphate is added to the system, the algal population density will eventually exceed the level that is controllable by the predator as shown in Figure 10.9. The significance of predation as a controlling factor of the algal population would be decreased greatly. If this is the case, then part of the eutrophication process in lakes may be the lessening of the role of predation as a controlling factor for the algal population.

CONCLUSION

It should be apparent from the preceding discussion that generalizations concerning predator-prey interactions must be treated cautiously and with an open mind. Clearly, the growth kinetics of the predator are of considerable importance in determining the outcome of a predator-prey interaction. Unfortunately, statistically representative experimental information is not yet available to verify completely any of the predator models presented. However, the theoretical observation, that increased predator effectiveness in grazing at low prey population densities tends to destabilize the predator-prey interaction, may be relatively important to consider.

It is also apparent that seemingly simple predator-prey interactions can turn out to be surprisingly complex. This was certainly the case with the *Colpoda-Anacystis* interaction that was described. Here in addition to the normal growth, consumption, and hydrodynamic washout considerations, the response of the predator to its own density, the encystment of the *Colpoda*, the attachment of the cysts to the wall of the vessel, and the subsequent excystment of the *Colpoda* cysts appear to have significant impact on this predator-prey interaction.

REFERENCES

Bader, F. G. "The Predator-Prey Interactions between a Protozoan and a Blue-Green Alga," Ph.D. thesis, University of Minnesota (1974).

Bartlett, M. S. "On Theoretical Models for Competitive and Predatory Biological Systems," *Biometrika* **44**, 27 (1957).

Blackman, F. F. "Optima and Limiting Factors," *Annals Bot.* **19**, 281 (1905).

Bracket, F. S. "Light Intensity and CO_2 as Factors in Photosynthesis of Wheat," *Cold Springs Harbour Symposium on Quant. Biol.* **3**, 117 (1935).

Canale, R. P., T. D. Lustig, P. M. Kehrberger, and J. E. Salo. "Experimental and Mathematical Modeling Studies of Protozoan Predation on Bacteria," *Biotech. Bioeng.* **15**, 707 (1973).

Canale, R. P. "An Analysis of Models Describing Predator-Prey Interactions," *Biotech. Bioengr.* **12**, 353 (1970).

Dabes, J. N., R. K. Finn, and C. R. Wilke. "Equations of Substrate-Limited Growth: The Case for Blackman Kinetics," *Biotech. Bioeng.* **15**, 1159 (1973).

Gause, G. F. *The Struggle for Existence.* (New York: Hafner Publ. Co., 1934), p. 133.

Jost, J. L. "Dynamics of a Symbiotic System. Interactions in a Microbial Food Web," Ph.D. thesis, University of Minnesota (1972).

Jost, J. L., J. F. Drake, A. G. Fredrickson, and H. M. Tsuchiya. "Interactions of *Tetrahymena pyriformis, Escherichia coli, Azotobacter vinelandii*, and Glucose in a Minimal Medium," *J. Bacteriol.* **113**, 834 (1973a).

Jost, J. L., J. F. Drake, H. M. Tsuchiya, and A. G. Fredrickson. "Microbial Food Chains and Food Webs," *J. Theor. Biol.* **41**, 461 (1973b).

Lotka, A. J. *Elements of Physical Biology.* (Baltimore, Md.: Williams and Wilkins Co., 1925), p. 148.

Monod, J. *Recherches sur la Croissance des Cultures Bacteriennes.* (Paris: Hermann et Cie, 1942).

Monod, J. "The Growth of Bacterial Cultures," *Ann. Rev. Microbiol.* **3**, 371 (1949).

Proper, G. and J. C. Garver. "Mass Culture of the Protozoa *Colpoda steinii*," *Biotech. Bioeng.* **8**, 287 (1966).

Salt, G. W. "Predation in an Experimental Protozoa Population (*Woodruffia-Paramecium*)," *Ecol. Monogr.* **37**, 113 (1967).

Steele, J. H. "Notes on Some Theoretical Problems in Production Ecology. Primary Production in Aquatic Environmenta," in *Me. Inst. Indrobiol.* 18 Supplement, C. R. Goldman, Ed. (Berkeley: University of California Press, 1965), p. 383.

Tamiya, H., E. Hase, K. Shibata, A. Mituya, T. Nihei, and T. Sasa. in *Algal Culture from Laboratory to Pilot Plant*, J. S. Burlew, Ed. (Washington, D.C.: Carnegie Inst. Washington, 1953), p. 204.

Tsuchiya, H. M., J. F. Drake, J. L. Jost, and A. G. Fredrickson. "Predator-Prey Interactions of *Dictyostelium discoideum* and *Escherichia coli* in Continuous Culture," *J. Bacteriol.* **110**, 1147 (1972).

van den Ende, P. "Predator-Prey Interactions in Continuous Culture," *Science* **181**, 562 (1973).

van Oorschot, J. L. P. "Effects of Light Intensity and Temperature upon Algal Growth. Mededelingen van De Landbourhogoschool," *Wageningen. Nederland. Verhardeling* **5**, 225 (1955).

11

APPLICATION OF A MODEL OF ZOOPLANKTON COMPOSITION TO PROBLEMS OF FISH INTRODUCTIONS TO THE GREAT LAKES

Donald C. McNaught and Donald Scavia[1]

INTRODUCTION

Size-selective predation by fishes on zooplankton, as well as the species composition of both herbivores and algae, has recently drawn the most attention in discussions of biological control of algal production. Current papers have regarded changes involving the succession toward smaller forms as an adjustment to fish predation (Allan 1974, Dodson 1974). Such a strategy includes high probability for small adult size at age of first reproduction as a developmental factor of major importance (McNaught 1975). However, little attention has been given to the seasonality of predatory intensity in determining the relative abundance of zooplankton prey over a growing season. The relative importance of size-selection by fishes as well as the effects of the nonuniform seasonal inshore distribution of fish predators   both upon zooplankton composition and abundance will be examined using a simple two-equation model.

Size-selective feeding by zooplankton upon phytoplankton, in contrast, may determine species success in an oligotrophic environment. In low nutrient situations where small nannoplankton dominate, those calanoid copepods that are small-particle specialists may likewise dominate (McNaught 1975), chiefly because they have a high efficiency for removing

[1] Department of Biological Sciences, State University of New York at Albany, Albany, New York 12222 and Visiting Scientist, Grosse Ile Laboratory, National Environmental Research Center-Corvallis, U.S. Environmental Protection Agency, Grosse Ile, Michigan 48138; Freshwater Institute, Rensselaer Polytechnic Institute, Troy, New York, present address: Great Lakes Environmental Research Lab., NOAA, 2300 Washtenaw, Ann Arbor, Michigan 48104.

small algal cells (Bogdan and McNaught 1975). Within such a genus (*Diaptomus*) of small-cell specialists characteristic of the Great Lakes, those which give birth at the smallest size (*Diaptomus minutus*) may be most successful. In some instances zooplankton production, but not species composition, may be correlated with food abundance, as in fishponds (Hall, *et al.* 1970). In other instances zooplankton production may be tied to fish predation, while species composition is related chiefly to fish predation and then to algal food composition, as we suspect to be true in the Great Lakes and in most aquatic environments.

THE MODEL

As we have suggested, the availability of phytoplankton and detrital food resources (R_i), the selectivity or preference of the zooplankton for these foods (WZ_{ji}), and size-selective predation by fishes (fish biomass times selectivity) are some of the inputs necessary to simulate zooplankton succession. Predation by fishes (Figure 11.1) dominates, especially in determining which species of zooplankton succeed, while algal productivity influences how many of each species are present. The real-time output of the model is easily used by the modeler to determine when the system has reached steady-state, the principal criterion in determining when to examine community composition (the output provides estimates of zooplankton and resource numbers at predetermined intervals), as readily observed in Figure 11.2. Both in nature, and in these mathematical simulations, steady-state conditions occur when the birth rate is equal to the death rate of a species, where both natural and predation related deaths are considered.

In our simulations, once steady-state has been reached, it persists indefinitely (Figure 11.2, day 121-day 601 for the copepod) because food resources and predation are fixed. In nature, changing food production and predatory pressures, as well as natural cycles of reproduction, limit steady-state conditions to brief periods. Likewise, in nature maximum standing crops of zooplankton usually occur at the time of steady-state. Thus we have used simulated densities at steady-state (Figure 11.2) to simulate the standing crop of a species.

The following equations have been used to model zooplankton dynamics (Equation 11.1) and the production of algal and detrital food resources (Equation 11.2).

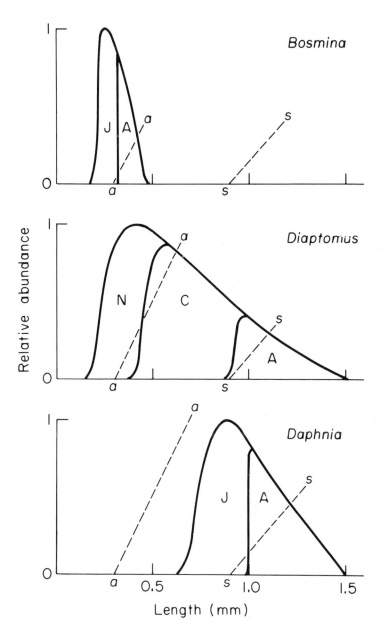

Figure 11.1   Pictorial representation of size structure of populations of *Bosmina longirostris, Diaptomus minutus* and *Daphnia galeata*, and relative number and size of each available to fine-filtering alewife (to right line a-a) and coarse filtering salmonids (right line s-s).   (J = juveniles, A = adult, N = nauplii, C = copepodites).

```
MODEL PARAMETERS

BZ=    .500    .500    .500    .500    .500    .500    .500    .500    .500
WZ=  .40  .33  .33  .17  .33  .33  .47  .01  .01 1.00 1.00 1.00
 E=    .400    .170    .470
DZ=    .140    .140    .140 PZ=    .13+002    .13+002    .13+002
BEZ=   .18-007    .36-008    .18-007    .36-008    .18-007    .36-008
PSZ=   .070    .030    .100
MDA=   .200+004    .200+004    .200+004  MR=    .560+009    .212+010    .230
+011
RMIN=  .15+005    .75+005    .75+005    .30+003    .15+004    .15+004
       .30+003    .30+003    .30+003    .00        .00        .00
OZ=  .100+008    .100+008    .100+008    .100+003    .100+003
PCMIN=  .140    .090    .900
CMAX=     7.    ALPHA= .999 RE=    .35000    .35000    .35000
INPUT:
>PRI
INITIAL CONDITIONS

BIOMIC=  .100+004    .100+004    .100+004
         .100+009    .100+009    .100+009    .165-001
TFIRST=  1.00  TLAST=601.00  STEP=30.000  ACCURC=.00500  IMAX=10
INPUT:
>INT
-------------------------------------------------------------------------
TIME    COPE       CLAD1      CLAD2      NANNO      NET        POM
-------------------------------------------------------------------------
   1   .100+004   .100+004   .100+004   .100+009   .100+009   .100+009
  31   .434+003   .252+003   .417+003   .358+009   .230+009   .185+010
  61   .416+003   .200+003   .357+003   .524+009   .488+009   .147+011
  91   .418+003   .202+003   .376+003   .555+009   .899+009   .224+011
 121   .420+003   .214+003   .420+003   .559+009   .137+010   .230+011
 151   .420+003   .227+003   .483+003   .560+009   .173+010   .230+011
 181   .421+003   .239+003   .564+003   .560+009   .194+010   .230+011
 211   .421+003   .248+003   .663+003   .560+009   .204+010   .230+011
 241   .421+003   .254+003   .782+003   .560+009   .209+010   .230+011
 271   .421+003   .259+003   .923+003   .560+009   .211+010   .230+011
 301   .421+003   .261+003   .109+004   .560+009   .211+010   .230+011
 331   .421+003   .263+003   .129+004   .560+009   .212+010   .230+011
 361   .421+003   .264+003   .152+004   .560+009   .212+010   .230+011
 391   .421+003   .265+003   .180+004   .560+009   .212+010   .230+011
 421   .421+003   .265+003   .181+004   .560+009   .212+010   .230+011
 451   .421+003   .265+003   .181+004   .560+009   .212+010   .230+011
 481   .421+003   .266+003   .181+004   .560+009   .212+010   .230+011
 511   .421+003   .266+003   .180+004   .560+009   .212+010   .230+011
 541   .421+003   .266+003   .191+004   .560+009   .212+010   .230+011
 571   .421+003   .266+003   .180+004   .560+009   .212+010   .230+011
 601   .421+003   .266+003   .181+004   .560+009   .212+010   .230+011
-------------------------------------------------------------------------
```

Figure 11.2 Simulation for Lake Michigan for inshore alewife biomass of 414 kg/ha during May 1972, producing simulations shown in Table 11.4.

Zooplankton Equations

$$\frac{dN_j}{dt} = \text{births}_j - \text{deaths}_j - \text{predatory deaths}_j - \text{metabolic costs}_j \tag{11.1}$$

where

$$\text{births}_j = \sum_i BZ_j \cdot N_j \left(\frac{WZ_{ji} (R_i - RMIN_{ji})}{QZ_j + \sum_i WZ_{ji} (R_i - RMIN_{ji})} \right) \tag{11.1a}$$

$$\text{deaths}_j = DZ_j \cdot N_j \tag{11.1b}$$

$$\text{deaths}_{\text{predatory}_j} = CMAX \cdot N_f \left[WZ_{fj}(N_j - NMIN_{fj}) \right] \cdot \text{size} \tag{11.1c}$$

$$\text{size} = \begin{cases} 1 \text{ where } N_j/MDA_j > 1 \\ 0 \text{ where } N_j/MDA_j < PCMIN_j \\ \text{or } 1 - \left[\dfrac{\alpha}{1 - PCMIN} \right] \left[1 - \dfrac{N_j}{MDA_j} \right], \text{ in all other cases} \end{cases} \tag{11.1d}$$

$$\text{metabolic costs} = (S_j + E_j)N_j \tag{11.1e}$$

Abbreviations for Zooplankton Equations

N_j = zooplankton standing crop (no/m^3)
N_f = fish standing crop (no/m^3)
$RMIN_{ji}$ = minimum level of resource i for feeding by consumer j (cells/m^3)
WZ_{ji} = selectivity index for consumer j upon food resource i
QZ_j = half-saturation constant for feeding by consumer j (cells/m^3)
BZ_j = intrinsic birth rate of zooplankter j (ind/ind/day)
DZ_j = intrinsic death rate of zooplankter j (ind/ind/day)
$CMAX$ = maximum foraging area of fish per day (m^3/fish/day)
$NMIN_{fj}$ = minimum level of zooplankter j for feeding by fish f (no/m^3)
MDA_j = density of zooplankton j population at which all organisms are mature (no/m^3)
$PCMIN$ = per cent of MDA necessary to originate feeding by fish f
S_j = respiratory rate for zooplankter j (l/day)
E_j = excretory rate for zooplankter j (l/day)
α = per cent of CMAX not achieved when fish consumption begins at PCMIN

Resource Equations

$$\frac{dR_i}{dt} = \text{growth}_i - (\text{grazing}_{\text{clad}} + \text{grazing}_{\text{cope}}) \tag{11.2}$$

where

$$\text{growth}_i = PSZ_i \cdot R_i \cdot [1 - \frac{R_i}{MR_i}] \tag{11.2a}$$

$$\text{grazing}_{\text{clad}} = \sum_j BEZ_j \cdot N_j \cdot (R_i - RMIN_{ji}) \tag{11.2b}$$

$$\text{grazing}_{\text{cope}} = \sum_j PZ_j \cdot N_j \cdot \frac{WZ_{ji}(R_i - RMIN_{ji})}{QZ_j + \sum_i WZ_{ji}(R_i - RMIN_{ji})} \tag{11.2c}$$

Abbreviations for Resource Equations

R_i = resource standing crop (cells/m^3)
N_j = zooplankton standing crop (no/m^3)
PSZ_i = turnover rate of resource i (l/days)
MR_i = maximum resource level in temperate eutrophic lake (cells/m^3)
BEZ_j = filtering rate of cladocerans (m^3/ind/day)
$RMIN_{ji}$ = minimum level of resource i for feeding by consumer j (cells/m^3)
PZ_j = feeding rate of copepods (cells/ind/day)
WZ_{ji} = selectivity index for consumer j upon food resource i
QZ_j = half-saturation constant for feeding by consumer j (cells/m^3)

Zooplankton population growth by species and thus relative community composition is controlled by instantaneous rates of birth, natural death, metabolic costs and above all by fish predation. Birth rates are maximum only with high food levels (R_i); at lower food concentrations food controls birthing. Below minimum food levels ($RMIN$), the individual species cannot harvest the relatively rare resource and birthing does not proceed. We have assumed that the relation between food and birth rate can be described by a modified Michaelis-Menton expression (Equation 11.1a). Food does not stimulate birthing unless it is plentiful (exceeds $RMIN_{ji}$), and is preferred by the grazer (high WZ_{ji}). The assumption that increased amounts of food stimulate birthing has been demonstrated for *Daphnia galeata* by Hall (1964) and others.

Fish predation is proportional to the population density of fishes (N_f), and a function of zooplankton species and density. Size selective predation is the major control factor (Equation 11.1d). This construct is based on the following assumptions: (1) the percentage of young in a population is proportional to the ratio of the observed density (N_j) to an all adult density (MDA_j), and (2) when the population reaches a certain percentage $(PCMIN)$ of MDA_j, the organisms are large enough for predation to begin at some low rate, where $N_j/MDA_j > PCMIN$, and increase until the zooplankton density reaches MDA_j and consumption reaches $CMAX$. Since size-selective fish predation is based upon the percentage $(PCMIN)$ of the ultimate adult population that must be achieved before predation commences, $PCMIN$ is a particular characteristic of each species of fish predator relative to zooplankton composition. Since the model, and hopefully the real aquatic world, is controlled chiefly by predation from the top of the trophic pyramid, this parameter deserves further attention.

The susceptibility of populations of zooplankton to predation can be visualized with regard to $PCMIN$ (Figure 11.1). This vital parameter, describing size-selection by alewife and trout, was determined by plotting the size distributions (McNaught, unpublished) of the three common crustacean zooplankters being simulated, considering the selectivity by alewife (Allan 1974) and trout (Galbraith 1967), and calculating graphically the *per cent* of the adult population of appropriate size-structure unavailable to alewife (14, 9 and 90%) and trout (94, 89 and 99%) of the three crustaceans, *Diaptomus, Daphnia* and *Bosmina*, respectively. As depicted in Figure 11.1, selectivity begins with zooplankters 0.3 mm in length and is maximum at lengths greater than 0.7 mm in the case of the alewife (Allan 1974), while the onset is at 0.9 mm and the maximum begins at 1.6 mm in the case of trout (Galbraith 1967). Thus the slope of lines a-a and s-s (Figure 11.1) indicates the increased efficiency at which both predators capture larger animals.

The resources, nannoplankton, netplankton and detritus, are controlled by turnover rates (PSZ), which drive these standing crops toward preset maximum levels (MR). In addition, these resources are controlled by grazing. Copepods (PZ) and cladoceran plankters (BEZ) have been modeled to filter and ingest food rather differently to simulate their described feeding behavior. Each has a food preference (WZ_{ji}) and a minimum food level $(RMIN_{ji})$ at which feeding commences. The term WZ_{ji} thus represents the palatability and susceptibility to capture of a particular resource by a consumer (O'Neill 1969). This concept is particularly useful in describing zooplankters that exhibit selective feeding, but could also be employed to provide a (mathematical) sanctuary for distasteful or otherwise unavailable prey from fish predators. In a similar fashion, fish-feeding commences at a minimal zooplankton level $(NMIN_{fj})$ and is linear with abundance. For further details of the constructs of this model consult the detailed description (McNaught and Scavia 1975).

Calibration of the Model

This resource allocation-predation model for zooplankton community composition has been calibrated (Table 11.1) for a large inland lake (McNaught and Scavia 1975). The details of the experiments and observations involved in calibration can be found elsewhere. However, it is vital for the reader to understand the extent to which we went to either find or make original determination of parameters describing zooplankton physiology and behavior. This initial calibration has not been altered for either the Lake Ontario or Lake Michigan simulations. Thus results of these simulations, compared to observed densities of zooplankton, have enabled us to *validate* a previously *calibrated* model.

The necessary parameters may be divided into two groups. The first includes system parameters, which describe the maximum biotic potential of a species. These are presented in Table 11.1; it must be understood that these are maximum estimates for the rate functions involved as well as the parameters describing submaximum behavior. The second group is the lake parameters, which vary by degree of eutrophication and include the turnover rate of the food resources (PSZ) and the abundance or biomass of fish predators ($BI\emptyset MIC$).

VALIDATION USING LAKE MICHIGAN PARAMETERS

Once the model had been calibrated to produce reasonable simulations of zooplankton density by species for a large inland lake (Lake George, New York) (McNaught and Scavia 1975), the organismic characteristics, such as maximum birth rate, ingestion rates, and natural death rates, were held constant. That is, once calibration was completed, these characteristics of the biotic potential of *Diaptomus, Daphnia galeata*, and *Bosmina longirostris* should describe these species under all environmental conditions. Simulations of zooplankton composition in various ecosystems thus require only a knowledge of the turnover times for food resources (nannoplankton, netplankton and detritus) and the biomass ($BI\emptyset MIC$) and characteristics of size selectivity ($PCMIN$) of the major fish predators.

Turnover Rates for Algal and Detrital Food Resources

Limited estimates for the rate of primary productivity ($mgCm^{-3} day^{-1}$) and the standing crop of algae in terms of carbon ($mgCm^{-3}$) are available for Lake Michigan. The most extensive seasonal study of primary productivity is that of Fee (1973), while the best seasonal estimates of algal carbon are those of Robertson, *et al.* (1971). From such shipboard estimates of primary productivity, made during 1970, and corresponding

Table 11.1 Values and Sources of Parameters used in Calibration of Zooplankton Model[a]

Parameter	Units	Genera of Zooplankton			Source
		Diaptomus	Daphnia	Bosmina	
BZ	ind ind⁻¹ day⁻¹	0.5	0.5	0.5	Hall 1964, McNaught unpublished
DZ	ind ind⁻¹ day⁻¹	0.14	0.14	0.14	Allan 1974
PZ	cells anim⁻¹ day⁻¹	0.13×10^2	—	—	Bogdan and McNaught 1975
BEZ	m³ anim⁻¹ day⁻¹	—	0.18×10^{-7}	0.36×10^{-8}	Bogdan and McNaught 1975
WZ	—	(nanno) 0.40	0.33	0.33	Bogdan and McNaught 1975
	—	(net) 0.17	0.33	0.33	Bogdan and McNaught 1975
	—	(detritus) 0.47	0.01	0.01	Bogdan and McNaught 1975
RE	ind ind⁻¹ day⁻¹	0.35	0.35	0.35	LaRow 1973
RMIN	cells m⁻³	(nanno) 0.15×10^5	0.75×10^5	0.75×10^5	Richman 1966, McMahon & Rigler 1963
	cells m⁻³	(net) 0.3×10^3	0.15×10^4	0.15×10^4	Richman 1966, McMahon & Rigler 1963
	cells m⁻³	(detritus) 0.7×10^3	0.3×10^3	0.3×10^3	Richman 1966, McMahon & Rigler 1963
NMIN	no. m⁻³	0	0	0	this chapter
PCMIN	%	(alewife) 14	09	90	this chapter
	%	(trout) 94	89	99	this chapter
CMAX	m³ anim⁻¹ day⁻¹	7	7	7	calculated from stock
QZ	cells m⁻³	1×10^7	1×10^7	1×10^7	Hall 1964
MDA	no. m⁻³	2000	2000	2000	from literature
MR	cells m⁻³	nannoplankton 5.6×10^8			from literature
	cells m⁻³	net plankton 2.1×10^9			from literature
	cells m⁻³	detritus 2.3×10^{10}			from literature

[a]For abbreviations of parameters see text.

standing crops of algal carbon (C) made in 1969, we have estimated resource turnover rates for the entire algal assemblage (Table 11.2). These
doubling-times ranged from 8 to 22 days during May and August. From
these estimates of turnover time (days) we calculated the instantaneous
rates (days^{-1}) shown in Table 11.2. In running all simulations, we then
assumed that nannoplankton turnover more rapidly than netplankton
(*ca.* 2:1). We have also assumed that rates of primary productivity in
Lake Michigan were unchanged between 1966 and 1972 and identical to
those measured by Fee (1973) in 1970.

Table 11.2   Estimates of Forcing Functions for Simulation of
Zooplankton Composition of Lake Michigan[a]

Month and Location	Rate of Primary Productivity		Algal Carbon	Turnover (T)	1/T
	(mgC m^{-2} d^{-1})	(mgC m^{-3} d^{-1})	(mgC m^{-3})	(days)	(day^{-1})
May 1970					
Inshore	1188[b]	29.7	222[c]	7.5	0.134
Offshore	450[b]	11.2	210[c]	18.8	0.053
August 1970					
Inshore	425[b]	10.6	181[c]	17.1	0.059
Offshore	317[b]	7.9	172[c]	21.8	0.046

[a]Turnover rates for phytoplankton and detrital resources (*PSZ*)
[b]Fee 1973.
[c]Robertson, *et al.* 1971.

Biomass of Predatory Alewife

The alewife, *Alosa pseudoharengus*, is the specific planktivore whose
effect upon zooplankton has been modeled simply because of its over-
whelming impact on the Lake Michigan ecosystem (Wells 1970). But
because of our general knowledge concerning the dominance of salmonids
and especially coregonines prior to 1954 (Beeton 1969), we have also
attempted to arrive at biomass estimates for salmonids characteristic of
pre-1954 levels. We have not attempted to calibrate the model for the
coregonidae, as the model will currently accept only one predatory fish.
However, a common coregonid, the bloater (*Coregonus hoyi*) is chiefly
a nonplanktonic feeder as an adult (Wells and Beeton 1963).

Fortunately the Great Lakes Fishery Laboratory of the U.S. Fish and Wildlife Service has made yearly estimates of alewife abundance, utilizing a standard series of trawls off Saugatuck, Michigan, to arrive at an estimate of lakewide biomass, which has ranged from 191 kg/ha at the peak abundance during the spring of 1967 (lakewide = 1.1 billion kg) to a low of 23 kg/ha following the massive die-off in late 1967 and 1969 (Brown 1972). In addition, the horizontal distribution of alewives is beginning to be understood. Age-classes III-VII move inshore in April (Wells 1968) in Lake Michigan, certainly inside the 54.7 m (30 fathom) contour, whereas age-classes I-II remain more uniformly distributed. Later in August, the younger alewife (I) prefers inshore waters, but the bulk of the population is more uniformly dispersed. Based on these observations of Wells (1968), we have made estimates of the biomass of alewives inshore (<54.7 m) and offshore (>54.7 m) during May and August, because such inshore spawning aggregations most likely have a profound effect on the relative numbers of zooplankton selectively cropped. Thus, as with zooplankton, standing crops of the alewife are well documented in Lake Michigan, especially with regard to inshore abundance, but with limited knowledge of open water populations. Unfortunately, we have only records of commercial catches for the salmonids.

Biomass of Predatory Salmonids Prior to 1954

Accurate lakewide estimates of salmonid standing stocks prior to 1954 are not currently available. Approximately 7.7 x 10^6 kg of lake herring, lake trout and whitefish were harvested commercially from Lake Michigan in 1949. By 1960, the commercial yield of these three salmonids had decreased to 0.08 x 10^6 kg (Beeton 1969). In simulating the predatory pressures of salmonids upon the zooplankton, we have arbitrarily assigned a biomass of 100-145 kg/ha, certainly crude at best. This is based upon the fact that standing crops of phytoplankton in 1954 were 43% of those in 1958 (Damann 1960) and 31% of those in 1963 (Damann, unpublished).

Results of Simulation of Species Composition for Lake Michigan

Using available estimates of algal productivity (Table 11.2) and predatory alewife biomass (Table 11.3), we have simulated the relative abundance of the dominant crustaceans *Diaptomus spp, Daphnia galeata* and *D. retrocurva,* and *Bosmina longirostris,* for the four major periods in which alewife abundance varied greatly. The alewife appeared in Lake Michigan in 1949 (Smith 1972). The pre-1954 simulation (Table 11.4) is characteristic of this period, during which salmonid predators prevailed.

Table 11.3   Estimates of Forcing Functions for Simulation of
Zooplankton Composition of Lake Michigan[a]

Month and Year	Biomass Inshore kg/ha	Biomass Offshore kg/ha	Mean Lakewide Biomass kg/ha
May 1966	423	93	220
August 1966	286	132	191[b]
May 1969	49	11	26
August 1969	34	16	23[b]
May 1972	462	100	240[c]
August 1972	313	145	210[c]

[a]Alewife biomass estimates (BIØMIC).
[b]Brown 1972.
[c]Edsall, et al., M.S.

The simulation for May and August of 1966 permits validation of the model and understanding of the importance of alewife predation, especially at the maximum period of alewife abundance in the 1960s. The 1968-69 simulation examines the effect of reduced alewife predation, and the 1972 simulation again verifies the general species composition expected under very heavy alewife predation.

Changes in Zooplankton Composition

For the Great Lakes the most comprehensive documentation of changes is available for Lake Michigan. The large cladocerans (*Daphnia retrocurva* and *D. galeata*) were abundant in 1927 (Brooks 1969) and 1954 (Wells 1960). They gave way to the smaller more predation-free *D. longiremis* by 1966, with a return by 1968 to the larger *D. retrocurva*, along with other larger cladocerans like *Leptodora kindtii* and the calanoids *Limnocalanus macrurus, Epischura lacustris* and *Diaptomus sicilis* (Wells 1960, Wells 1970). The abrupt return of larger zooplankton during 1968 has been linked with the alewife die-off in 1966 (Wells 1970).

As of 1972, the zooplankton composition of Lake Michigan is similar to that of Lake Ontario, with *Bosmina longirostris* and *Cyclops bicuspidatus* most abundant (Roth and Stewart 1973). Fortunately, this careful historical documentation of changes in species composition of zooplankton communities of the Great Lakes will enable us to perform a variety of validations of our model.

Table 11.4 Results of Simulation of Zooplankton Standing Crops (no/m³) in Lake Michigan, Compared to Observed Densities for 1954, 1966, 1968, 1972[a]

Month and Location	Fish Biomass (kg/ha)	Zooplankton Standing Crop (no/m³)					
		Diaptomus spp (4)		Daphnia galeata + retrocurva		Bosmina longirostris	
		Predicted	Observed	Predicted	Observed	Predicted	Observed
Pre 1954 - Salmonid Community							
May-June – lakewide	100	1930	2534	1830	430	1980	250
August – lakewide	145	1910	2282	1810	2600	1980	26
1966 – Alewife Maximum							
May-June – inshore	423	433	1167	273	8	1810	320
– offshore	93	991	–	615	–	1850	–
August – inshore	286	509	579	319	16	1810	100
– offshore	132	779	–	483	–	1830	–
1968-69 Alewife Crash							
May-June – inshore	49	1580	–	913	–	1890	–
– offshore	11	52400	–	4370	–	5500	–
August – inshore	34	2840	3206	1220	2100	1930	32
– offshore	16	20700	–	2100	–	2730	–
1972 Alewife Buildup							
May-June – inshore	462	421	7650	266	45	1810	7500
– offshore	100	931	–	579	–	1840	–
August – inshore	313	492	8360	309	1700	1810	2500
– offshore	145	731	–	.454	–	1840	–

[a]Observed densities: 1954 (Wells 1960), 1966 (Wells 1970), 1968 (Wells 1970), 1972 (Roth and Stewart 1973).

Comparison Between Predicted (Simulated)
and Observed Changes

Results of simulations for four years are summarized with respect to observed values of relative zooplankton abundance (Table 11.4). As McNaught (1975) suggested, the large calanoid *Diaptomus* is especially sensitive to predation, while the small, early reproducing cladoceran *Bosmina* is least sensitive both in results of simulations and as observed in nature. Under salmonid predation before 1954, we have predicted (1910-1930 m^{-3}) only slightly fewer *Diaptomus* than observed (2282-2534 m^{-3}). Under intense predation by alewife (1966), we have predicted 433 and 509 m^{-3} in inshore waters, with 1167 and 579 m^{-3} observed. The model properly simulated decreases in *Diaptomus* under heavy (93-423 kg/ha) alewife predation in 1966. Following the alewife crash in 1967 (Brown 1972), we simulated increases in *Diaptomus* to about 2840 m^{-3} inshore during August. Approximately 3206 *Diaptomus* m^{-3} were observed (Wells 1970). Again, with the alewife increase on the upswing in 1972 (Edsall, *et al.* 1972), the abundance of *Diaptomus* decreased (Table 11.4).

Daphnia (*galeata* and *retrocurva*) also responded in the results of simulations as expected. Predictions of abundance in pre-1954 years and 1968-69 were confirmed by field observations (Wells 1970), as we have summarized (Table 11.4). Generally, the least impressive relationship between predicted and observed densities occurred in the case of *Bosmina*. Only in 1972 were the high densities observed (Roth and Stewart 1973) that would be expected under heavy alewife predation (Table 11.4).

Since this model has been constructed to simulate interrelationships between the herbivorous zooplankton and their food resources and predators, we have not attempted to simulate the abundance of *Cyclops bicuspidatus*. This common cyclopoid copepod is chiefly predatory as an adult, and even the copepodites exhibit low feeding rates on algae in the Great Lakes (McNaught, unpublished). Thus this model is not appropriate for predicting the abundance of *Cyclops*.

Clearly, our simulations of zooplankton abundance based on available foods and predatory pressure by the alewife are of the proper order of magnitude and in proper relationship to changes in predatory pressure (Figure 11.3). While the model is useful for careful investigation of the interactions between fish predator and zooplankton prey, more detail could be developed to focus upon the relationship between species composition of zooplankton, especially *Bosmina*, and algal species diversity. Alternatively, it may be that *Bosmina* abundance is not a function of predator density because it is essentially predator-free in the Great Lakes.

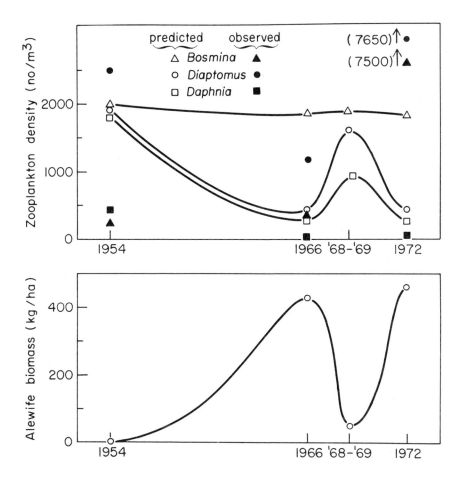

Figure 11.3a Simulated and observed changes in abundance of *Bosmina longirostris* (simulated △, observed ▲), *Daphnia* (□, ■), and *Diaptomus* spp. (○, ●) with regard to biomass of alewife, during May-June in inshore waters of Lake Michigan.

VALIDATION FOR LAKE ONTARIO PARAMETERS

Intensive study of Lake Ontario during the IFYGL program has provided excellent information on primary productivity (Stadelmann, *et al.* 1974) and zooplankton abundance (McNaught, *et al.* 1975), although in contrast to Lake Michigan little is known of the alewife populations. Generally turnover rates during May of the primary producers in Lake Ontario (0.10) are higher than in Lake Michigan (0.09) (Tables 11.2 and 11.5). Inshore phytoplankton populations doubled in about 10 days

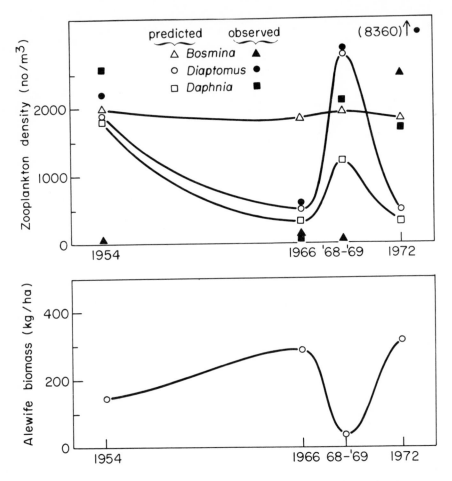

Figure 11.3b   Simulated and observed changes (as in Figure 11.3a) during August
in inshore waters of Lake Michigan.

during 1972. The alewife was less abundant inshore and generally less
abundant lakewide than in Lake Michigan, probably because it has been
in equilibrium with its predator for much longer.

Detailed Observations of Historical Abundance
of Alewife and Zooplankton

The alewife has had access to Lake Ontario since the end of the
Pleistocene. However, it did not become abundant there until 1873, following the loss of its major fish predator, the Atlantic salmon (Smith
1972). Currently, the spring months (April-June) find the alewife inshore

spawning, at which times their population density is 100-268 kg/ha (personal communication, A. Larsen, Bureau of Sports Fisheries). While detailed estimates of these populations after they have moved offshore (July) are not available, a simple calculation based on the surface area of Lake Ontario 'outside of the 40 meter contour (70% area) would indicate open water populations of approximately 11 kg/ha (Table 11.5).

In Lake Ontario rather limited evidence has indicated that zooplankters of the genera *Diaptomus* and *Daphnia* were abundant during the summer of 1939; *Bosmina* was clearly not observed. By 1969, *Bosmina longirostris* and *Cyclops bicuspidatus* were most abundant (McNaught and Buzzard 1973). *Bosmina longirostris* is especially abundant in nearshore waters.

Comparison Between Predicted and Observed Zooplankton Composition

Since our model is most sensitive to fish predation upon zooplankton, it is not surprising that our results for Lake Ontario, where little is known of alewife abundance, are less precise than those for Lake Michigan. Where estimates are available for alewife biomass (May-June inshore, Larsen, personal communication) we have predicted similar zooplankton composition as we ourselves observed. For May-June inshore, with an alewife biomass of 185 kg/ha, we predicted 420 *Diaptomus* m^{-3}, 405 *Daphnia* m^{-3}, and 4020 *Bosmina* m^{-3} and observed 719, 300 and 6800 m^{-3} respectively, a very good relationship indeed. Calculation of an offshore biomass of alewife of 11 kg/ha from crude inshore densities did not provide enough predation to get our simulated abundance down to observed levels (Table 11.6). The model does suggest that openwater alewife biomass is greater than 11 kg/ha. Possibly simulations, using known zooplankton composition, could be used to predict fish biomass.

USE OF MODEL IN MANAGEMENT OF GREAT LAKES

The idea that aquatic ecosystems are controlled in a large degree from atop the pyramid—the predatory fishes controlling the abundance of planktivorous alewives, these small fishes in turn controlling the turnover and standing crop ($\sim$ grazing) of herbivorous zooplankton, and these grazers the type of algae—is not a new idea. However, simulations presented for Lake Michigan and Lake Ontario are strong evidence for such control.

Table 11.5   Estimates of Forcing Functions for Simulation of Zooplankton Composition of Lake Ontario

| Month and Location | Rate of Primary Productivity (mg Cm^{-3} d^{-1}) | Turnover Rates for Phytoplankton Resources (PSZ) | | | Alewife Biomass Estimates (BIØMIC) (kg/ha) |
		Algal Carbon (mg Cm^{-3})	Turnover (T) (days)	1/T (days^{-1})	
May-June					
Inshore	54.2[a]	550[a]	10.1	0.10	185[b]
Offshore	18.5[a]	183[a]	9.9	0.10	11

[a]Stadelmann, et al. 1974.
[b]Larsen, personal communication.

Table 11.6   Results of Simulation of Zooplankton Standing Crops (no/m^3) in Lake Ontario

| Month and Location | Fish Biomass (kg/ha) | Zooplankton Standing Crop (no/m^3) | | | | | |
| | | Diaptomus spp. | | Daphnia galeata + retrocurva | | Bosmina longirostris | |
		Predicted	Observed	Predicted	Observed	Predicted	Observed
May-June							
Inshore	185	420	719[a]	405	300[a]	4020	6800[a]
Offshore	11	56,000	697[a,b]	5670	32[a]	8420	290[a]

[a]McNaught, et al. 1975.
[b]Does not include abundant nauplii.

Suppose that a commission with power to regulate these magnificent bodies of water decided that we should attempt to restore the oligotrophic zooplankton fauna as a first step in returning the lakes to an earlier trophic state, or at least controlling the increased rate of eutrophication. This is not ecologically a bad idea, since most herbivores associated historically with oligotrophic periods in the Great Lakes were large. Larger herbivores are more efficient filtrators (filtering capacity is proportional to the cube of the length $(L^{3.02})$(Burns and Rigler 1967). A large calanoid 1.5 mm in length as an adult should filter about 8.1 times as much as a smaller calanoid species of 0.75 mm, although for populations of equal biomass the difference will be less. One method would be to add a large salmonid predator like the Coho salmon, *Oncorhynchus kisutch*, or the Chinook salmon, *Oncorhynchus tschawytscha*, as has been done in Lake Michigan. As these salmonid predators reduced the alewife standing stock, the percentage of composition of the zooplankton would shift toward dominance by the calanoids (Figure 11.4).

A detailed examination of this series of simulations for Lake Ontario is of interest from a management point of view. They were run using constant turnover rates as one forcing function. The instantaneous rate of resource turnover was held at a constant characteristic of offshore waters > 40 m) during August [1/T (nannoplankton) = 0.15, 1/T (netplankton) = 0.05], as in Table 11.5. At an alewife biomass of 11 kg/ha, *Diaptomus* (52,000 m^{-3}) is dominant, followed distantly by the more eutrophic *Bosmina longirostris* (7,200 m^{-3}) and *Daphnia* (4,700 m^{-3}). With increased predation by the alewife, which as we remember becomes progressively more efficient for zooplankters between 0.3 and 0.7 mm in length, the standing crops of the larger *Diaptomus* and *Daphnia* drop rapidly, reaching a sharp break to an asymptote at approximately 33 kg/ha. Thereafter increased alewife predation, equated to increased standing crop of these fishes, has little effect upon zooplankton composition. Thus, to increase the proportion of oligotrophic *Diaptomus* in the community requires holding the alewife *lakewide* below 33 kg/ha.

Unfortunately, we do not presently have a good offshore estimate of alewife stocks for the open waters of Lake Ontario. We realize that the alewife has been in this lowest of the Great Lakes for a long time, and in abundance since 1873 (Smith 1972). Thus alewife stocks may be in equilibrium and likely lower than in Lake Michigan. Yet changes in the historical past point to an increased incidence of *Bosmina longirostris* (McNaught and Buzzard 1973), itself evidence for increased fish predation in the system. One way to control alewife stocks is by introducing predatory salmon. The introduction of salmonids has an additional ecological advantage, since they are coarse filter feeders

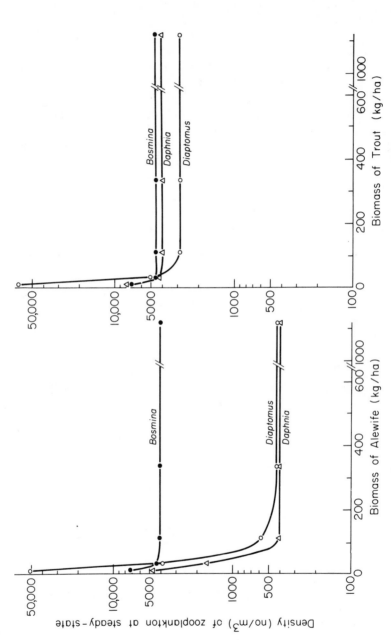

Figure 11.4 Simulated densities (no/m³) of *Diaptomus* (○), *Daphnia* (△) and *Bosmina longirostris* (●), with varying biomass of (left) alewife (kg/ha) and (right) trout, in Lake Ontario at levels of primary productivity corresponding to inshore waters during May.

(*PCMIN* is large). Thus the young salmon should act more like the oligotrophic planktonphage fishes (coregonines) that originally occupied this lake in large numbers. The commercial yield of coregonines, especially the lake herring and whitefish, did not drop markedly until 1945-1955 in Lake Ontario (Beeton 1969).

Salmonids like the rainbow trout, *Salmo gairdneri*, about which a good deal is known concerning size-selective feeding (Galbraith 1967) would permit the original oligotrophic zooplankton fauna to persist. Remnants of relict populations still exist in deeper water, such as small populations of *Limnocalanus macrurus* and *Diaptomus sicilis*, but as relatively low (0.2 and 0.1%) components of the crustacean zooplankton in 1972 (McNaught and Buzzard 1973). With the reestablishment of salmonid-like planktivorous fishes, standing crops of *Bosmina longirostris* would be expected to increase slightly (8.8%), but those of the large calanoid *Diaptomus* (384%) and the large cladoceran *Daphnia galeata* (885%) would increase markedly at a salmonid biomass of 100 kg/ha, according to our simulation (Figure 11.4). In terms of simulated composition of zooplankton community, the dominant calanoid (*Diaptomus*) would comprise 13% of the standing crop (numbers) under predation by 100 kg/ha of alewives, but would constitute a larger fraction of the standing crop (34%) under the same biomass of salmonids. From the standpoint of zooplankton biomass, these differences would be even larger, since *Diaptomus* is one of the largest genera.

From a management viewpoint, the introduction of salmonid predators upon the exotic alewife, concomitant with the reestablishment of coregonine populations, makes considerable ecological sense. But we are dealing with simulations. Whether they can be directly converted to management recommendations is unknown and untested. The challenge, however, is clear; biological control of the alewife in Lake Michigan is certainly desirable.

THEORETICAL IMPLICATIONS

Two basic theoretical implications of these simulations are evident: the first suggests that fish predation is the principal factor controlling zooplankton composition. Second, these simulations clearly indicate that the often-used eutrophic indicator, *Bosmina longirostris*, is an indicator of *predatory pressure* and not solely of advanced eutrophication or shifts to larger algal species. In suggesting that selective fish predation controls zooplankton species composition, we agree with recent conclusions of others (Allan 1974, Dodson 1974).

The conclusion that inshore *spawning* populations of the alewife during May and June in effect largely determine summertime zooplankton composition is both stimulating and surprising. This new and basic observation may negate previous statements concerning interpretation of shifts from *Eubosmina coregoni* to *Bosmina longirostris* during cultural eutrophication. This conclusion was drawn from the observations that inshore spawning runs of alewives are followed by rapid growth of *Bosmina longirostris* populations and the independent simulation of large *Bosmina* populations with high predatory density. Thus the seasonality of predation related to spawning may influence zooplankton composition. We suggest that an exotic predator, like the alewife, in addition to nutrient enhancement and concomitant changes in algal composition, may be responsible for such shifts from larger to smaller *congeneric* species.

ACKNOWLEDGMENTS

Development of the resource-predation model was supported in part by The Eastern Deciduous Forest Biome, U.S. International Biological Program, funded by The National Science Foundation under Interagency Agreement AG-199, BMS 69-01147-A09, with The Energy Research and Development Administration, Oak Ridge National Laboratory. This chapter is Contribution No. 211 from The Eastern Deciduous Forest Biome (Lake George site).

Extensive data collections on Lake Ontario were made with support of the Environmental Protection Agency (Grant 800536) and logistic support by the crew of the *RV Researcher*, a vessel of the National Oceanic and Atmospheric Administration. The chapter was prepared during the tenure of the senior author as Visiting Scientist at the Grosse Ile Laboratory, National Environmental Research Center, U.S. Environmental Protection Agency. Support of the Director, Dr. Tudor Davies, and Project Officer, Mr. Nelson Thomas, is acknowledged.

Both Edward Brown and LaRue Wells of the Great Lakes Fisheries Laboratory, Department of Interior, Ann Arbor, were helpful in determining alewife biomass and zooplankton composition for Lake Michigan. Professor Frank DiCesare provided a helpful critique of the manuscript, as did an anonymous reviewer and the Editor, Raymond Canale.

REFERENCES

Allan, J. D. "Balancing Predation and Competition in Cladocerans," *Ecol.* **55**, 622 (1974).

Beeton, A. M. "Changes in the Environment and Biota of the Great Lakes," in *Eutrophication: Causes, Consequences, Correctives.* (Washington, D.C.: National Academy of Science, 1969), pp. 150-187.

Bogdan, K. G. and D. C. McNaught. "Selective Feeding by *Diaptomus* and *Daphnia*," *Verh. Internat. Verein. Limnol.* **19** (1975).

Brooks, J. L. "Eutrophication and Changes in the Composition of the Zooplankton," in *Eutrophication: Causes, Consequences, Correctives.* (Washington, D.C.: National Academy of Sciences, 1969), pp. 236-255.

Brown, E. H. "Population Biology of Alewives, *Alosa pseudoharengus*, in Lake Michigan, 1949-70," *J. Fish. Res. Bd. Canada* **29**, 477 (1972).

Burns, C. W. and F. H. Rigler. "Comparison of Filtering Rates of *Daphnia rosea* in Lake Water and in Suspensions of Yeast," *Limnol. Oceanogr.* **12**, 492 (1967).

Damann, K. E. "Plankton Studies of Lake Michigan. II. Thirty-Three Years of Continuous Plankton and Coliform Bacteria Data Collected from Lake Michigan at Chicago, Illinois," *Trans. Amer. Micro. Soc.* **74**, 397 (1960).

Dodson, S. I. "Zooplankton Competition and Predation: An Experimental Test of the Size-Efficiency Hypothesis," *Ecol.* **55**, 605 (1974).

Edsall, T. A., E. H. Brown, T. G. Yocom, and R. S. C. Wolcott. "Utilization of Alewives by Coho Salmon in Lake Michigan," M.S., Great Lakes Fishery Lab., Ann Arbor, Michigan (1972).

Fee, E. J. "A Numerical Model for Determining Integral Primary Production and its Application to Lake Michigan," *J. Fish. Res. Bd. Canada* **30**, 1447 (1973).

Galbraith, M. G. "Size-Selective Predation on *Daphnia* by Rainbow Trout and Yellow Perch," *Trans. Am. Fish. Soc.* **96**, 1 (1967).

Hall, D. J. "An Experimental Approach to the Dynamics of a Natural Population of *Daphnia galeata* Mendotae," *Ecol.* **45**, 94 (1964).

Hall, D. J., W. E. Cooper, and E. E. Werner. "An Experimental Approach to the Population Dynamics of Freshwater Animal Communities," *Limnol. Oceanogr.* **15**, 839 (1970).

LaRow, E. J. "Effect of Food Concentration on Respiration and Excretion in Herbivorous Zooplankton," U.S. Int. Biol. Prog. Memo Rept. 73-69:1-11 (mimeo) (1973).

McMahon, J. W. and F. H. Rigler. "Mechanisms Regulating the Feeding Rate of *Daphnia magna* Straus," *Can. J. Zool.* **41**, 321 (1963).

McNaught, D. C. "A Hypothesis to Explain the Succession from Calanoids to Cladocerans During Eutrophication," *Verh. Internat. Verein. Limnol.* **19**, 724 (1975).

McNaught, D. C., M. Buzzard and S. Levine. "Zooplankton Production in Lake Ontario as Influenced by Environmental Perturbations," Ecological Research Series, U.S. Environmental Protection Agency, Rept. EPA-660/3-75-021 (Washington, D.C.: U.S. Government Printing Office, 1975).

McNaught, D. C. and M. Buzzard. "Changes in Zooplankton Populations in Lake Ontario (1939-1972)," *Proceedings 16th Conference Great Lakes Research* (1973), pp. 76-86.

McNaught, D. C. and D. Scavia. "Simulation of Changes in Zooplankton Production During Eutrophication," unpublished manuscript (1975).

O'Neill, R. V. "Indirect Estimation of Energy Fluxes in Animal Food Webs," *J. Theoret. Biol.* **22**, 284 (1969).

Richman, S. "The Effect of Phytoplankton Concentration on the Feeding Rate of *Diaptomus oregonensis*," *Verh. Internat. Verein. Limnol.* **16**, 392 (1966).

Robertson, A., C. F. Powers and J. Rose. "Distribution of Chlorophyll and its Relation to Particulate Organic Matter in the Offshore Waters of Lake Michigan," *Proceedings 14th Conference Great Lakes Research* (1971), pp. 90-101.

Roth, J. C. and J. A. Stewart. "Nearshore Zooplankton of Southeastern Lake Michigan," *Proceedings 15th Conference Great Lakes Research* (1973), pp. 132-142.

Schelske, C. L., E. F. Stoermer and L. E. Feldt. "Nutrients, Phytoplankton Productivity and Species Composition as Influenced by Upwelling in Lake Michigan," *Proceedings 14th Conference Great Lakes Research* (1971), pp. 102-113.

Smith, S. "Factors of Ecologic Succession in Oligotrophic Fish Communities of the Laurentian Great Lakes," *J. Fish. Res. Bd. Canada* **29**, 717 (1972).

Stadelmann, P., J. E. Moore and E. Pickett. "Primary Production in Relation to Temperature Structure, Biomass Concentration and Light Conditions at an Inshore and Offshore Station in Lake Ontario," *J. Fish. Res. Bd. Canada* **31**, 1215 (1974).

Wells, L. "Seasonal Abundance and Vertical Movements of Planktonic Crustacea in Lake Michigan," *U.S. Fish Wild. Serv., Fish. Bull.* **60**, 343 (1960).

Wells, L. "Seasonal Depth Distribution of Fish in Southeastern Lake Michigan," *U.S. Fish. Wild. Serv., Fish. Bull.* **67**(1), 1 (1968).

Wells, L. "Effects of Alewife Predation on Zooplankton Populations in Lake Michigan," *Limnol. Oceanogr.* **15**, 556 (1970).

Wells, L. and A. M. Beeton. "Food of the Bloater, *Coregonus hoyi*, in Lake Michigan," *Trans. Amer. Fish. Soc.* **92**, 245 (1963).

12

EFFECTS OF A CHLORINATED HYDROCARBON POLLUTANT ON THE GROWTH KINETICS OF A MARINE DIATOM

Nicholas S. Fisher, Robert R. L. Guillard, and Charles F. Wurster[1]

INTRODUCTION

Recent reports suggest that several widespread aquatic pollutants have deleterious effects on different species of planktonic algae. Chlorinated hydrocarbon pollutants in particular have been investigated and have proven toxic to many species. Furthermore, various physical, chemical, and biological characteristics of the environment can influence the toxicity of these pollutants to phytoplankton. Thus, temperature (Fisher and Wurster 1973), salinity (Batterton, *et al.* 1972), light intensity (MacFarlane, *et al.* 1972), evolutionary history (Fisher, *et al.* 1973), other pollutants (Mosser, *et al.*, 1974), and interspecific competition (Mosser, *et al.* 1972b; Fisher, *et al.* 1974) all affect the toxicity to algae of DDT [1,1,1-trichloro-2,2-bis(*p*-chlorophenyl)ethane] or PCB (polychlorinated biphenyls), two of the more common and persistent aquatic pollutants. While this research has helped identify several environmental factors that can influence a pollutant's toxicity, it has only tangentially touched upon the consequences of the myriad of possible interactions that may occur between pollutants and various nutrient

[1]Woods Hole Oceanographic Institution, Woods Hole, Massachusetts; Woods Hole Oceanographic Institution, Woods Hole, Massachusetts; Marine Sciences Research Center, State University of New York, Stony Brook, New York, respectively.

concentrations. Basic information concerning nutrient-inhibitor interactions is lacking, though it is fundamental to understanding and successfully predicting pollutant effects in nature.

As a first step toward assessing the significance of these interactions, we investigated the effects of PCB on the growth of an organochlorine-sensitive marine diatom, *Thalassiosira pseudonana* Hasle and Heimdahl, in media containing different nitrate concentrations. The cells were grown under a wide range of nitrate concentrations to determine whether PCB inhibition, if it occurred, would be greater in high or low nitrogen media, thereby allowing a distinction between competitive and noncompetitive inhibition (Sizer 1957). Thus, the experiment was not designed to be a detailed study of the kinetic constants of *T. pseudonana*, but rather an attempt to describe the pattern of PCB's effects on the diatom's growth kinetics.

MATERIALS AND METHODS

The diatoms were grown in axenic batch cultures with relatively low cell densities, thus simulating the steady-state environment of a chemostat for the first few cell divisions. This method was found to be satisfactory for determining specific growth rates of algae at different nutrient concentrations (Droop 1973; Guillard *et al.* 1973; Swift and Taylor 1974; Thomas and Dodson 1974).

An artificial seawater medium was prepared by dissolving reagent-grade salts in double-distilled water. It was chemically equivalent to Instant Ocean (Aquarium Systems, Inc., Eastlake, Ohio) with f/2 nutrient enrichment (Guillard and Ryther 1962), with the exception that no inorganic nitrogen in any form was added. The medium was sterilized by autoclaving and inoculated with *Thalassiosira pseudonana* (formerly *Cyclotella nana*), Woods Hole clone 3H, a small, centric diatom of estuarine origin. The inoculum came from a population of cells, growing in f/2 medium but with an initial nitrate concentration of 88.3 μM (f/20), that were just at the end of their exponential growth phase. Experiments showed that growth of these cells could not continue without further nitrogen enrichment, suggesting that the cells' internal nitrogen pools were insufficient to support further cell division.

Initial experimental cell densities were set at 10^4/ml, and the cell suspension was then divided into eleven equal fractions. To each fraction was added a different amount of sterile-filtered $NaNO_3$ (through 0.22-μm Millipore filters) so that the cells were suspended in 11 different media. The nitrate additions to the media yielded added nitrogen concentrations of 0.2, 0.4, 5, 59, 82, 118, 177, 235, 294, 529, and 835 μM.

Fifteen ml of each cell suspension was then asceptically dispensed into each culture tube (15-ml capacity). For each medium, three tubes received methanolic solutions of PCB (Aroclor 1254) at 1 $\mu g/l$ (1 ppb) and three received 1.5 μl of methanol (controls). The procedure for addition of the PCB has been described by Mosser, *et al.* (1972a). Methanol at a 10^{-4} dilution was previously shown to have no effect on algal growth (Fisher 1974). The culture tubes were then sealed with Teflon-coated screw caps and incubated at 23.5 ± 0.5°C under 7500 lumens/m^2 of continuous light from cool white fluorescent bulbs. Samples of each cell suspension were taken at 24 hr, fixed with Lugol's solution, and counted with a Speirs-Levy eosinophil counter.

Growth rates (μ) were calculated according to Equation (12.1),

$$\mu = (35)(\ln n_t - \ln n_{t_o}) / (t - t_o) \qquad (12.1)$$

where ($t - t_o$) equals elapsed time in hours, and n_t and n_{t_o} denote the population densities at times t and t_o, respectively (Eppley and Strickland 1968).

RESULTS

Table 12.1 presents the mean growth rates with their 95% confidence intervals (Sokal and Rohlf 1969) for *T. pseudonana*. *T. pseudonana*'s growth rate was unaffected by the PCB in high nitrate media but was

Table 12.1   Growth, at 24 h, of *T. pseudonana* at Different Nitrate Concentrations with or without 1 ppb of PCB[a]

Added Nitrate Concentration	Growth Rate	
	0 ppb PCB	1 ppb PCB
0.2	1.73 ± 0.11	0.23 ± 0.19
0.4	2.80 ± 0.05	0.41 ± 0.40
5	3.47 ± 0.09	3.15 ± 0.17
59	3.48 ± 0.05	3.30 ± 0.09
82	3.47 ± 0.07	3.37 ± 0.12
118	3.56 ± 0.02	3.40 ± 1.88
177	3.66 ± 0.64	3.48 ± 0.09
235	3.70 ± 0.06	3.55 ± 0.0
294	3.70 ± 0.06	3.64 ± 0.45
529	3.73 ± 0.07	3.70 ± 0.25
835	3.76 ± 0.18	3.71 ± 0.05

[a]Nitrate concentrations are in μM and growth rates (calculated by Equation 12.1) are given as divisions per day. Data are means of triplicate cultures and are given with their 95% confidence intervals.

substantially reduced in the media containing the lower nitrate levels. A two-way analysis of variance (Sokal and Rohlf 1969) of the growth rates, given in Table 12.2, showed that growth rate was significantly dependent on PCB concentration (P < 0.001) and on nitrate concentration (P < 0.001), and that the PCB effect was significantly dependent on the nitrate concentration (P < 0.001).

Table 12.2 Two-Way Analysis of Variance of *T. pseudonana*'s Growth Rates Showing the Effects of PCB Treatment (1 ppb), Nitrate Concentration, and PCB-Nitrate Interaction

Source of Variation	df	SS	Fs
PCB treatment	1	3.575	895.99 (P < 0.001)
Nitrate concentration	10	52.157	1307.19 (P < 0.001)
PCB x nitrate interaction	10	8.716	218.46 (P < 0.001)
Error	44	0.176	

Figures 12.1 and 12.2 present the data as growth rates *vs.* initial nitrate concentrations with and without 1 ppb of PCB, respectively. The lines drawn by a Hewlett-Packard Plotter through the data points on these graphs are hypothetical. The presumption was made, based on others' observations (see Discussion), that growth rate and nutrient concentration are hyperbolically related. The lines are based, according to Equation (12.2), on the empirically derived Ks and μ_{max} values.

$$\mu = (\mu_{max}) (S) / (Ks + S) \tag{12.2}$$

where μ equals the rate of growth, μ_{max} equals the maximum specific growth rate, S equals the limiting nutrient concentration, and Ks represents the half-saturation constant, or that nutrient concentration at which growth is at a half maximal rate. This equation to describe microbial growth kinetics has been adapted by Monod from the Michaelis-Menten equation (Equation 12.3) used in the study of enzyme kinetics (see Eppley and Strickland 1968).

$$v = (V_{max}) (S) / (Ks + S) \tag{12.3}$$

where v equals the velocity of a reaction, V_{max} equals the maximum velocity at which the reaction can occur, S equals the substrate concentration, and Ks equals the dissociation constant of the enzyme-substrate complex (Lehninger 1970).

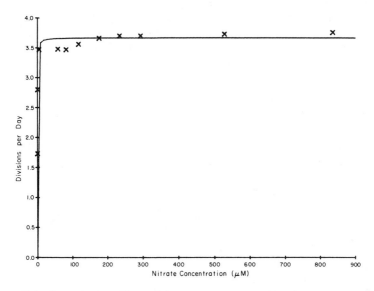

Figure 12.1 Control culture *T. pseudonana* growth rates (divisions/day) *vs.* added nitrate concentrations (μM). Data points are means of triplicate cultures. Error bars are not graphed because, generally, they would not exceed the size of the symbols. The variability of the data is presented in Table 12.1.

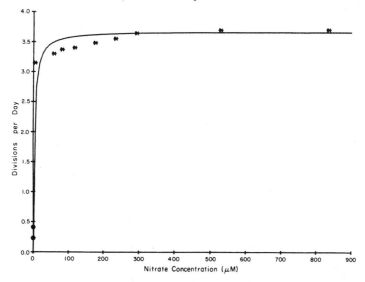

Figure 12.2 PCB (1 ppb)-treated culture *T. pseudonana* growth rates (divisions/day) *vs.* added nitrate concentrations (μM). Data points are means of triplicate cultures. Error bars are not graphed because, generally, they would not exceed the size of the symbols. The variability of the data is presented in Table 12.1.

A Lineweaver-Burk plot (Lehninger 1970) of the data was drawn by a Hewlett-Packard Plotter to determine the half-saturation constant (Ks) and the maximum growth rate (μ_{max}) by finding the negative reciprocal of the abcissa intersect and the reciprocal of the ordinate intersect, respectively. The plot, presented in Figure 12.3, shows an excellent fit of the converted data to a straight line, indicating a close fit to Michaelis-Menten kinetics. (Figure 12.3 does not show all of the data points for lack of space, but the regression lines drawn took into account all the data.) *T. pseudonana*'s Ks value was found to be 0.21 μM nitrogen in control cultures but 3.04 μM nitrogen in PCB-treated cultures. The PCB, however, did not affect *T. pseudonana*'s μ_{max} (3.67 divisions/day in control cultures, 3.68 divisions/day in treated cultures).

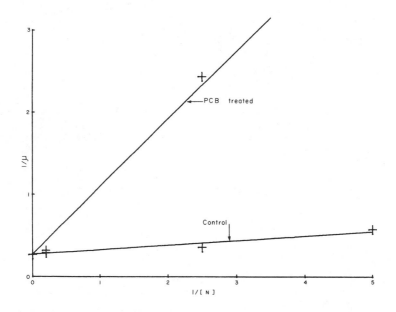

Figure 12.3 Lineweaver-Burk plot of *T. pseudonana* growth rates[-1] (divisions/day)[-1] *vs.* added nitrate[-1] (μM nitrate)[-1] with or without 1 ppb of PCB. Data points are means of triplicate cultures. The coefficient of determination (Sokal and Rohlf 1969) for control culture data is 0.95 and for PCB-treated culture data is 0.99.

DISCUSSION

The hyperbolic relationship observed between growth rate and nutrient concentration is in agreement with the data of many other investigations of microorganisms (Droop 1973, Meers 1973). These data indicate a clarification of the term "limiting" is warranted in that a nutrient can be yield-limiting, as expressed by Liebig (1847), or it can be growth rate-limiting, as shown here and elsewhere. Mathematically, these hyperbolic curves conform with the Michaelis-Menten expression. Since growth is a result of many enzyme reactions, the rate of growth is limited by the slowest one or set of these reactions, and a growth rate *vs.* nutrient concentration curve takes the shape of an enzyme activity *vs.* substrate concentration curve (hence, Equation 12.2 is derived from Equation 12.3).

Hyperbolic relationships conforming to Michaelis-Menten kinetics have been obtained for algal uptake of such nutrients as glucose (Wright and Hobbie 1965) nitrate (Caperon and Meyer 1972b; Eppley and Coatsworth 1968; Eppley and Thomas 1969; Eppley, *et al.* 1969), silicate (Paasche 1973; Goering, *et al.* 1973), and phosphate (Fuhs 1969). Many of these adsorption systems have been shown to be enzyme-regulated. Under certain environmental conditions, there also appears to be a hyperbolic relationship (described by Equation 12.2) between algal growth rate and external (extracellular) nutrient concentration for such nutrients as nitrogen (Eppley and Thomas 1969; Qasim, *et al.* 1973), silicate (Guillard, *et al.* 1973), phosphorus (Qasim, *et al.* 1973), and carbon (Goldman 1972), indicating a tight coupling of nutrient uptake and growth. However, a number of reports have shown growth rates to be more closely coupled to internal nutrient pools than external concentrations. Hence, growth rate was hyperbolically related to internal concentrations of vitamin B_{12} (Droop 1968), phosphate (Fuhs 1969), iron (Davies 1970), and nitrate (Caperon and Meyer 1972a), and these relationships generally followed Michaelis-Menten kinetics (Droop 1973).

While the results reported here agree favorably with others, the exact value of the Ks is questionable. Nitrogen contamination of the medium before the nitrate additions were made was presumed to be negligible; any nitrogen contamination would raise the absolute value of the Ks reported above. Organic nitrogen was present in the medium (23.4 μM of nitrogen) in the form of the metal chelator EDTA (ethylenediamine-tetraacetic acid), but it is not available for metabolism by algae (Hutner, *et al.* 1950). Subsequent tests, involving algal bioassays similar to those used for determining vitamin requirements of diatoms (Guillard and Cassie 1963), revealed that biologically utilizable nitrogen contaminants were present at a concentration of 0.26 μM of nitrogen. The control Ks is, therefore, raised from 0.21 μM to 0.42 μM under the experimental conditions, though the shape of the curve is not appreciably changed, and the conclusion that the PCB increased the Ks for growth remains valid. It should be noted that chemostats generally allow for greater certainty in studies of growth rate *vs.* substrate concentration because they provide both greater control over the physiological state of the inoculum and a steady-state environment in which limiting nutrients are maintained at a constant concentration. Thus, while batch cultures are suitable for determining the general effects of the PCB (or similar pollutant), chemostats may be employed to estimate more accurately the "true" Ks for a given species growing under a given set of environmental conditions.

The Ks for growth of *T. pseudonana* (clone 3H) in control cultures is lower than its Ks for nitrate uptake (1.87 μM of nitrogen) as reported by Carpenter and Guillard (1970). This difference between growth and uptake half-saturation constants is consistent with many earlier reports (see Eppley and Renger 1974, for discussion) and supports the theory that growth rate is directly limited by internal nutrient pools, whose size is governed by the nutrient uptake rate, while uptake rate is hyperbolically related to external nutrient concentrations.

The data suggest that the PCB acted as a competitive inhibitor of one or more growth rate-limiting enzymes involved in nitrogen uptake or metabolism. Competitive inhibition is characterized by an increase in Ks but no effect on μ_{max}, presumably due to blockage of an enzyme's active site by the inhibitor (Lehninger 1970). Thus, a competitive inhibitor's effects can be reversed by excessive substrate concentrations, as shown in Figure 12.4. It is quite possible that the inhibited system is membrane-bound, for the hydrophobic PCB would be likely to localize in the lipid portion of cell membranes. Recent experiments indicate that glutamic dehydrogenase is competitively inhibited by PCB (Fisher and Murphy, unpublished). The PCB could also come into contact with ATPases, which are often found in lipoprotein membranes (Risebrough, *et al.* 1970) and may interfere with their function in active transport across membranes. Batterton, *et al.* (1972) similarly suggested that DDT inhibited Na^+, K^+-activated ATPase in the blue-green alga *Anacystis nidulans*.

Reports have described similar effects of other pollutants upon *T. pseudonana*'s growth. Hannan and Patouillet (1972) found that mercury toxicity to three algal species, including *T. pseudonana*, increased with decreasing nutrient concentrations, and Stoll and Guillard (unpublished data) found naphthalene toxicity to *T. pseudonana* increased with decreasing phosphorus levels.

The calculated chlorinated hydrocarbon concentration per cell in this experiment (about 50 $\mu g/g$ or 50 ppm wet weight) exceeds levels generally found in natural phytoplankton samples. However, in nitrogen-limited chemostats (Fisher 1974) it was shown that *T. pseudonana*'s growth was diminished when cellular PCB concentrations (0.2-5 ppm) were well within the range of chlorinated hydrocarbon concentrations found in marine phytoplankton samples (Harvey, *et al.* 1974; Williams and Holden 1973; Giam, *et al.* 1973; Ware and Addison 1973).

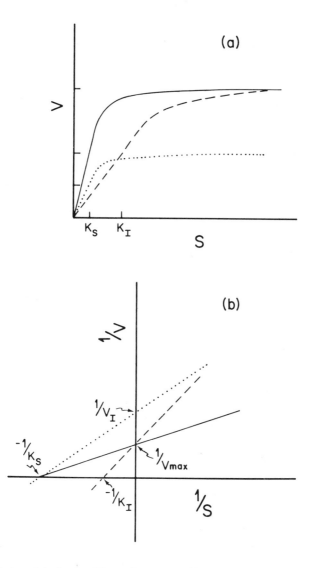

Figure 12.4 A model of competitive and noncompetitive enzyme inhibition. S represents the substrate concentration, V represents the velocity of the reaction, V_{max} equals the maximum velocity of the reaction catalyzed by the uninhibited and competitively inhibited enzymes, V_I equals the maximum velocity of the reaction catalyzed by the noncompetitively inhibited enzyme, Ks equals the half-saturation constant of the uninhibited and the noncompetitively inhibited enzymes, and K_I equals the half-saturation constant of the competitively inhibited enzyme. Solid lines represent uninhibited enzyme, dashed lines represent competitively inhibited enzyme, and dotted lines represent noncompetitively inhibited enzyme. (a) Velocity vs. substrate concentration. (b) Lineweaver-Burk plot of enzyme activity.

The dependence on nitrate concentration of algal sensitivity to PCB is noteworthy in that marine phytoplankton are often nitrogen-limited in nature (Ryther and Dunstan 1971). The vulnerability of phytoplankton to pollutants may therefore vary with the season, being greatest during nutrient deprivation and minimal during bloom conditions. This interaction is suggestive of the great number of interactions possible between nutrient abundance and aquatic pollutants, and illustrates the difficulty in determining safe environmental levels of persistent waste products.

It is of ecological significance that PCB and similar pollutants can reduce algal growth rates. It is possible that, owing to the impermanence of conditions in aquatic environments and to the relatively low phytoplankton cell densities normally occurring in nature, many phytoplankton species would adopt a competitive strategy of rapid resource exploitation (*i.e.*, high growth rate) rather than a density-dependent strategy involving such mechanisms as allelochemics. Experimental evidence for this, however, is lacking. By diminishing growth rates and thereby interfering with the competitive abilities of some phytoplankton, chlorinated hydrocarbons could disrupt successional patterns and alter species composition in algal communities in nature.

Growth kinetics studies have been used to explain observed PCB-induced disturbances of continuously cultured phytoplankton communities in which direct antibiosis via allelochemics did not occur (Fisher 1974). Such studies can facilitate predictions on the outcome of competitive interactions and may aid in the mass culturing of select species of microorganisms for industrial and aquaculture projects and in the assessment of pollution hazards in nature.

ACKNOWLEDGMENTS

We thank Drs. Vaughan T. Bowen and George R. Harvey for critically reading the manuscript. This research was supported by U.S. National Science Foundation Grant GB-11902 and by a grant from the Sarah Mellon Scaife Foundation to the Woods Hole Oceanographic Institution. This chapter was Contribution No. 3496 from the Woods Hole Oceanographic Institution and Contribution No. 112 from the Marine Sciences Research Center (MSRC), State University of New York at Stony Brook.

REFERENCES

Batterton, J. C., G. M. Boush and F. Matsumura. "DDT: Inhibition of Sodium Chloride Tolerance by the Blue-Green Alga *Anacystis nidulans*," *Science* **176**, 1141 (1972).
Caperon, J. and J. Meyer. "Nitrogen-Limited Growth of Marine Phytoplankton. I. Changes in Population Characteristics with Steady-State Growth Rate," *Deep-Sea Res.* **19**, 601 (1972a).

Caperon, J. and J. Meyer. "Nitrogen-Limited Growth of Marine Phyto-plankton. II. Uptake Kinetics and Their Role in Nutrient-Limited Growth of Phytoplankton," *Deep-Sea Res.* **19**, 619 (1972b).

Carpenter, E. J. and R. R. L. Guillard. "Intraspecific Differences in Nitrate Half-Saturation Constants for Three Species of Marine Phyto-plankton," *Ecol.* **52**, 183 (1970).

Davies, A. G. "Iron, Chelation and the Growth of Marine Phytoplankton. I. Growth Kinetics and Chlorophyll Production in Cultures of the Euryhaline Flagellate *Dunaliella tertiolecta* under Iron-Limiting Condi-tions," *J. Mar. Biol. Assoc. U.K.* **50**, 65 (1970).

Droop, M. R. "Vitamin B_{12} and Marine Ecology. IV. The Kinetics of Uptake, Growth and Inhibition in *Monochrysis lutheri*," *J. Mar. Biol. Assoc. U.K.* **48**, 689 (1968).

Droop, M. R. "Some Thoughts on Nutrient Limitation in Algae," *J. Phycol.* **9**, 264 (1973).

Eppley, R. W. and J. L. Coatsworth. "Uptake of Nitrate and Nitrite by *Ditylum brightwellii*—Kinetics and Mechanisms," *J. Phycol.* **4**, 151 (1968).

Eppley, R. W. and E. H. Renger. "Nitrogen Assimilation of an Oceanic Diatom in Nitrogen-Limited Continuous Culture," *J. Phycol.* **10**, 15 (1974).

Eppley, R. W. and J. D. H. Strickland. "Kinetics of Marine Phytoplankton Growth," in *Advances in Microbiology of the Sea*, Vol. I, M. R. Droop and E. J. F. Wood, Eds. (London: Academic Press, 1968), pp. 23-62.

Eppley, R. W. and W. H. Thomas. "Comparison of Half-Saturation Con-stants for Growth and Nitrate Uptake of Marine Phytoplankton," *J. Phycol.* **5**, 375 (1969).

Eppley, R. W., J. N. Rogers and J. J. McCarthy. "Half-Saturation Con-stants for Uptake of Nitrate and Ammonium by Marine Phytoplankton," *Limnol. Oceanogr.* **14**, 912 (1969).

Fisher, N. S. "Effects of Chlorinated Hydrocarbon Pollutants on Growth of Marine Phytoplankton in Culture," Ph.D. thesis, State University of New York at Stony Brook (1974).

Fisher, N. S. and C. F. Wurster. "Individual and Combined Effects of Temperature and Polychlorinated Biphenyls on the Growth of Three Species of Phytoplankton," *Environ. Pollution* **5**, 205 (1973).

Fisher, N. S., L. B. Graham, E. J. Carpenter and C. F. Wurster. "Geo-graphic Differences in Phytoplankton Sensitivity to PCBs," *Nature* **241**, 548 (1973).

Fisher, N. S., E. J. Carpenter, C. C. Remsen and C. F. Wurster. "Effects of PCB on Interspecific Competition in Natural and Gnotobiotic Phytoplankton Communities in Continuous and Batch Cultures," *Microb. Ecol.* **1**, 39 (1974).

Fuhs, G. W. "Phosphorus Content and Rate of Growth in the Diatoms *Cyclotella nana* and *Thalassiosira fluviatilis*," *J. Phycol.* **5**, 312 (1969).

Giam, C. S., M. K. Wong, A. R. Hanks, W. M. Sackett and R. L. Richardson. "Chlorinated Hydrocarbons in Plankton from the Gulf of Mexico and Northern Caribbean," *Bull. Environ. Contam. Toxicol.* **9**, 376 (1973).

Goering, J. J., D. M. Nelson and J. A. Carter. "Silicic Acid Uptake by Natural Populations of Marine Phytoplankton," *Deep-Sea Res.* **20**, 777 (1973).

Goldman, J. C., W. J. Oswald and V. Jenkins. "The Kinetics of Inorganic Carbon-Limited Algal Growth," *J. Water Poll. Control Fed.* **46**, 554 (1974).

Guillard, R. R. L. and V. Cassie. "Minimum Cyanocobalamin Requirements of Some Marine Centric Diatoms," *Limnol. Oceanogr.* **8**, 161 (1963).

Guillard, R. R. L. and J. H. Ryther. "Studies of Marine Planktonic Diatoms. I. *Cyclotella nana* Hustedt, and *Detonula confervacea* (Cleve) Gran.," *Can. J. Microbiol.* **8**, 229 (1962).

Guillard, R. R. L., P. Kilham, and T. A. Jackson. "Kinetics of Silicon-Limited Growth in the Marine Diatom *Thalassiosira pseudonana* Hasle and Heimdahl (=*Cyclotella nana* Hustedt)," *J. Phycol.* **9**, 233 (1973).

Hannan, P. J. and C. Patouillet. "Effect of Mercury on Algal Growth Rates," *Biotech. Bioeng.* **14**, 93 (1972).

Harvey, G. R., H. P. Miklas, V. T. Bowen and W. G. Steinhauer. "Observations on the Distribution of Chlorinated Hydrocarbons in Atlantic Ocean Organisms," *J. Mar. Res.* **32**, 102 (1974).

Hutner, S. H., L. Provasoli, A. Schatz and C. P. Haskins. "Some Approaches to the Role of Metals in the Metabolism of Microorganisms," *Proc. Am. Phil. Soc.* **94**, 152 (1950).

Lehninger, A. L. *Biochemistry.* (New York: Worth Publishers, 1970).

Liebig, J. *Chemistry and Its Adaptation to Agriculture and Physiology,* 4th ed. (London: Taylor and Walton, 1847).

MacFarlane, R. B., W. A. Glooschenko and R. C. Harriss. "The Interaction of Light Intensity and DDT Concentration upon the Marine Diatom *Nitzschia delicatissima* Cleve," *Hydrobiol.* **39**, 373 (1972).

Meers, J. L. "Growth of Bacteria in Mixed Cultures," *Critic. Rev. Microbiol.* **3**, 139 (1973).

Mosser, J. L., N. S. Fisher, T.-C. Teng and C. F. Wurster. "Polychlorinated Biphenyls: Toxicity to Certain Phytoplankters," *Science* **175**, 191 (1972a).

Mosser, J. L., N. S. Fisher and C. F. Wurster. "Polychlorinated Biphenyls and DDT Alter Species Composition in Mixed Cultures of Algae," *Science* **176**, 533 (1972b).

Mosser, J. L., T.-C. Teng, W. G. Walther and C. F. Wurster. "Interactions of PCBs, DDT, and DDE in a Marine Diatom," *Bull. Environ. Contam. Toxicol.* **12**, 665 (1974).

Paasche, E. "Silicon and the Ecology of Marine Plankton Diatoms. II. Silicate-Uptake Kinetics in Five Diatom Species," *Mar. Biol.* **19**, 262 (1973).

Qasim, S. Z., P. M. A. Bhattathiri and V. P. Devassy. "Growth Kinetics and Nutrient Requirements of Two Tropical Marine Phytoplankters," *Mar. Biol.* **21**, 299 (1973).

Risebrough, R. W., J. Davis and D. W. Anderson. "Effects of Chlorinated Hydrocarbons," in *The Biological Impact of Pesticides in the Environment,* J. W. Gillett, Ed. (Corvallis, Oregon: Environmental Health Sciences Center, Oregon State University, 1970).

Ryther, J. H. and W. M. Dunstan. "Nitrogen, Phosphorus, and Eutrophication in the Coastal Marine Environment," *Science* **171**, 1008 (1971).

Sizer, I. W. "Chemical Aspects of Enzyme Inhibition," *Science* **125**, 34 (1957).

Sokal, R. R. and F. J. Rohlf. *Biometry*. (San Francisco: Freeman, 1969).

Swift, D. G. and W. R. Taylor. "Growth of Vitamin B_{12}-Limited Cultures of *Thalassiosira pseudonana, Monochrysis lutheri*, and *Isochrysis galbana*," *J. Phycol.* **10**, 385 (1974).

Thomas, W. H. and A. N. Dodson. "Effect of Interactions Between Temperature and Nitrate Supply on the Cell-Division Rates of Two Marine Phytoflagellates," *Mar. Biol.* **24**, 213 (1974).

Ware, D. M. and R. F. Addison. "PCB Residues in Plankton from the Gulf of St. Lawrence," *Nature* **246**, 519 (1973).

Williams, A. and A. V. Holden. "Organochlorine Residues from Plankton," *Mar. Pollut. Bull.* **4**, 109 (1973).

Wright, R. T. and J. E. Hobbie. "The Uptake of Organic Solutes in Lake Water," *Limnol. Oceanogr.* **10**, 22 (1965).

13

A STEADY-STATE MODEL OF LIGHT-, TEMPERATURE-, AND CARBON-LIMITED GROWTH OF PHYTOPLANKTON

Dale A. Kiefer and Theodore Enns[1]

INTRODUCTION

An examination of charts of intermediary metabolism indicates that the biochemical system that operates during the growth of phytoplankton is extremely complex. Pathways such as those responsible for the assimilation of carbon and nitrogen interact both directly, by sharing common intermediates, and indirectly, by sharing pools of energized compounds such as the reduced pyridine nucleotides (NADPH) and adenosine triphosphate (ATP). Such complexity hinders the plant physiologist and ecologist who wish to describe the performance of the entire biochemical system. Thus, a description or a model of the growth response of phytoplankton to limiting light intensities, temperature, and nutrient concentrations must be based upon simplifications that do not violate the basic features of metabolic regulation.

Most ecological and physiological descriptions of the growth of phytoplankton have used Monod's (1942) kinetic model for the steady-state growth of microbes:

$$\mu = \frac{\mu_{max}\,(S)}{K_s + (S)} \tag{13.1}$$

where μ is the specific growth rate at limiting substrate concentration S, constant μ_{max} is the maximum specific growth rate at concentrations of

[1]Visibility Laboratory of the Scripps Institution of Oceanography, and Marine Biology Research Division, University of California, San Diego, LaJolla, California, 92093, respectively.

substrate that are saturating, and K_S is the concentration of substrate at which the specific growth rate is one-half its maximal value. Although Monod's model has been applied usefully to a number of descriptions of nutrient-limited growth (*e.g.*, Williams 1965, Caperon 1965, Dugdale 1967, Droop 1968), simplifications inherent in Equation (13.1) have severely limited its usefulness in describing more complex features of metabolic regulation (*e.g.*, Caperon and Meyer 1972a, 1972b; Droop 1973; Eppley and Renger 1974).

ASSUMPTIONS OF THERMODYNAMIC MODELS

Recognizing the limitations of such descriptive models for microbial growth, we have developed a mechanistic model for phytoplankton growth. This model describes the conversions of energy and the flows of elements such as carbon, hydrogen, and oxygen according to principles of non-equilibrium thermodynamics. The application of such principles requires two basic assumptions. First, we must assume that the biochemical system reaches a steady state. This means that the state variables of the system remain constant with time (Prigogine 1967). Such an assumption may at least be partially justified by studies of the properties of phytoplankton grown in continuous culture. These studies (*e.g.*, Myers and Graham 1971, Caperon and Meyer 1972a) indicate that unique physiological and biochemical properties are associated with a given rate of steady–state growth.

Second, we must assume that the biochemical system can be described mathematically as a system that is near equilibrium and reversible. Such an assumption allows one to use phenomenological equations that are the key to our model. This assumption is difficult to verify, but it does appear that such complex biochemical pathways as the photosynthetic and oxidative electron transport systems may be characterized as reversible and near equilibrium (*e.g.*, Mitchell 1970, Rottenberg *et al.* 1970). We will present the model by first describing the system and the relevant parameters. We will then consider the series of equations that determine the behavior of the system, and finally, we will apply the model to carbon-, light-, and temperature-limited growth of phytoplankton.

A QUALITATIVE DESCRIPTION

In Figure 13.1 we have combined all reactions of photosynthetic growth into two pairs of coupled biochemical pathways. Primary photochemical reactions occur at coupling site 3-4, where the free energy released in the absorption of light, reaction 4, establishes the electrochemical potential

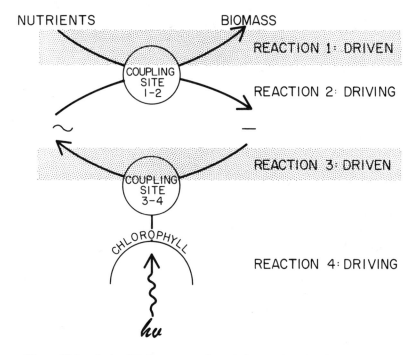

NUTRIENTS

BIOMASS

REACTION 1: DRIVEN

COUPLING SITE 1-2

REACTION 2: DRIVING

REACTION 3: DRIVEN

COUPLING SITE 3-4

CHLOROPHYLL

REACTION 4: DRIVING

$h\nu$

Figure 13.1   A simplified representation of the reactions leading to growth in phytoplankton. The driving reactions release free energy while the driven reactions consume free energy. Energy conversion occurs at the two coupling sites: at coupling site 3-4 the energy of absorbed quanta is converted into energy available for photosynthetic electron transfer, and at coupling site 1-2 the energy released in photosynthetic electron transport is converted into the energy consumed in the assimilation of nutrients.

required for photosynthetic electron transport, reaction 3. Secondary biochemical reactions involving oxidation-reduction, group transfer, and rearrangements occur at coupling site 1-2. Here the free energy released in photosynthetic electron transport, reaction 2, drives the dark reactions of photosynthesis.

Reaction 1 can be represented as (Kok 1960)

$$6.14 \ CO_2 + 3.7 \ H_2O + NH_3 = C_{6.14} H_{10.4} O_{2.24} N + 6.9 \ O_2 \qquad (13.2)$$

Since 4 electrons are transferred per molecule of O_2, the equation indicates that 4.5 electrons are transferred per molecule of CO_2 fixed. We will define standard conditions for phytoplankton growth by a temperature of $293°K$, an incident irradiance of 1.9×10^{-2} m einsteins cm^{-2} hr^{-1}, and a nutrient solution containing 10 μM CO_2 and NH_3 and 280 μM O_2.

The two coupling sites are functional rather than structural; the sites consist of the collection of enzymes catalyzing the four pathways of Figure 13.1. We characterize each coupling site by two features, conductivity and coupling, whose values are determined by the enzymes of the respective pathways. The conductivity at a site will be high if both the *in vivo* activity of enzymes and the concentration of substrates are high; conversely, if either the activity of enzymes or substrate concentrations are low, conductivity will be low. Thus, conductivity may be considered as a measure of the catalytic capability of the cell. The coupling at a site is an intrinsic feature of coupled pathways in the steady state and is independent of either *in vivo* enzymatic activity or substrate concentration. Coupling describes the extent to which coupled reactions "drag" each other; thus it determines the efficiency with which a site functions (Kedem and Caplan 1965). Our model assumes variable conductivity but constant coupling. Both features will be defined mathematically by phenomenological coefficients.

A QUANTITATIVE DESCRIPTION

The behavior of the system shown in Figure 13.1 is described by the 11 equations given below. They describe light absorption, rates of biochemical reactions, and steady-state conditions.

$$J_4 = J_i \, \text{Chl a*} \, (A \cdot L) \tag{13.3a}$$

$$J_4 = L_{43}A_3 + L_{44}A_4 \tag{13.4a}$$

$$J_3 = L_{33}A_3 + L_{34}A_4 \tag{13.4b}$$

$$J_2 = L_{21}A_1 + L_{22}A_2 \tag{13.4c}$$

$$J_1 = L_{11}A_1 + L_{12}A_2 \tag{13.4d}$$

$$q_{34} = \frac{L_{34}}{\sqrt{L_{33} \, L_{44}}} = 0.96 \tag{13.5a}$$

$$z_{34} = \frac{\sqrt{L_{33}}}{\sqrt{L_{44}}} = 1.00 \tag{13.5b}$$

$$q_{12} = \frac{L_{12}}{\sqrt{L_{11} \, L_{22}}} = 0.91 \tag{13.5c}$$

$$z_{12} = \frac{\sqrt{L_{11}}}{\sqrt{L_{22}}} = 0.95 \tag{13.5d}$$

$$J_2 = J_3 \tag{13.6a}$$

$$A_2 = f(a_1, A_4, q_{12}, q_{34}, z_{12}, z_{34}) \tag{13.6b}$$

Equation (13.3a) is an approximation of Beer's Law describing the flow of quanta absorbed by a lamina of cells. J_4, the flux of absorbed quanta, is a function of incident irradiance (J_i), the chlorophyll concentration of the cell suspension (Chl), the molar absorption coefficient for cellular chlorophyll (a^*), and a dimensional parameter ($A{\cdot}L$). The product $A{\cdot}L$ relates the flux of absorbed quanta in a one-liter suspension of cells to the incident flux of quanta. Since the incident flux has units of einsteins $cm^{-2} \cdot hr^{-1}$, $A{\cdot}L$ is equal to 1×10^3 cm^3 as shown in Table 13.1. The molar absorption coefficient includes all photosynthetic pigments normalized to chlorophyll.

The value for a^* was determined by measuring the mean (400-700 nm) diffuse absorption coefficient for a suspension of cells that contains a given concentration of chlorophyll a (Duntley et al. 1974). The coefficient was then normalized by dividing the coefficient by the chlorophyll a concentration. Use of such a normalized coefficient in Equation (13.3a) is based upon the assumption that the relative concentrations of chlorophyll a and accessory pigments remain constant at all growth rates.

Equations (13.4a) and (13.4b) are phenomenological equations describing fluxes at coupling site 3-4 as functions of conductivity coefficients (the four L terms) and chemical affinities (the two A terms). L_{34} and L_{43} are the coupling coefficients that relate the rates of biochemical reactions 3 and 4 to their nonconjugate forces A_4 and A_3, respectively. L_{33} and L_{44} are straight coefficients relating the rates of reactions 3 and 4 to their conjugate forces A_3 and A_4, respectively. We invoke Onsager's law of reciprocality for the model:

$$L_{34} = L_{43} \tag{13.7}$$

Likewise, at coupling site 1-2:

$$L_{12} = L_{21} \tag{13.8}$$

The chemical affinity for reaction 4, the de-excitation of reaction center chlorophyll, is a constant estimated to be +40 kcal per einstein of quanta for photosystems I and II (e.g., Rabinowitch and Govindjee

Table 13.1. Constants for Equations Describing Limited Growth of Phytoplankton[a]

Equations	Parameter	Value
13.3a	$a*$, molar coefficient for cellular chlorophyll	1.3×10^2 cm^{-1} ·mM Chl^{-1}
13.3a	$A \cdot L$, dimension parameter	1.0×10^3 cm^3
13.4a 13.4b	A_4, chemical affinity of reaction 4	40 cal m einstein^{-1}
13.4c 13.4d	A_1, chemical affinity of reaction 1	-18 cal m mole^{-1}
13.14	b, slope $\dfrac{\Delta Chl}{\Delta J_i}$	6.0×10^{-4} m mole Chl m einstein^{-1} cm^2 hr
13.14	a, intercept	-1.8×10^{-2} m mole Chl mg-at C^{-1}
13.16a	c, constant for mass balance equation	6.5×10^{-2} cal m mole^{-1}
13.16a	d, constant for mass balance equation	-17 cal m mole^{-1}
13.16b	L_{12}^{max}, maximal cross coefficient for coupling site 12	5.2×10^{-2} m moles^2 cal hr^{-1}
13.16b	K_s, half saturation constant	5.0×10^{-3} m M CO$_2$
13.17	L_{12}^{T1}, cross coefficient for coupling site 12 at 293°K	3.47×10^{-2} m moles^2 cal hr^{-1}
13.17	E_a, activation energy	11.4 cal m mole^{-1}
13.17	R, gas constant	1.99×10^{-3} cal °K^{-1} m mole^{-1}
13.17	T_1, temperature at standard growth	293°K

[a]Standard growth conditions are 293°K, 10 μM CO$_2$, 10 μM NH$_3$, 1.9×10^{-2} m einsteins · cm^{-2} · hr^{-1}.

1970). Since chemical affinity is defined as the negative of the partial molar free energy change for a reaction, reaction 4 releases 40 kcal of free energy. Reaction 3, which is the flux of electrons from primary donors to acceptors across reaction centers, consumes free energy. At the steady state its chemical affinity has been estimated at -28 kcal per mole of electrons for reaction centers of both photosystems (Crofts *et al.* 1971, Ross and Calvin 1967).

The two remaining phenomenological equations describe flows at coupling site 1-2. The chemical affinity for reaction 2, the flow of electrons down the gradient established by primary photochemistry is +28 kcal/ mole e⁻ at steady state. Thus,

$$A_2 = - A_3 \qquad (13.9)$$

The chemical affinity of reaction 1, shown in Equation (13.2), is -18 kcal/mole e^- (Kok 1965). While A_1 and A_4 are constants, A_2, A_3, and the phenomenological coefficients are variables that must be solved for any steady-state rate of growth.

Equations (13.4a-13.4d) define relationships between phenomenological coefficients. The ratios defined by q and z were introduced by Kedem and Caplan (1965), who showed that such ratios were useful in describing energy conversions for coupled reactions. q_{12} and q_{34} are called the degrees of coupling, and z_{12} and z_{34} are ratios of straight coefficients. Values for q may range from 0 for uncoupled reactions to 1 for completely coupled reactions. Values for z will be positive in our model since all phenomenological coefficients are limited to positive values. From data describing the optimal efficiency of growth in *Chlorella* (Kok 1960), we have guessed q_{12} and z_{12} to be 0.91 and 0.95, respectively, and q_{34} and z_{34} to be 0.96 and 1.00, respectively (Kiefer *et al.* 1975).

Although the coupling parameters q and z are guessed, errors that may be introduced will have no effect on the qualitative response of the model to variations in light, carbon, or temperature. However, the efficiency of energy conversion by the model cells will depend upon the combined effect of the two pairs of coupling parameters. The four equations restrict the coupling characteristics at the two sites to a constant value at all steady-state rates of growth. Since for each coupling site there are three phenomenological coefficients and two equations, the coefficients have only one degree of freedom.

Equations (13.6a) and (13.6b) are conditions for steady-state growth. Equation (13.6a) constrains the photosynthetic electron transport system to equal rates of input and output of electrons. Equation (13.6b) is a function that fixes A_2, the chemical affinity of the electron transport system, to a value that optimizes the overall efficiency of energy conversion by the system. As indicated, this value will depend upon the chemical affinities of reactions 1 and 4 and upon the coupling characteristics at both coupling sites.

In coupled reactions the efficiency of energy conversion is the ratio of the rate of free energy consumed by the driven reaction to the rate of free energy released by the driving reaction. Thus, the efficiencies of energy conversion at coupling sites 1-2 and 3-4 are

$$\eta_{12} = -\frac{J_1 A_1}{J_2 A_2} \tag{13.10a}$$

$$\eta_{34} = -\frac{J_3 A_3}{J_4 A_4} \tag{13.10b}$$

The efficiency of overall energy conversion, the conversion of light energy into chemical bond energy, is

$$\eta_{14} = -\frac{J_1 A_1}{J_4 A_4} \qquad (13.10c)$$

Since $A_2 = -A_3$ (Equation 13.9), and $J_2 = J_3$,

$$\eta_{14} = \eta_{12} \cdot \eta_{34} \qquad (13.11)$$

As shown by Kedem and Caplan (1965), the efficiencies n_{12}, n_{34}, and n_{14} are described conveniently as functions of the coupling parameters q and z and the ratio of chemical affinities, $\frac{\text{A driven}}{\text{A driving}}$. In addition, they showed that for given values of q and z there are unique values for the ratio of chemical affinities that allow optimal efficiency in energy conversion. For example,

$$\frac{A_1}{A_2} = \frac{-q_{12} \; z_{12}}{1 + \sqrt{1 - q_{12}^2}} \qquad (13.12)$$

Equation (13.6b) is a similar, although more complex, function that determines the value of A_2, which, in turn, yields an optimal value for n_{14}. Figure 13.2 is a graph of the dependence of efficiencies upon the chemical affinity of reaction 2. n_{14}, the overall efficiency, is 0.23 when the chemical affinity of reaction 2 equals 28 kcal/mole e⁻. If A_2 increases, then n_{14} decreases, primarily because of decreases in n_{34}; if A_2 decreases, then n_{14} decreases, primarily because of decreases in n_{12}. Values for coupling parameters and chemical affinities of reaction 1 and 4 are shown in the figure. Equation (13.6b) was obtained by optimizing n_{14} with respect to A_2 according to Equation (13.11).

$$\frac{\partial \eta_{14}}{\partial A_2} = \frac{\partial (\eta_{12} \cdot \eta_{34})}{\partial A_2} = 0 \qquad (13.13)$$

We note that if the coupling parameters of sites 1-2 and 3-4 were equal, the optimal value for A_2 merely would be the mean value of A_1 and A_4.

These 11 equations are basic to the thermodynamic model. We will consider now the response of this system of equations to variations in J_i, in the case of light-limited growth of phytoplankton, and variations in L_{12} for carbon- and temperature-limited growth. Solution of these equations for growth-limiting conditions will require one or two additional equations, which unfortunately must be derived empirically.

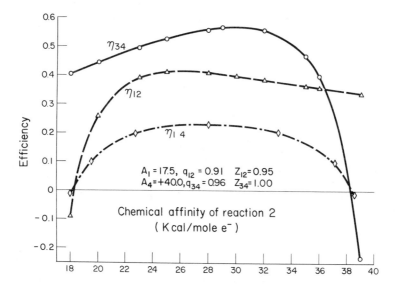

Figure 13.2   Changes in the efficiency of energy conversion at coupling site 1-2, n_{12}, and site 3-4, n_{34}, and overall conversion, n_{14}, with changes in the chemical affinity of photosynthetic electron transport. The values for chemical affinities and degrees of coupling are shown in the figure.

LIGHT-LIMITED GROWTH

Solution of the thermodynamic model for light-limited growth requires an additional equation:

$$Chl = a - b \cdot J_i \qquad (13.14)$$

Cellular chlorophyll a, Chl [in moles of chlorophyll a · (g-at. of cell carbon)$^{-1}$] is described as a function of incident irradiance, J_i (in einsteins cm^{-2} hr^{-1}). The slope of this linear equation comes from the observations of Myers and Graham (1971) on steady-state, light-limited growth of *Chlorella pyrenoidosa.* They noted that at subsaturating levels of irradiance, $\Delta CHl/\Delta J_i$ is constant and equals the value of b given in Table 13.1. The intercept of the equation is obtained by solving the 12 equations of the model (13.3a-13.6b and 13.14) for conditions of optimal growth of *C. pyrenoidosa.* Under such conditions, Myers and Graham noted that the cell's specific growth rate is approximately 0.075 hr^{-1}, and the chlorophyll (a and b) content of the cells is 3×10^{-4} moles/g- at carbon. The value for the intercept, a, is given in Table 13.1.

Figure 13.3 summarizes the response of the model to subsaturating levels of irradiance. The variations in parameters are obtained by solving the 12 equations, given A_1 and A_4 and variable values for J_i. According to the figure, the specific growth rate of the phytoplankton decreases with decreasing light levels, the response being a quadratic function of irradiance. This decrease in growth rate is paralleled by decreases in the cross coefficients, L_{12} and L_{34}, and the straight coefficients for both coupling sites. Such a decrease in conductivity is interpreted as a decrease in the catalytic capability of cells, or more precisely a decrease in either enzyme concentration or specific activity. As a direct consequence of Equation (13.14), the chlorophyll content of model cells increases linearly with decreases in light. Despite a decrease in specific growth rate, the quantum efficiency of photosynthesis remains constant in the steady state:

$$\phi = \frac{J_{CO_2}}{J_q} = 0.11$$

A value of 0.11 approaches the maximum value of 0.125 for an eight quanta process and is probably too high. The model also predicts that the driving potential for electron transport, A_2, is maintained constant at all light intensities. This is, of course, a consequence of the optimization principle and constant coupling parameters q and z.

These predictions can be compared with the steady-state growth of *Chlorella pyrenoidosa* shown in Figure 13.4 (Myers and Graham 1971). The growth response of *Chorella* is similar to the model cells except for two features. First, the curve for *Chorella* does not pass through the origin, indicating the importance of energy-consuming processes unrelated to growth (basal metabolism). Second, experimentally grown *Chlorella* requires over twice the irradiance for growth rates comparable to those of model cells.[2] Two additional features of steady-state growth of *Chlorella* are a constant quantum efficiency of growth and a decreased rate of turnover for photosynthetic units, τ. Both features are consistent with the predictions of constant quantum efficiency and decreased enzymatic activity for model cells.

---

[2] This discrepancy is caused primarily by an overestimate of the chlorophyll *a* content of model cells at a given light level. The overestimate comes from the use of Equation (13.14), an empirical relationship between light and the cellular content of both chlorophyll *a* and *b* in *Chlorella*. Since the light absorption of model cells is normalized to chlorophyll *a* concentration alone, we overestimate the flux of absorbed quanta for model cells. This does not affect the behavior of the model.

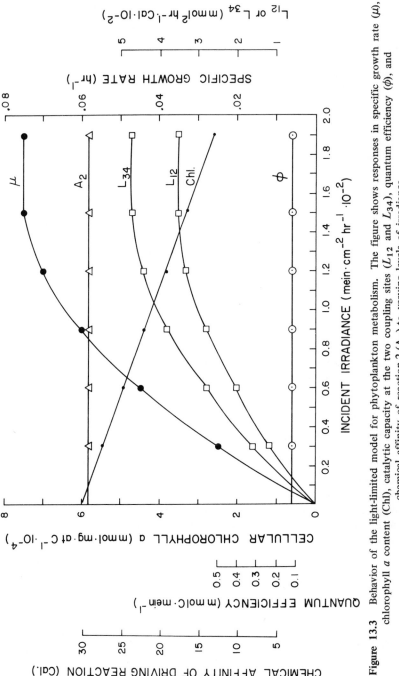

Figure 13.3 Behavior of the light-limited model for phytoplankton metabolism. The figure shows responses in specific growth rate (μ), chlorophyll a content (Chl), catalytic capacity at the two coupling sites (L_{12} and L_{34}), quantum efficiency (ϕ), and chemical affinity of reaction 2 (A_2) to varying levels of irradiance.

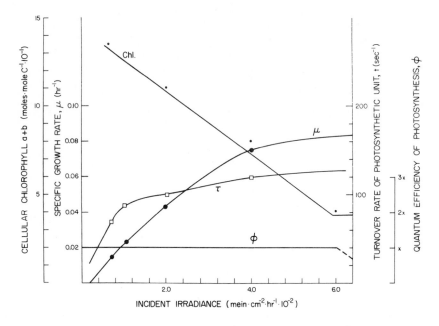

Figure 13.4 Changes in photosynthetic characteristics of *Chlorella pyrenoidosa* grown at varying levels of irradiance. The figure shows changes in specific growth rate (μ), cellular chlorophyll $a + b$ (Chl), quantum efficiency (ϕ), and maximum turnover rate for photosynthetic units (τ). The data were taken from a table by Myers and Graham (1971).

CARBON-LIMITED GROWTH

Referring to Figure 13.1 and Equations (13.3a-13.6b), one may consider that reduced concentrations of carbon dioxide affect the thermodynamic system by way of two parameters at coupling site 1-2. First, diminishing concentrations of carbon dioxide cause decreases in the chemical affinity of reaction 1 as predicted by Equations (13.2) and (13.14):

$$A_1 = -\Delta\overline{G} = - \left[\Delta\overline{G}^\circ + RT \ln \frac{(C_{6.14}H_{10.4}O_{2.24}N)(O_2)^{6.9}}{(CO_2)^{6.14}(NH_3)} \right] \qquad (13.15)$$

Since A_1 is kcal/mole e^- at standard growth conditions (*e.g.*, $CO_2 = 10 \ \mu M$), we can calculate changes in A_1 with changes in the concentration of carbon dioxide from Equation (13.15). Second, diminishing concentrations of carbon dioxide will affect decreases in the conductivity at the coupling site. Due to a limited affinity of the enzyme ribulose 1,5-diphosphate carboxylase for carbon dioxide, decreases in carbon dioxide will cause decreases in the phenomenological coefficients L_{11}, L_{22}, and L_{12}.

Thus, the series of equations describing carbon-limited growth will include Equations (13.3a-13.6b) and two additional equations that describe the above mentioned changes in A_1 and L_{12}:

$$A_1 = c \ln(CO_2) + d \qquad (13.16a)$$

$$L_{12} = \frac{L_{12}^{max} (CO_2)}{K_s + (CO_2)} \qquad (13.16b)$$

The constants of Equation (13.16a) shown in Table 13.1 were obtained by solving Equation (13.15) for conditions of standard growth and rearranging. Equation (13.16b) is based upon the assumption that the enzyme ribulose 1,5-diphosphate carboxylase is the rate-limiting enzyme in carbon-limited growth. The equation is of the form of the Michaelis-Menten equation for enzyme kinetics with L_{12} substituted for V, the velocity of the catalyzed reaction. The half saturation constant for the carboxylase is assigned a value of 5.0×10^{-3} mM. The maximum value for the cross coefficient, L_{12}^{max}, at saturating concentrations of carbon dioxide was obtained by solving Equation (13.16b) for standard conditions of growth for which L_{12} and CO_2 are known.

Parameters for the solutions to the 13 equations of the carbon-limited model are shown in Figure 13.5. The decrease in L_{12} caused by sub-saturating concentrations of carbon dioxide is paralleled by decreases in specific growth rate, cellular chlorophyll concentration, and L_{34}. The responses of all three parameters are approximated by rectangular hyperbolas. The decrease in chlorophyll concentration is of some interest, since the loss of cellular chlorophyll in nutrient-starved cells (called chlorosis) often has been ascribed to an inability to synthesize chlorophyll rather than to a consequence of metabolic regulation (e.g., Devlin 1966).

The model also predicts small decreases in the quantum efficiency of steady-state photosynthesis, ϕ, and small increases in the chemical affinity of reaction 2, A_2. Both changes are caused by small increases in the chemical affinity of reaction 1, and both are probably too small to be measured experimentally. The dominant factor in determining the response of the biochemical model is the decrease in conductivity at coupling site 1-2 rather than the decreased chemical affinity of reaction 1.

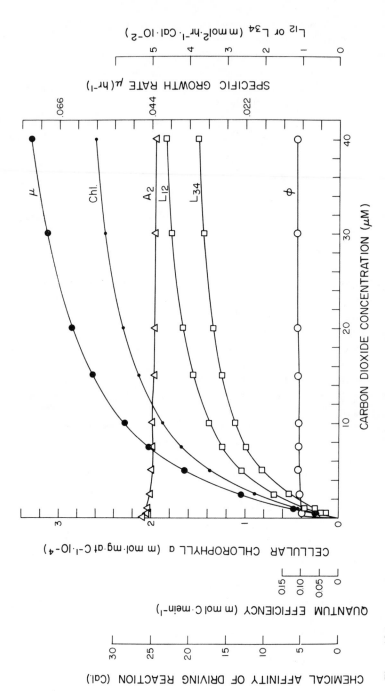

Figure 13.5 Behavior of the carbon dioxide-limited model for phytoplankton metabolism. The figure shows predicted responses in specific growth rate (μ), chlorophyll content (Chl), catalytic capacity at the two coupling sites (L_{12} and L_{34}), quantum efficiency (ϕ), and chemical affinity of reaction 2 (A_2) to varying concentrations of carbon dioxide.

TEMPERATURE-LIMITED GROWTH

Just as insufficient concentrations of carbon dioxide limit growth by decreasing the conductivity at coupling site 1-2, so low temperatures also limit growth by decreasing the conductivity at this site. Primary photochemistry, which occurs at coupling site 3-4, is insensitive to variations in temperature. Our description of temperature-limited growth draws upon the 11 equations, (13.3a-13.6b), and a single equation describing L_{12} as a function of temperature:

$$L_{12} = L_{12}^{T_1} \exp \left\{ \frac{Ea}{R} (\frac{1}{T_1} - \frac{1}{T_2}) \right\}$$ (13.17)

This is an adaptation of the Arrhenius equation in which phenomenological cross coefficients are substituted for the rate constants of chemical reactions. L_{12} is the cross coefficient at temperature T_2. $L_{12}^{T_1}$, the cross coefficient under standard growth conditions, is a constant whose value is given in Table 13.1. Ea is the activation energy for the dark reactions of photosynthesis; its value is also shown in Table 13.1 and corresponds to a Q_{10} of 2.0. R is the universal gas constant, T_1 is 293°K, and T_2 is the temperature input to the model. Use of Equation (13.17) assumes both a constant concentration of the rate-limiting enzyme and no thermal denaturation of the enzyme. The model was applied to a temperature range of 273-303°K.

Figure 13.7 shows the response of the system to such a range in temperature. The decrease in L_{12} with temperature is paralleled by a decrease in specific growth rate, cellular chlorophyll, and L_{34}. As is the case for light-limited growth, the quantum efficiency of photosynthesis and the chemical affinity of reaction 2 remain constant at all growth rates. The changes in specific growth rate and cellular chlorophyll predicted by the model agree with those observed for temperature-limited growth of the green algae *Dunaliella tertiolecta* (Eppley and Sloan 1966).

SUMMARY

We summarize the predictions of the thermodynamic model as follows. Light-limited cells will have high concentrations of chlorophyll with low concentrations of enzymes; carbon-limited cells will have low concentrations of chlorophyll with low concentrations of enzymes, with the exception of the carboxylase; and temperature-limited cells will have low concentrations of chlorophyll with high concentrations of enzymes. In all three cases the model predicts a relatively constant quantum efficiency of photosynthesis in the steady state.

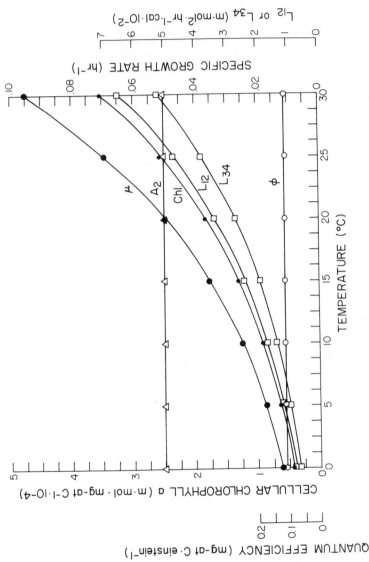

Figure 13.6 Behavior of temperature-limited model for phytoplankton metabolism. The figure shows predicted responses in specific growth rate (μ), chlorophyll content (Chl), catalytic capacity at the two coupling sites (L_{12} and L_{34}), quantum efficiency (ϕ), and chemical affinity of reaction 2 (A_2) to varying temperatures.

The thermodynamic model presented has several advantages over previous models of phytoplankton growth. First, the model is a more complete description of metabolic regulation since it describes the relationship between the flow of energy and matter within the cell. Second, the model should offer insights into the nature of multiple limitation of growth such as limiting light and carbon concentrations or limiting temperature and carbon concentrations. Since limiting temperature and carbon concentrations decrease the conductivity at coupling site 1-2, decreases in both temperature and carbon may decrease growth rate multiplicatively. On the other hand, since limiting light does not act directly upon the conductivity at coupling site 1-2, decreases in light and carbon concentration may decrease growth rate according to the principle of a single limiting factor. Third, the model is easily extended to nitrogen- or phosphate-limited growth. This is done by introducing two driven flows at coupling site 1-2, one for carbon and the other for either nitrogen or phosphorus. By appropriate choice of coupling parameters, it may be possible to describe nutrient-induced changes in the carbon, nitrogen or phosphorus content of cells as well as changes in chlorophyll content and enzymatic activity.

ACKNOWLEDGMENTS

We wish to thank Elisabeth Stewart for solving the simultaneous equations and Dr. Osmund Holm-Hansen for encouragement. This work was supported in part by the National Science Foundation Grants No. GA-36511 and DES 76-00405, and in part by the U.S. Atomic Energy Commission Contract AT (11-1)GEN 10, P.A. 20.

REFERENCES

Bannister, T. T. "Production Equations in Terms of Chlorophyll Concentration, Quantum Yield, and Upper Limit to Production," *Limnol. Oceanogr.* **19**, 1 (1974).

Caperon, J. "The Dynamics of Nitrate-Limited Growth of *Isochrysis galbana* Populations," Ph.D. Thesis, University of California, San Diego (1965).

Caperon, J. and J. Meyer. "Nitrogen-Limited Growth of Marine Phytoplankton. I. Changes in Population Characteristics with Steady-State Growth Rate," *Deep-Sea Res.* **19**, 601 (1972a).

Caperon, J. and J. Meyer. "Nitrogen-Limited Growth of Marine Phytoplankton. II. Uptake Kinetics and Their Role in Nutrient-Limited Growth of Phytoplankton," *Deep-Sea Res.* **19**, 619 (1972b).

Crofts, A. R., C. R. Wraight, and D. E. Fleischman. "Energy Conservation in the Photochemical Reactions of Photosynthesis and its Relation to Delayed Fluorescence," *F.E.B.S. Letters* **15**, 89 (1971).

Devlin, R. M. *Plant Physiology*. (New York: Reinhold Publishing Co., 1966).

Droop, M. R. "Some Thoughts on Nutrient Limitations in Algae," *J. Phycol.* **9**, 264 (1973).

Droop, M. R. "Vitamin B-12 and Marine Ecology. IV. The Kinetics of Uptake, Growth, and Inhibition in *Monochrysis lutheri*," *J. Mar. Biol. Assoc. U.K.* **48**, 689 (1968).

Dugdale, R. C. "Nutrient Limitation in the Sea: Dynamics, Identification, and Significance," *Limnol. Oceanogr.* **12**, 685 (1967).

Duntley, S. Q., R. W. Austin, W. H. Wilson, C. F. Edgerton, and S. E. Moran. "Ocean Color Analysis," Final Technical Report S.I.O. Ref. 74-10, Visibility Laboratory, San Diego, California (1974).

Eppley, R. W. and P. R. Sloan. "Growth Rates of Marine Phytoplankton: Correlation with Light Absorption by Cell Chlorophyll *a*," *Physiol. Plant* **19**, 47 (1966).

Eppley, R. W. and E. H. Renger. "Nitrogen Assimilation of an Oceanic Diatom in Nitrogen-Limited Continuous Culture," *J. Phycol.* **10**, 15 (1974).

Kedem, O. and S. R. Caplan. "Degree of Coupling and its Relation to Efficiency of Energy Conversion," *Trans. Faraday Soc.* **61**, 1897 (1965).

Kiefer, D. A., E. Stewart and T. Enns. "A Thermodynamic Model of Phytoplankton Metabolism. Part I. Theory," unpublished manuscript (1975).

Kiefer, D. A. "A Thermodynamic Model of Phytoplankton Metabolism. Part II. Light and CO_2-Limited Growth," unpublished manuscript (1975).

Kok, B. "On the Efficiency of *Chlorella* Growth," *Acta Botanica Neerl* **I**, 445 (1952).

Kok, B. "Photosynthesis: The Path of Energy," in *Plant Biochemistry*, J. Bonner and J. E. Varner, Eds. (New York: Academic Press, 1965).

Mitchell, P. "Reversible Coupling Between Transport and Chemical Reactions," in *Membranes and Ion Transport*, Vol. I. E. E. Bittar, Ed. (New York: Wiley-Interscience, 1970).

Monod, J. *La Croissance des Cultures Bacteriennes*. (Paris: Herman, 1942).

Myers, J. and J. Graham. "The Photosynthetic Unit in *Chlorella* Measured by Repetitive Short Flashes," *Plant Physiol.* **48**, 282 (1971).

Prigogine, I. *Introduction to Irreversible Thermodynamics*, 3rd ed. (New York: Wiley-Interscience, 1967).

Rabinowitch, E. and Govindjee. *Photosynthesis*. (New York: John E Wiley & Sons, Inc., 1969), Chapter 5.

Ross, R. T. and M. Calvin. "Thermodynamics of Light Emission and Free Energy Storage in Photosynthesis," *Biophys. J.* **7**, 595 (1967).

Rottenberg, H., S. R. Caplan and A. Esig. "A Thermodynamic Appraisal of Oxidative Phosphorylation with Special Reference to Ion Transport by Mitochondria," in *Membranes and Ion Transport*, Vol. I. E. E. Bittar, Ed. (New York: Wiley-Interscience, 1970), p. 165.

Williams, F. M. "Population Growth and Regulation in Continuously Cultured Algae," Ph.D. Thesis, Yale University (1965).

14

SIMULATION OF ALGAL GROWTH AND COMPETITION IN A PHOSPHATE-LIMITED CYCLOSTAT

S. W. Chisholm, and Paul A. Nobbs[1]

INTRODUCTION

Numerous physiological and developmental processes in algae exhibit daily rhythmicity. Among these processes are photosynthesis (Hastings *et al.* 1960; Senger 1970), luminescence (Sweeney and Hastings 1958), phototaxis (Bruce and Pittendrigh 1956), sporulation (Bühnemann 1955), nitrogen uptake (Eppley *et al.* 1971), enzyme levels (Eppley *et al.* 1971; Sulzman and Edmunds 1972), and cell division (see Pirson and Lorenzen 1966; Lorenzen 1970). The endogenous basis for many of these rhythmic functions is suspected or established.

The growing evidence for endogenous rhythmicity in all eucaryotic organisms and the fact that the rhythms have period lengths that correspond to those of environmental oscillations strongly suggest that daily rhythms have adaptive functions (Enright 1970; Cloudsley-Thompson 1970; Remmert 1969). The adaptive role of daily rhythms in the dynamics of algal growth and regulation is not clear, but it is expected that the rhythms do confer a selective advantage on the organisms in some way. Rhythmicity may serve to time and coordinate intracellular processes, or to "tune" the organism to daily oscillations in the physical environment. Daily rhythms may also function adaptively by optimally timing the daily activities of individuals of a given species with respect to those of another species, either competitors or predators (Bünning

[1] Institute of Marine Resources, University of California San Diego, LaJolla, California 92092; and Department of Biology, California State University at San Diego, San Diego, California 92115, respectively.

1967). In short, individuals whose rhythm in a certain critical process (such as uptake of a limiting nutrient) is optimally phased with respect to environmental oscillations, rhythms in other species, and internal metabolic rhythms, should be selected for.

In previous work examining phosphate uptake and P-limited growth in *Euglena gracilis* cultures (Chisholm 1974; Chisholm *et al.* 1975) it was shown that phosphate uptake is rhythmic in synchronously dividing populations, and that this rhythm persists and maintains amplitude in nondividing, phosphate-limited cells, suggesting a possible endogenous basis. Cell division in populations grown in a P-limited cyclostat (continuous culture on a light/dark cycle) was phased, and the V_{max} for phosphate uptake oscillated with a 24-hour periodicity, even though only part of the population divided each day. The half-saturation constant for uptake remained constant throughout the daily cycle. The cyclostats were run with a variety of influent phosphate concentrations and dilution rates, and growth rate was found to be a hyperbolic function of cellular phosphorus level.

In an attempt to examine the potential selective advantage of a rhythm in the rate of uptake of a limiting nutrient, a computer model was developed to simulate the growth of *Euglena* in a cyclostat, using the rate functions and parameters resulting from the experiments described above. The model was then modified to simulate the simultaneous growth of two species to examine the competitive interactions resulting from selected rhythmic characteristics of the two species and their environment. It was demonstrated that proper phasing of rhythms in phosphate uptake and phosphate supply can permit competitive coexistence of two species in a cyclostat.

METHODS AND RESULTS

The Model

The model is a simulation model composed of a system of nonlinear differential equations. There are three major compartments in the system: dissolved phosphate concentration, intracellular phosphorus concentration, and cell density, which contains a subcompartment of only those cells capable of dividing (*i.e.,* not newly divided cells). The subcompartment keeps newly divided cells distinct from the other cells so that no cell can divide more than once during the division gate. Phosphorus concentration is the main forcing variable in the model. It is delivered to the system as dissolved P and lost as both dissolved and intracellular P at a constant dilution rate (D). Uptake of phosphate by the cells follows

Michaelis-Menten kinetics with V_{max} oscillating with 24-hour periodicity (Chisholm 1974) and decreasing with increased growth rate (Rhee 1973; Chisholm 1974). Growth rate is a hyperbolic function of the phosphorus content of the cells, and the division process, expressed as an increase in cell density, is restricted to a certain phase (gate) of the 24-hour cycle (Chisholm et al. 1975).

The design of the model obviously represents an oversimplification of the events that regulate the growth of algae. For example, treating intra-cellular phosphorus as a single compartment in the model grossly under-estimates the complexity of the system in view of known facts concerning various phosphorus pools and their relationship to growth (Fuhs 1969; Rhee 1973). These oversimplifications were intentional, however, because an effort was made to avoid complexity, which would require parameter values other than those that had been experimentally determined for *Euglena*. The objective was not to simulate perfectly the characteristics of *Euglena* growth by fitting parameters or calibrating the model, but to compare a simulation using known parameters with the actual behavior of *Euglena* in a cyclostat. Since the ultimate goal was to simulate com-petition between hypothetical species, perfect agreement between simula-tion and the actual *Euglena* system was not necessary as long as the two behaviors were qualitatively similar and the simulation responded as theory predicted.

The heart of the model is three differential equations:

$$dP/dt = D\ P_i - D\ P - X\ V \tag{14.1}$$

$$dX/dt = U_e X_p - D\ X \tag{14.2}$$

$$dQ/dt = V - D\ Q \tag{14.3}$$

where

P_i = influent phosphorus concentration (μM ml^{-1})
Q = intracellular phosphorus concentration (μM cell^{-1})
P = residual dissolved phosphorus concentration (μM ml^{-1})
D = dilution rate of cyclostat (day^{-1})
X = cell density (cells ml^{-1})
V = velocity of phosphate uptake (μM cell^{-1} day^{-1})
U_e = growth rate, expressed as proportion of cells dividing (day^{-1})
X_p = subcompartment of X, excluding newly divided cells (cells ml^{-1})

X_p is set equal to X at the beginning of the division gate. As cells divide, they are lost from the X_p compartment and the newly divided cells are placed into compartment X. This ensures that newly divided cells cannot divide again during a given 24-hour period. Some of the cells in X_p are washed out of the system according to the equation

$$dX_p/dt = - D \ X_p \qquad (14.4)$$

Renewal of the X_p population occurs instantaneously by definition when it is set equal to X at the beginning of the division gate.

Velocity of phosphate uptake (V) is calculated as follows:

$$V = V_{max} \ P/ \ (K_s + P) \qquad (14.5)$$

where

$$V_{max} = R \ V_{maxm}/ \ (V_{maxm} \ M \ U_e + 1) \qquad (14.6)$$

and

$$R = V_a \ \sin \ 2\pi[(t\text{-}V_p)/24 + 1.25] \ + V_m \qquad (14.7)$$

where

V_{max} = asymptote of uptake velocity for a given growth rate and time of day (μM cell^{-1} day^{-1})

V_{maxm} = maximum possible V_{max} (μM cell^{-1} day^{-1})

K_s = half-saturation constant for uptake as a function of residual phosphorus concentration (μM)

M = slope of the linear function: $1/V_{max} = M \ U_e + 1/V_{maxm}$ which rearranged becomes $V_{max} = V_{maxm} / (V_{maxm} M \ U_e + 1)$. The units of M are cell day^2 μM^{-1}

R = time-dependent coefficient causing sinusoidal oscillation in V_{max} (dimensionless)

V_a = amplitude of oscillation coefficient (dimensionless)

V_p = time of peak of oscillation coefficient (hour)

V_m = mean value of oscillation coefficient (dimensionless)

t = time of day (fraction of a day)

Equation (14.6) is taken from Rhee (1973) who has shown that $1/V_{max}$ is a direct function of growth rate in *Scenedesmus* sp., which also appears to be the case for *Euglena*. Although the data regarding this relationship in *Euglena* is not extensive enough to confirm the linear relationship, V_{max} does decrease with increased growth rate in this species (Chisholm 1974).

Growth is a hyperbolic function of cellular phosphorus level in the model, and is described by the equation

$$U_e = U_{e \ max} \ (Q - Q_o) / Q \quad \text{(Droop 1973)} \qquad (14.8)$$

where

U_e = growth rate, restricted to division gate (day^{-1})

$U_{e\ max}$ = asymptote of growth rate (day^{-1})

Q = cellular phosphorus concentration (μM cell^{-1})

Q_o = cellular phosphorus concentration when $U_e = 0$ (μM cell^{-1})

As mentioned above, cell division was restricted to a defined period in the 24-hour cycle, which is referred to as the division gate. The times of the beginning and end of the gate are designated by G_O and G_c, respectively. At G_O, the proportion of the population that can divide is assessed according to Equation (14.8) and this daily growth rate is adjusted so the necessary number of divisions can be accomplished during the division gate.

In some simulations the influent phosphate concentration (P_i) was oscillated in a sinusoidal manner, similar to Equation (14.7), so that the maximum influent concentration was twice the mean concentration. The model was programmed in FORTRAN IV and run on a Univac 1108 computer.

Simulation of *Euglena* Growth in a Cyclostat

Numerous cyclostat runs were simulated with the model using the parameter values shown in Table 14.1 and the dilution rates and influent phosphorus concentrations used in the actual experiments (Chisholm, *et al.* 1975). All parameter values were determined directly from the

Table 14.1   Parameter Values used for Simulating *Euglena* Growth in a Cyclostat[a]

Parameter	Value	Units
$U_{e\ max}$	1.068	day^{-1}
Q_o	0.96×10^{-7}	μM cell^{-1}
V_{maxm}	3.60×10^{-6}	μM cell^{-1} day^{-1}
M	1.46×10^{-6}	cell day^2 μM^{-1}
V_a	0.25	relative
V_m	0.75	relative
V_p	2100	hours
K_s	1.50	μM
G_c	0800	hours
G_o	2200	hours

[a]Data from Chisholm (1974).

experiments (Chisholm 1974; Chisholm, *et al.* 1975). The mean steady-state values of residual phosphate P, intracellular phosphorus Q, and cell density X were calculated for each simulated run and compared with those actually measured in the *Euglena* cyclostat (Table 14.2). In general, agreement between the simulated and experimental values was good, although in the case of residual phosphate only rough comparisons could be made because the levels were undetectable when the cyclostat was run at the dilution rates simulated. The only significant differences between simulated and experimental runs were the mean values of Q and X for D = 0.47 day^{-1}, where the experimental value for X was unusually high (thus Q was unusually low), and the values of P for D = 0.60 and 0.52 day^{-1} where residual phosphate should have been detectable in the cyclostat if the values predicted by the model are correct.

The relationship between residual phosphate concentration and growth rate in the simulations is a rectangular hyperbola (Figure 14.1) as predicted by chemostat theory (Monod 1950). The half-saturation constant for growth as a function of residual phosphate (K) was calculated to be 0.112 μM PO$_4$ from these data, which agrees well with the value of K (0.151 μM PO$_4$) calculated indirectly from the cyclostat data (Chisholm 1974). It is noteworthy that this constant is an order of magnitude lower than the half-saturation constant for phosphate uptake (K_s) used in the simulations. As Droop (1973) has pointed out, the difference between these two constants is a necessary outcome of the variability in Q.

$U'_{e\ max}$, the growth rate predicted at infinite residual phosphate concentration, was calculated to be 0.854 day^{-1} from the simulations (Figure 14.1). This is significantly less than the $U_{e\ max}$ (maximum growth rate predicted at infinite Q) value (1.068 day^{-1}) determined experimentally and used as a parameter in the simulations (see Table 14.1). This result was surprising at first because intuitively it would be expected that there can be only one maximum growth rate for a population. It must be recognized, however, that these maximum values are asymptotes and do not represent achievable growth rates. Droop (1973) has analyzed this in detail, and has shown that the specific equations and assumptions from which the two constants are derived demand that $U'_{e\ max}$ be less than $U_{e\ max}$, which was found to be the case in the cyclostat system.

Two experimental cyclostat runs in which transients were created by changing dilution rate and influent phosphate concentration (Chisholm, *et al.* 1975) were simulated with the model (Figure 14.2). The mean steady-state values of Q and X before and after the change in conditions agree reasonably well with the experimental data in both cases (see data for D = 0.41, 0.29 and 0.49 in Table 14.2). The duration of the

Table 14.2 Comparison of Steady-State Mean Values of State Variables Obtained from Simulated and Experimental Examination of *Euglena* Growth in a Cyclostat[a]

D (day⁻¹)	$U = e^D - 1$ [b] (day⁻¹)	P_i (μM/l)	P (μM)		Q (μM/10⁷ cells)		X (cells/ml)	
			Exp.	Sim.	Exp.	Sim.	Exp.	Sim.
0.28	0.32	9.6	<0.30	0.06	1.38	1.38	69528	69225
0.29	0.34	4.4	<0.30	0.07	1.85	1.40	23822	30935
0.30	0.35	7.2	<0.30	0.08	1.38	1.42	52162	49985
0.32	0.38	3.8	<0.30	0.09	1.46	1.48	26000	25105
0.41	0.51	3.2	<0.30	0.17	1.96	1.79	16302	17370
0.41	0.51	5.6	<0.30	0.18	1.89	1.83	29586	29820
0.47	0.60	8.3	<0.30	0.29	1.54	2.17	53738	37160
0.49	0.63	4.4	<0.30	0.34	2.60	2.33	16866	17490
0.52	0.68	8.8	<0.30	0.48	3.03	2.65	28994	31655
0.60	0.82	8.3	<0.30	1.32	4.66	3.96	17770	17840

[a] Experimental data taken from Chisholm (1974).

[b] For derivation and explanation see Chisholm, *et al.* (1975).

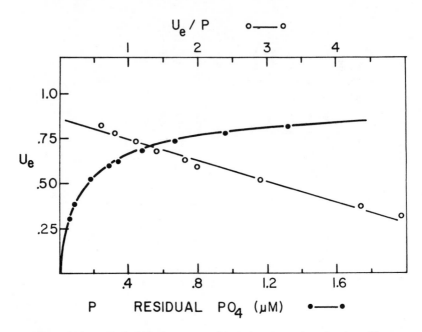

Figure 14.1 Relationship between residual phosphate concentration (P) and growth rate (U_e) generated from the simulations of *Euglena* growth in a cyclostat (●). The Hofstee (1952) linearization of the data is also shown (○).

transition from one steady-state to another in the simulations, however, was three times longer than those in the real system. The functional relationship between V_{max} and U_e (via Q) in the model is suspected as responsible for this discrepancy. It appears that intracellular P levels do not adjust fast enough in the model to provide the rapid transitions seen in the experimental data. It is likely also that the treatment of intracellular phosphorus as a single pool impairs transient flexibility in the model. These shortcomings, however, do not detract from the usefulness of the model for simulating competition between hypothetical species since the general responsiveness of the model is consistent with the cyclostat theory.

Simulation of Competition in a Cyclostat

The model was modified to simulate the growth of two species in the cyclostat. The functional relationships describing the growth of each species were identical with those used to simulate *Euglena* growth. Parameter values were varied to simulate different competitive relationships between the two species, with the goal of finding combinations of species characteristics permitting coexistence. Parameter values are the same as those listed in Table 14.1 unless otherwise indicated.

In the first simulations of competition the timing of the phosphate uptake peaks of the two species was shifted out of phase and influent phosphate was pulsed, with maximum levels coinciding with the uptake peak of one species. The uptake peak for species A was at 0900 hr and for species B at 2100 hr. The peak in phosphate supply was at 0900 hr and was double the mean influent concentration. Since species A had maximal uptake capabilities at the time of maximum phosphate availability, it had a distinct advantage over species B as indicated by the divergent mean cell densities over the first 14 days of the simulation (Figure 14.3a). On day 15 the competitive advantage of species A was neutralized by artificially lowering the K_s for nutrient uptake of species B from 1.50 μM PO_4 to 1.26 μM PO_4. A stable coexistence between the two species immediately resulted because the optimized timing of nutrient uptake in species A was counteracted with an overall increased effectiveness for uptake in species B.

Given this combination of nutrient uptake abilities of species A and B, coexistence between the two species will result regardless of the initial conditions of the simulation. For example, if the cyclostat was inoculated with low numbers of both species at different densities, stable coexistence was ultimately achieved (Figure 14.3b). The mean cell densities of the two species in the steady state was a function of initial conditions (compare Figures 14.3a and 14.3b) but coexistence was not affected, and the combined cell densities of species A and B was a constant number.

Coexistence of species A and B persisted through changes in influent phosphate concentration (Figure 14.4a), but not through changes in dilution rate (Figure 14.4b). This is to be expected, since a change in dilution rate forces changes in growth rate and residual phosphate concentration, whereas a change in influent phosphate concentration only affects mean cell densities in the steady state. The change in residual phosphate concentration caused by changing the dilution rate destroys coexistence by forcing uptake velocities into different regions in the V vs. P curves.

DISCUSSION

The results of the simulations show that properly phased daily rhythms in nutrient uptake coupled with a pulsed nutrient supply can create conditions permitting coexistence of algal species by restricting the competitive advantage of each species to a part of the daily cycle. Only one coexistence

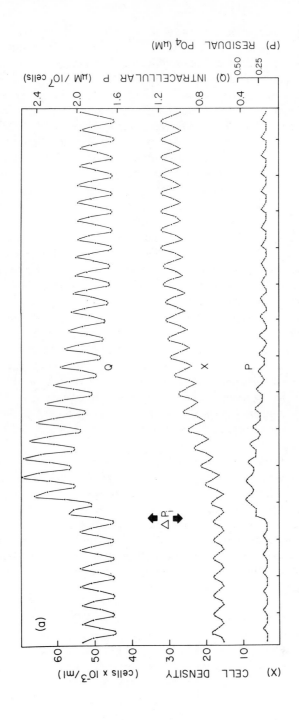

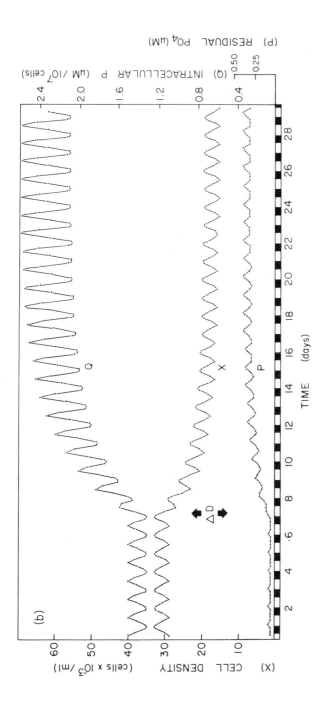

Figure 14.2 Simulation of *Euglena* growth in a cyclostat. Cellular phosphorus concentrations (Q), residual phosphate concentration (P), and cell density (X) are shown as a function of time. (a) Dilution rate was 0.41 day^{-1} throughout and influent phosphate concentration was changed from 3.2 to 5.6 μM PO$_4$ at the arrow. (b) Influent phosphate concentration was 4.4 μM PO$_4$ throughout and dilution rate was changed from 0.29 to 0.49 day^{-1} at the time indicated by the arrow.

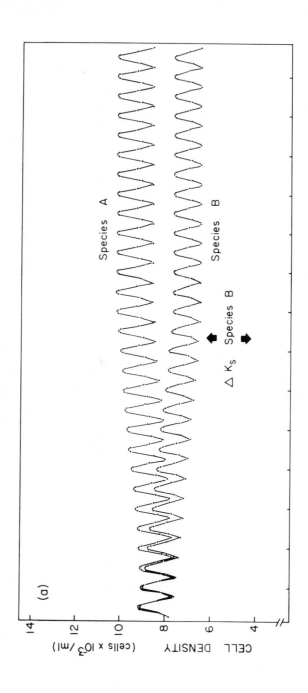

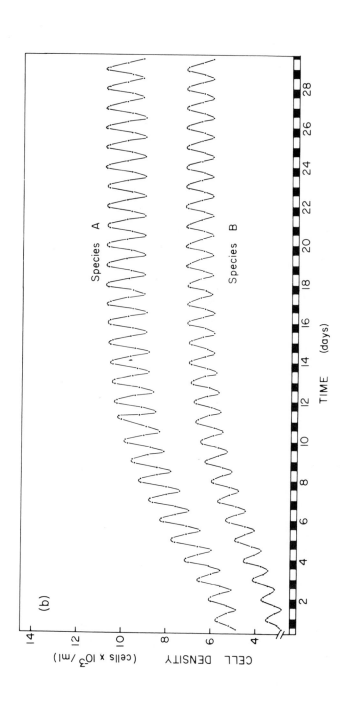

Figure 14.3 Simulation of competition between two species in a cyclostat. Dilution rate was 0.40 day^{-1} and mean influent phosphate concentration was 3.2 μM PO_4. (a) All parameters were initially the same for the two species (see Table 14.1) except that species B had maximal uptake capacity at 2100 hr and species A had its maximum at 0900 hr. Influent phosphate concentration was pulsed sinusoidally with the peak at 0900 hr. On day 15 the K_S of species B was lowered from 1.50 to 1.26 μM PO_4. (b) Conditions were the same as the final conditions in (a). The simulation was initiated with reduced cell densities and allowed to run to steady-state.

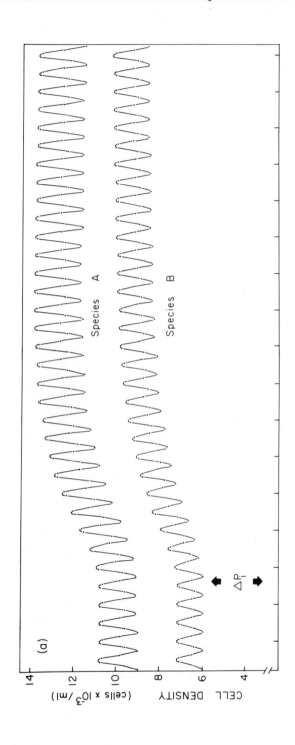

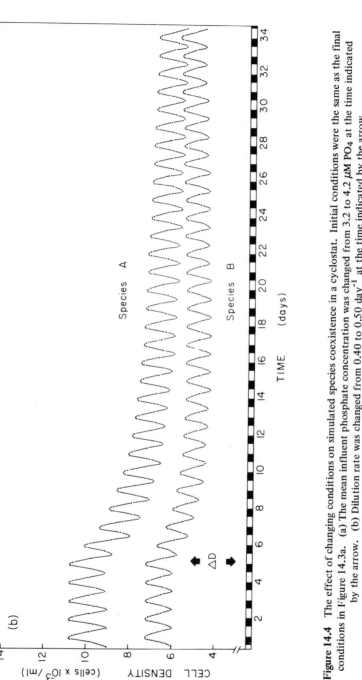

Figure 14.4 The effect of changing conditions on simulated species coexistence in a cyclostat. Initial conditions were the same as the final conditions in Figure 14.3a. (a) The mean influent phosphate concentration was changed from 3.2 to 4.2 μM PO_4 at the time indicated by the arrow. (b) Dilution rate was changed from 0.40 to 0.50 day^{-1} at the time indicated by the arrow.

condition was presented here, but there are numerous other combinations of species and environmental characteristics that permit coexistence.

The influence of oscillations in nutrient supply on phytoplankton competition has been described by Grenney, et al. (1973) and Doyle and Poore (1974). Model simulations were used in both of these studies. Grenney, et al. (1973) were able to generate stable coexistence between species in a nitrogen-limited system by introducing a pulse in the nitrogen supply with a period length greater than 10 days. Doyle and Poore (1974) have shown that nutrient pulses with 24-hour periodicities could be responsible for division synchrony in natural populations by selecting against those individuals not optimally phased with the nutrient pulse.

In both of these models, all periodicity in the system is a result of direct responses of organismic functions to fluctuations in the environment, with no inherent periodicity in the species themselves. In regular, predictable environments, responses are best governed by indirect means such as endogenous rhythmicity, which has the "power of anticipation [which] nullifies time lag, thereby increasing general stability (Margelef 1968)." Eppley, et al. (1971) and Williams (1971) have also recognized the stabilization power of inherent daily rhythmicity, particularly in nutrient uptake by algal species in stable environments. Williams' model, however, only deals with populations in which all the cells divide once each day, and it considers nutrient uptake a direct function of cellular growth. The rhythm in phosphate uptake in *Euglena* is independent of the cell division cycle, which results in a regular daily oscillation in nutrient uptake in the population regardless of how many cells divide each day. This type of independent, predictable oscillation is desirable over growth-related oscillations in that its constancy is favorable for the operation of selective processes; that is, individuals could undergo selection based on their temporal profile of nutrient uptake, and it is conceivable that a natural assemblage of algae consists of a group of species having a minimally overlapping combination of temporal profiles.

Although there is only suggestive evidence that the activities of algal species are temporally stratified in nature (Stross and Pemrick 1974), it is known that individual species divide at different times of the day when entrained to a light/dark cycle in culture, and the period over which division occurs is variable. For example, when grown in *L*:D *12*:12, *Chlamydomonas reinhardti* divides only during the last two hours of the dark period (Bernstein 1960, Bruce 1970), whereas *Nitzschia turgidula* (Paasche 1968), *Skeletonema costatum* (Jorgensen 1966) and *Dunaliella tertiolecta* (Eppley and Coatsworth 1966) divide during a 9-15-hour interval that spans the light and dark periods. *Ditylum brightwellii* divides only during the light phase of the light/dark cycle under most circumstances (Paasche 1968), whereas division in *Coccolithus huxleyi* is restricted to the dark period (Paasche 1967). Among the dinoflagellates, stratified timing of cell division is quite pronounced. When grown on $L:D$ 12:12, cell division in *Gymnodinium splendens, Gonyaulax sphaeroidea, Gonyaulax polyedra* and *Prorocentrum micans* is concentrated at 0100, 1100, 1200 and 1900 hr respectively, when the dark period begins at 2400 hr (Sweeney and Hastings 1962). Assuming that the phase angle between the division process and the imposed light/dark cycle reflects characteristics of an underlying oscillatory mechanism (Aschoff 1965), it could be proposed that the mechanism has species-specific properties that affect the relative fitness of individuals of a given species in a given environment.

One of the basic adaptive functions of endogenous rhythmicity is believed to be its effectiveness in adjusting the daily timing of organisms' activities to compensate for changes in day length (Cloudsley-Thompson 1970). Some algae, such as *Dunaliella tertiolecta* and *Nitzschia turgidula*, can compensate for changes in day length in their cell division patterns (Eppley and Coatsworth 1966, Paasche 1968). Others (*e.g., Euglena gracilis*) seem to begin the division process at a fixed time relative to a transition point in the light/dark cycle, regardless of the photoperiod (Edmunds 1965, Edmunds and Funch 1969). Elbrächter (1973) has observed that natural populations of *Ceratium* divide at the same time of day regardless of the photoperiod, or time of year. Differences among species in their response to changes in photoperiod could play a role in the seasonal succession of phytoplankton communities by continuously altering the competitive advantage of various species.

There is an obvious need for field studies designed to examine the performance of individual species under a variety of selective regimes to determine if species or clones are indeed selectively phased in their activities with respect to environmental oscillations and the rhythms of cooccurring species. The problem to date has been a lack of proper technology for the examination of species-specific processes, but some helpful techniques are slowly becoming available.

It is likely that the adaptive function of daily rhythms in algae will ultimately be shown to be complex and variable from system to system. Even more important than the elucidation of this function, however, is the recognition by ecologists that time is a very important dimension of the niche and should be considered an integral component of community diversity and stability.

ACKNOWLEDGMENTS

We thank Dr. Charles Cooper for providing the computer time and facilities at California State University, San Diego, for the development of the model, and Dr. R. G. Stross for his support and guidance throughout the work. Research was supported in part with grants from NSF GV 29347 and with funds from NSF and AEC through interagency agreement AG-199, 40-193-69.

REFERENCES

Aschoff, J. "The Phase Angle Difference in Circadian Periodicity," in *Circadian Clocks*, J. Aschoff, Ed. (Amsterdam: North Holland Publishing Co., 1965).

Bernstein, E. "Synchronous Division in *Chlamydomonas moewussi*," *Science* **131**, 1528 (1960).

Bruce, V. G. "The Biological Clock in *Chlamydomonas reinhardti*," *J. Protozool.* **17**, 328 (1970).

Bruce, V. G. and C. S. Pittendrigh. "Temperature Independence in a Unicellular 'Clock,' " *Proc. Natl. Acad. Sci., U.S.* **42**, 676 (1956).

Bühnemann, B. "Die rhythmische Sporenbildung von *Oedogonium cardiacium*," *Wittr. Biol. Zbl.* **74**, 1 (1955).

Bünning, E. *The Physiological Clock*. (New York: Springer-Verlag, 1967).

Chisholm, S. W. "Studies on Daily Rhythms of Phosphate Uptake in *Euglena* and Their Potential Ecological Significance," Ph.D. thesis, State University of New York at Albany (1974).

Chisholm, S. W., R. G. Stross, and P. A. Nobbs. "Light/Dark-Phased Cell Division in *Euglena gracilis* (Z) (Euglenophyceae) in PO_4-Limited Continuous Culture," *J. Phycol.* **11**(4) (1975).

Cloudsley-Thompson, J. L. "Recent Work on the Adaptive Functions of Circadian and Seasonal Rhythms in Animals," *J. Interdisc. Cycle Res.* **1**, 5 (1970).

Doyle, R. and R. V. Poore. "Nutrient Competition and Division Synchrony in Phytoplankton," *J. Exp. Mar. Biol. Ecol.* **14**, 201 (1974).

Droop, M. R. "Some Thoughts on Nutrient Limitation in Algae," *J. Phycol.* **9**, 264 (1973).

Edmunds, L. N. "Studies on Synchronously Dividing Cultures of *Euglena gracilis* Klebs (strain Z). I. Attainment and Characterization of Rhythmic Cell Division," *J. Cell. Comp. Physiol.* **66**, 147 (1965).

Edmunds, L. N. and R. Funch. "Effects of 'Skeleton' Photoperiods and High Frequency Light/Dark Cycles on the Rhythm of Cell Division in Synchronized Cultures of *Euglena*," *Planta* **87**, 134 (1969).

Elbrächter, M. "Population Dynamics of *Ceratium* in Coastal Waters of the Kiel Bay," *Oikos* (Suppl.) **15**, 43 (1973).

Enright, J. T. "Ecological Aspects of Endogenous Rhythmicity," *Ann. Rev. Ecol. Syst.* **1**, 221 (1970).

Eppley, R. W. and J. L. Coatsworth. "Culture of the Marine Phytoplankter *Dunaliella tertiolecta* with Light/Dark Cycles," *Archiv. Mikrobiol.* **55**, 66 (1966).

Eppley, R. W., J. N. Rogers, J. J. McCarthy, and A. Sournia. "Light/Dark Periodicity in Nitrogen Assimilation of the Marine Phytoplankters *Skeletonema costatum* and *Coccolithus huxleyi*," *J. Phycol.* **7**, 150 (1971).

Fuhs, G. W. "Phosphorus Content and Rate of Growth in the Diatoms *Cyclotella nana* and *Thalassiosira fluviatilis*," *J. Phycol.* **5**, 312 (1969).

Grenney, W. J., D. A. Bella, and H. C. Curl. "A Theoretical Approach to Interspecific Competition in Phytoplankton Communities," *Amer. Natur.* **107**, 405 (1973).

Hastings, J. W., L. Astrachan, and B. M. Sweeney. "A Persistent Daily Rhythm in Photosynthesis," *J. Gen. Physiol.* **45**, 69 (1960).

Hofstee, B. H. J. "On the Evaluation of the Constants V_m and K_m in Enzyme Reactions," *Science* **116**, 329 (1952).

Jorgensen, E. G. "Photosynthetic Activity During the Life Cycle of Synchronous Skeletonema Cells," *Physiol. Plant.* **19**, 789 (1966).

Lorenzen, H. "Synchronous Cultures," in *Photobiology of Microorganisms*, Per Halldal, Ed. (London: Wiley Interscience, 1970), pp. 187-212.

Margelef, R. *Perspectives in Ecological Theory.* (Chicago: U. of Chicago Press, 1968).

Monod, J. "La Technique de Culture Continue; Theorie et Applications," *Annls. Inst. Pasteur, Paris* **79**, 390 (1950).

Paasche, E. "Marine Plankton Algae Grown with Light/Dark Cycles. I. *Coccolithus huxleyi*," *Physiol. Plant.* **20**, 946 (1967).

Paasche, E. "Marine Plankton Algae Growth with Light/Dark Cycles. II. *Ditylum brightwellii* and *Nitzschia turgidula*," *Physiol. Plant.* **21**, 66 (1968).

Pirson, A. and H. Lorenzen. "Synchronized Dividing Algae," *Ann. Rev. Plant Physiol.* **17**, 440 (1966).

Remmert, H. "Tageszeitliche Verzahnung der Aktivität Verschiedner Organismen," *Oecologia* **3**, 214 (1969).

Rhee, G-Yull. "A Continuous Culture Study of Phosphate Uptake, Growth Rate and Polyphosphate in *Scenedesmus* sp.," *J. Phycol.* **9**, 495 (1973).

Senger, H. "Charakterisierung einer Synchronkultur von *Scenedesmus obliquus*, ihrer Potentiellen Photosyntheseleistung und des Photosynthese-Quotienten Während des Entwicklungscyclus," *Planta* **90**, 243 (1970).

Stross, R. G. and S. M. Pemrick. "Nutrient Uptake Kinetics in Phytoplankton: A Basis for Niche Separation," *J. Phycol.* **10**, 164 (1974).

Sulzman, F. M. and L. N. Edmunds. "Persisting Circadian Oscillations in Enzyme Activity in Nondividing Cultures of *Euglena*," *Biochem. Biophys. Res. Comm.* **47**, 1338 (1972).

Sweeney, B. M. and J. W. Hastings. "Rhythmic Cell Division in Populations of *Gonyaulax polyedra*," *J. Protozool.* **5**, 217 (1958).

Sweeney, B. M. and J. W. Hastings. "Rhythms," in *Physiology and Biochemistry of the Algae*, R. A. Lewin, Ed. (New York: Academic Press, 1962), pp. 687-700.

Williams, F. M. "Dynamics of Microbial Populations," in *Systems Analysis and Simulation in Ecology*, Vol. I, B. C. Patten, Ed. (New York: Academic Press, 1971).

15

COMPONENT MODELING: A DIFFERENT APPROACH TO REPRESENT BIOLOGICAL GROWTH DYNAMICS

W. J. Grenney and D. B. Porcella[1]

INTRODUCTION

The adoption of federal and state water quality standards and other regulations has created a demand for engineers to predict the impact of pollution loads on water quality during periods of water scarcity and for future user demands. Environmental systems are extremely complex. Therefore the prediction of system responses to external perturbations is difficult and full of uncertainty. The indirect effects of some materials injected into the environment may trigger reverberations through the entire system, which can ultimately lead to an unexpected state of degradation. Mathematical models provide an excellent means of bringing together the state-of-the-art knowledge from a variety of disciplines into a form that can be readily applied to practical problems. Models are being used more and more to represent the dynamic responses of environmental systems.

Models have been used for numerous purposes: to gain additional insight about the mechanisms influencing a particular system, to predict the impact of an expected future disturbance to a system, to evaluate alternative methods for improving existing problems, and as part of optimization programs to determine the most economical method of avoiding or alleviating problems in a particular area. It is important, therefore, that the ability of a particular model selected to represent a system be suited to the purposes for which the model will ultimately be used.

[1] Utah State University, Utah Water Research Laboratory, Logan, Utah.

Environmental models vary considerably in the degree of resolution (refinement) with which they represent the physical world. Low resolution models represent general trends for a few linked dependent variables over a limited set of boundary conditions. High resolution models are much more flexible and may represent the responses of a large number of linked dependent variables for a much wider set of boundary conditions. As the order of resolution of a model increases so does the difficulty and cost of application. The differential equations become complex, and usually nonlinear and time-consuming numerical techniques are required to attain solutions. The number of coefficients in the model increases and the estimation of coefficient values from observed data is complicated by the nonlinearities in the equation. Relatively large amounts of field and laboratory data must be collected because, obviously, the realism of model responses cannot exceed the accuracy and precision of the data used to validate the model. The development of a model for a particular situation, therefore, requires a great deal of engineering judgment with a trade-off between the practicality and economy of model application and the amount and refinement of information to be provided by the model responses.

MODEL DESCRIPTION

Model Resolution

In this chapter we attempt a different conceptual approach to biological modeling. It is our hypothesis that certain biological growth phenomena can be represented in terms of macroscopic energy forms as defined in the physical sciences. We suggest further that the components and coefficients used in this type of model have intuitive correspondence to the microscopic properties of biological systems. An illustration is presented in which phytoplankton growth dynamics are modeled. The model formulation and application promotes conceptual insight, provides an unusually good fit to observed data, and yet results in relatively simple linear mathematical equations. Although a linear equation approach was taken in this paper, nonlinear models could have been used if appropriate.

A component is a mathematical model of a physical process involving energy flow or transformation. The mathematical model relating the complementary variables of potential and flux for a component may be either linear or nonlinear in form. Linear components are used in approximations of nonlinear processes because of the greatly simplified nature of the mathematics and stability of computer simulations of the system behavior. The error due to the linear approximation may be acceptable if the system state does not vary too far from the state for which the approximation is made and if the state for which the approximation is made is not too far removed from equilibrium (Boudart 1968, DeGroot 1963, Gallen 1960, Prigogine 1961).

Orders of model resolution in the physical sciences are frequently referred to as either macroscopic or microscopic. Many of the macroscopic quantities, such as pressure and temperature, are directly associated with our ability to measure them. On the microscopic scale, quantities that describe the atoms and molecules that make up the system (such as their speeds, energies and masses) are considered. Mathematical formulations based on microscopic quantities form the basis for the science of statistical mechanics. The microscopic properties are usually not directly measurable.

For any system the macroscopic and the microscopic quantities must be related because they are simply different ways of describing the same property. No existing electronic computer could solve the problem of applying the laws of mechanics individually to every atom in a bottle of oxygen. And even if it could, the results of such calculations would be too unwieldy to be useful. Fortunately, however, the detailed life histories of individual atoms in a gas are not important in order to calculate only the macroscopic behavior of the gas. The laws of mechanics are applied statistically and all of the thermodynamic variables can be expressed as certain averages of microscopic properties. For example, the pressure of the gas, viewed macroscopically, is measured operationally using a pressure gauge. Viewed microscopically, it is related to the average rate per unit area at which the molecules of the gas deliver momentum to the pressure gauge as they strike its surface.

Component Structure

Physical systems have been successfully modeled for years by mentally distinguishing two classes of complementary variables to represent the state of a system: propensity (potential) and flux. Propensity may be defined as a displacement from some implied reference state and is frequently referred to as the "across" variable because it is associated with a two-point measurement. Table 15.1 shows how variables have been defined for various types of energy systems.

For example, the across variable is measured as the relative displacement of a spring in a mechanical system and as the pressure head in a hydraulic system. The flux may be defined as a velocity, that is the rate change in displacement, and is frequently referred to as the "through" variable because it is a one-point measurement. Referring to Table 15.1, the through variable is velocity for both the mechanical and the hydraulic

Table 15.1 Components Defined for Various Types of Energy Systems.

System	Potential Unit	Flux Unit	Inertance	Capacitance	Resistance	Kinetic Energy (K)	Potential Energy (U)	Energy Dissipation
Mechanical (spring-weight) system $F = -\beta_1 x$	Distance (x)	Velocity (v)	$F = m\dfrac{dv}{dt} = m\dfrac{dx^2}{dt^2}$ $x = -\dfrac{m}{\beta_1}\dfrac{dv}{dt}$	$v = \dfrac{dx}{dt}$	$F_f = mG\beta_2$	$K = m\displaystyle\int^v vdv$ $= \tfrac{1}{2}mv^2$	$U = \displaystyle\int^x F\,dx$ $= \tfrac{1}{2}\beta_1 x^2$	$\dfrac{dE}{dt} = -mG\beta_2 v$
Hydraulic (pipe) system	Head (x) (cm)	Velocity (v) (cm/sec)	$x = -\dfrac{\ell}{G}\dfrac{dv}{dt}$	$v = -\dfrac{\beta_3}{A}\dfrac{dx}{dt}$	$F_f = \rho QD\beta_4 v^2$	$K = m\displaystyle\int vdv$ $= \tfrac{1}{2}\rho v^2$ $= \tfrac{1}{2}\rho kA v^2$	$U = \displaystyle\int^h G\rho x\,dx$ $= \tfrac{1}{2}G\rho x^2 \rho$	$\dfrac{dE}{dt} = -\dfrac{\rho QD A\beta_4}{\beta_3}v^3$
Electrical (circuit) system	Electrical potential (e)	Current (i)	$e_L = L\dfrac{di}{dt}$	$i = C\dfrac{de_c}{dt}$	$e_R = Ri$	$K = \displaystyle\int e_c C\dfrac{de_c}{dt}$ $= \tfrac{1}{2}\dfrac{q^2}{C}$	$U = \displaystyle\int Li\dfrac{di}{dt}$ $= \tfrac{1}{2}Li^2$	$\dfrac{dE}{dt} = Ri^2$

A = Pipe x-sectional area (cm²)
C = Capacitance = (farad = Coulomb/Volt)
D = Pipe diameter (cm)
E = Total energy = U + K (ergs = dyne·cm = g cm²/sec²)
e = Electrical potential (volts = Joule/Coulomb)
F = Force (dynes = g·cm/sec²)
F_f = Friction force (dynes)
G = Gravity coefficient = (cm/sec²)

i = Current (ampere = coulomb/sec)
L = Inertance (henerys = amps/sec)
ℓ = Length of pipe (cm)
M = Mass (grams)
q = Charge (coulombs)
R = Resistance (ohms = volts/amps)
t = Time (sec)

v = Velocity (cm/sec)
x = Distance (cm)
β_1 = Spring coefficient (dynes/cm = g/sec²)
β_2 = Friction coefficient (dimensionless)
β_3 = Storage coefficient (cm²)
β_4 = Friction coefficient (dimensionless)
ρ = Density (g/cm³)

system. Other energy states within the system can be represented as functions of these two fundamental variables. The mathematical expressions for several important energy states (including capacitance, inductance, and resistance) are shown in Table 15.1. These processes incorporate the phenomenon of kinetic energy storage, potential energy storage, and energy dissipation.

In continuous flow systems it is usually convenient to select slightly different definitions for the complementary variables so that the across and through variables represent "specific energy" and "mass flow" respectively. Specific energy is defined as the available work per unit mass. In a hydraulic system the across variable may be represented by ergs per gram of water and in an electrical system by joules/coulomb (volts). The complementary mass flow is grams of water per second for the hydraulic system and coulombs per second (amperes) for the electrical system. The modeling of matter and energy flows simultaneously may be accomplished conveniently by this approach because matter transport, storage and transformation processes are driven by energy intensity (specific energy) imbalances (Smerage 1975).

In component modeling, a homogeneous system is subdivided mentally into homogeneous subsystems called components. Several classes of components have been defined in the physical sciences and each class has a specific physical phenomenon associated with it. For example, an inertance component represents processes associated with the concept of momentum. Therefore, a model for any system possessing the properties of momentum would include at least one inertance component. It should be emphasized that a component is a mental invention; it has meaning as a macroscopic property of the system and says nothing about the individual microscopic quantities lumped to form the variable. However, as explained in the preceding section, the expected value from a statistical analysis of the microscopic properties should converge to the macroscopic expression.

Table 15.1 contains three classes of components that are frequently found in physical systems: inertance, capacitance, and resistance. Although the mathematical expressions vary slightly from system to system, the conceptual property represented by a particular class of component is the same for all systems. Components may also be classified as reversible and irreversible; those that store energy without loss are referred to as "reversible" components and those that dissipate energy without storage are "irreversible" components.

Biological Model

Biological systems are made up of microscopic quantities that collectively exhibit macroscopic properties. Since engineering models usually deal with macroscopic variables, it would seem appropriate to model biological aspects of aquatic systems at that level even though components in terms of macroscopic properties say nothing about the individual microscopic quantities involved. It is the purpose of this chapter to demonstrate that certain modeling techniques and concepts developed for the physical sciences can be applied to the macroscopic properties of biological systems.

The first step in constructing the biological model is the selection of the across variable and the through variable analogous to pressure head and velocity in a hydraulic system. Since the model will be applied to a dynamic system of phytoplankton, the most appropriate across variable (potential) would be the propensity for phytoplankton population. Therefore, the across variable for this model is defined as "cell number potential" (CNP). The through variable then, analogous to the hydraulic system, can be defined as the flux of CNP (CNP per day).

Next, the system to be modeled is represented as a network of components. Therefore the initial component description of the system is formulated in graphical symbolism (Figure 15.1). The bond graph representation is convenient and the application of this technique is summarized by Hill and Porcella (1974). The graphical representation is the heart of the component description of the system because the applicability of the ensuing analysis is limited by the ability of the investigator to represent his conceptual model of the process in graphic form. Each component is represented by a mathematical model (a differential equation or an algebraic expression) characterizing the behavior of that component of the system as an entity independent of other interconnected components of the system. This implies that the various components can be removed either experimentally or conceptually from each other and studied in isolation to establish models of their characteristics.

Five classes of components are of interest to us in this chapter. Three are functional components representing the properties inertance, capacitance, and resistance. Two are connective components (flux and potential junctions) used to couple the functional components. The connective components describe energy pathways for the interaction of the components within an energy form. A flux junction (JF) represents a series pathway; that is, the flux is the same to all connecting components:

$$f_i = f_j \qquad i = 1, 2, ..., n$$
$$\Sigma P_i = 0 \qquad j = 1, 2, ..., n \qquad (15.1)$$

where f_i = flux (the through variable), P_i = potential (the across variable) and n = the number of paths intersecting at a junction. A potential junction (JP) represents a parallel pathway; that is, the potential is the same to all connecting components:

$$\Sigma f_i = 0 \qquad i = 1, 2, ..., n$$

$$P_i = P_j \qquad j = 1, 2, ..., n \qquad (15.2)$$

Capacitance may be defined as the component used to represent the storage of potential. For example, in a hydraulic system, a water tank that stores hydraulic pressure can be modeled as a fluid capacitance (see Table 15.1). If we assume linear capacitance in our biological system (similar to the capacitance in the electrical system in Table 15.1), the component can be represented mathematically as

$$f = C \frac{dP}{dt} \qquad (15.3)$$

where C is a capacitance coefficient.

Just as capacitance represents the ability of a system to do additional work due to a potential displacement, inertance represents the system's ability to do additional work associated with the flux. The energy of flow that is stored in the inertia of the mass fluid moving through a pipe can be modeled as a fluid inertance. In physical systems inertance is the component inducing overshoot when sudden changes are exerted on the system. In a hydraulic system, for example, it is this stored energy that gives rise to the water hammer effect when a valve is closed quickly. The mathematical representation of a linear inertance component is

$$P = L \frac{df}{dt} \qquad (15.4)$$

where L is an inertance coefficient.

Resistance is defined as a component used to represent the dissipation of energy. In a hydraulic system for instance it represents the head loss due to friction through a fitting. A linear resistance is the potential difference necessary to move a unit flow through the component in a unit time and may be expressed as

$$P = Rf \qquad (15.5)$$

where R is the "resistance" coefficient.

Figure 15.1 represents the component model conceptualized for a continuous culture of phytoplankton under conditions of constant light and

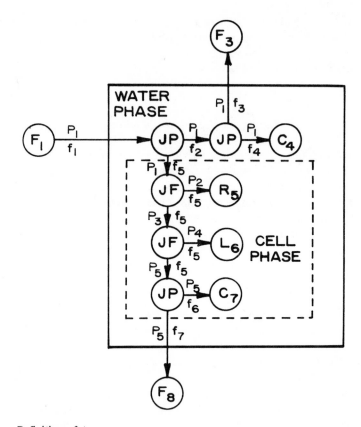

Definition of terms

C	capacitance component
F	flux source
f	flux
JF	flux junction ($f_i = f_j$; $\Sigma_{P_i} = 0$) $i = 1, 2, ..., n$
	$j = 1, 2, ..., n$
JP	potential junction ($p_i = p_j$; $\Sigma_{f_i} = 0$) $i = 1, 2, ..., n$
	$j = 1, 2, ..., n$
L	inertance component
n	number of paths intersecting at a junction
P	potential
R	resistance component

Figure 15.1   Bond graph representation for a continuous culture of phytoplankton.

temperature. The flux entering the system is a controlled source and can be represented by QS_OY_O, where Q = flow into the culture (liters per day), S_O = concentration of nutrient in the flow (microgram atoms per liter), and Y_O = the specific energy of the nutrients (CNP per microgram atom). This flux is split as it enters the culture, part of it going into the medium (f_2) and part being assimilated by the phytoplankton population (f_5). The flux entering the medium is again split at the next potential junction, part leaving the system with the effluent (f_3), and part being stored in the medium represented by the capacitance component C_4. The magnitude of the storage is equal to the potential at the component at any time (P_1).

The flux entering the cell phase (f_5) is hypothesized to possess inertia and to encounter resistance (energy dissipation). The flux, therefore, enters flux junctions that connect inertance (L_6) and resistance (R_5) to the system.

The flux is split at the next potential junction, part remaining in the culture as cell biomass and part being washed out in the effluent from the culture. The potential (P_5) stored at the capacitance component C_7 represents the size of the cell population in the culture at any time.

In summary, Figure 15.1 represents the component structure for a biological system in a continuous culture. Although the system is made up of microscopic quantities, it demonstrates macroscopic properties that can be modeled by components analogous to those in the physical sciences. A controllable flux enters the system and is split between the medium and the cell population. The flux entering the cell phase is subjected to resistance and inertance and is eventually stored as a potential (total number of cells in the culture at any time). The model is represented by the first system of equations shown in Table 15.2 where Θ is equal to the dilution rate.

Example of Model Responses

Data from a continuous culture were found in the literature from an experiment reported by Caperon (1968, 1969) using *Isochrysis galbana*, a small littoral flagellate, in a nitrate-limited environment. The influent nitrate nitrogen concentration (S_O) was 8.65 microgram atoms per liter. During the experiment steady-state populations were attained at several flow rates (Q). Cell numbers were measured and plotted *vs.* time for transient and steady-state conditions. These data are plotted as dots in the graphs in Figure 15.2 for dilution rates of 0.10, 0.24, and 0.41 per day. Flows were changed at approximately 19 and 26 days. The nitrate concentration in the reactor was very low at all flow rates.

Table 15.2   Equations and Coefficient Values for Several Types
of Models Compared in Figure 15.2

Model Equations	Parameters
$\dfrac{dP_1}{dt} = (QS_0Y_0 - f_5 - \Theta P_1)/C_4$	Y_0 = 0.29 (CNP x $10^8/\mu g$ - at N)
	R_5 = 1.6 (days)
	R'_5 = - 6.4 (CNP x 10^8)
$\dfrac{df_5}{dt} = (P_1 - R'_5 - R_5f_5 - P_5)/L_6$	C_4 = 0.1 (dimensionless)
	L_6 = 0.2 /days^2)
$\dfrac{dP_5}{dt} = (f_5 - \Theta P_5)/C_7$	C_7 = 0.04 (dimensionless)
$\dfrac{dX}{dt} = \hat{\mu} \dfrac{S_1X}{K_S + S_1} - \Theta X$	$\tilde{\mu}$ = 14 (per day)
	K_S = 1.0 (μg - at N/l)
$\dfrac{dS_1}{dt} = -\dfrac{\hat{\mu}}{Y} \dfrac{S_1X}{K_S + S_1} + \Theta(S_0 - S_1)$	Y = 3.0 (cells x $10^8/\mu g$ - at N)
$\dfrac{dX}{dt} = KYS_1 - \Theta X$	K = 14 (per day)
	Y = 3.0 (cells x $10^8/\mu g$ - at N)
$\dfrac{dS_1}{dt} = -KS_1 + \Theta(S_0 - S_1)$	

The components represented by the first set of differential equations
in Table 15.2 can be reduced to a steady state system by allowing the
time derivatives to approach zero. The resulting steady-state equations
may be expressed as

$$QS_0Y_0 - f_5 - P_1\Theta = 0 \tag{15.6a}$$

$$P_1 - R'_5 - R_5f_5 - P_5 = 0 \tag{15.6b}$$

$$f_5 - P_5\Theta = 0 \tag{15.6c}$$

It should be noted at this point that the only unknown coefficients remaining in the steady-state equations are Y_0 and R (R_5f_5 was replaced
by $R'_5 + R_5f_5$ to account for constant energy input as explained in the

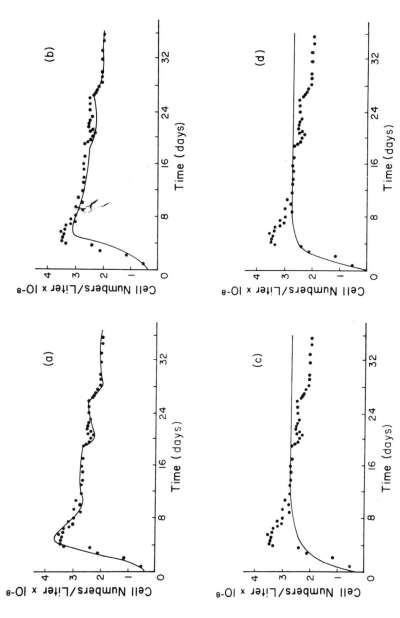

Figure 15.2 Comparison of responses from four models for the same set of observed data: (a) component model, (b) multicompartment model, (c) Monod model, and (d) first order model.

following paragraph); the coefficients C_4, L_6, and C_6 are associated only with the dynamic responses of the system. Since the equations are linear, it is a simple matter to express them in a form that is convenient for fitting observed data as follows:

$$X\Theta = Y_0 (QS_0 - S_1\Theta) \tag{15.7a}$$

$$S_1Y_0 - X = R'_5 + R_5X\Theta \tag{15.7b}$$

where Q and Θ are the known flow rate and the dilution rate, respectively, S_1 (P_1/Y_0 in the model) is the mass of the nutrient in the medium (which for Caperon's experiment was approximately zero), S_0 is the influent nitrogen concentration, and X is the steady-state cell population (P_5 in the model). Figure 15.3a is a plot of Caperon's data using Equation (15.7a) so that the slope of a straight line through the points represents Y_0. Since the data points do not fall on a straight line, it is evident that we are attempting to model a nonlinear system using linear equations. The model responses, therefore, could be expected to diverge from the measured data; thus, only the first three flow conditions out of a total of six presented in Caperon's paper are plotted in Figure 15.2 (further work on developing nonlinear components is continuing in order to represent the extreme range of flow conditions).

Figure 15.3b shows the graph of Equation (15.7b) used for estimating the resistance coefficient. Theoretically, for a linear resistance component, the curve-fit line should pass through the origin and the slope of the line should be equal to the resistance; however, the curve-fit line intercepts the ordinate at approximately -6.4. This negative resistance would indicate that there is a constant energy (light) input into the system. In a hydraulic system this would be analogous to having a pump in the line that increases the head potential. For a photosynthetic system this energy input is not surprising since light energy is required for the growth of the microorganisms.

Once the coefficients R'_5, R_5, and Y_0 were estimated from steady-state conditions, the dynamic coefficients C_4, L_6, and C_7 were estimated by trial and error curve fitting the model responses plotted in Figure 15.3a.

Figure 15.2a shows the responses of the component model (the first set of equations in Table 15.2). The initial small population grows very rapidly in an unlimited nutrient condition for approximately the first four days. The population continues to grow after the exogenous nitrate is depleted because of the momentum of the rapidly changing population. This overshoot is represented by the inertance component in the model. During the overshoot process the yield coefficient (Y_0) increases to a

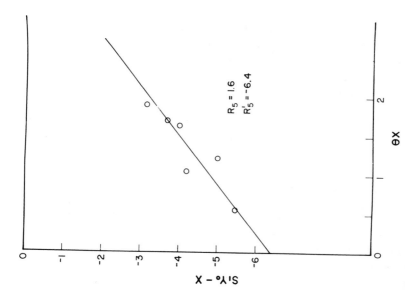

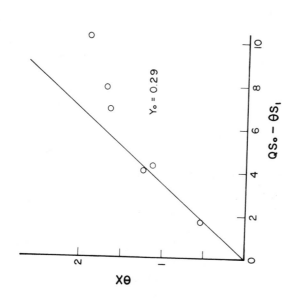

Figure 15.3 Technique used for estimating coefficients for the component model.

level that is suboptimal for the existing exogenous conditions. This yield coefficient is analogous to specific energy in a physical system. After the dynamic responses have dampened out, the population converges to a steady-state at an optimal nitrogen content. Then the flow rate increases at days 19 and 26, and the population converges to new steady-state conditions.

Figure 15.2b shows the responses of a multicompartment storage model (Grenney, *et al.* 1973). In this model nitrogen was stored in intercellular pools; thus, the amount of nitrogen per cell (the yield coefficient) could change under different exogenous conditions. This model was able to represent the steady-state population quite well; however, it was inadequate to simulate the initial surge observed in the population.

Figure 15.2c shows the responses using the second system of equations in Table 15.2. Growth dynamics are based on saturation kinetics, a technique which is most often referred to as the Monod Model (Monod 1949). The model completely misses the population surge and is quite insensitive to variations in the dilution rate. As indicated by Figure 15.2c, even the steady-state populations at various dilution rates are not represented well by the model.

Figure 15.2d shows the response of a first-order model using the third system of equations in Table 15.2. The response is very similar to that of the Monod Model and is not capable of representing the dynamic surge or the steady-state populations at various dilution rates.

A sensitivity analysis was conducted on the coefficients C_7, L_6, and R_5. Results of the analysis are shown in Figure 15.4. The abscissa indicates the value of these specific coefficients being varied while all other coefficients are held at the values indicated in Table 15.2. The ordinate (ϵ) is the percentage difference between the maximum model response during the first dilution rate and the maximum population observed in the chemostat (approximately 3.7×10^8 cells per liter). Note that the abscissa is plotted on a log scale. The figure indicates that for the particular set of coefficients used in this demonstration the model responses are very sensitive to variations in coefficient values in the range between 0.1 and 10.

DISCUSSION

Biological systems are made up of microscopic quantities that exhibit macroscopic properties. It is therefore consistent to model these properties using techniques developed by the physical sciences to represent certain observed phenomena such as capacitance, inertance, and resistance. This chapter describes the development of a model that incorporates

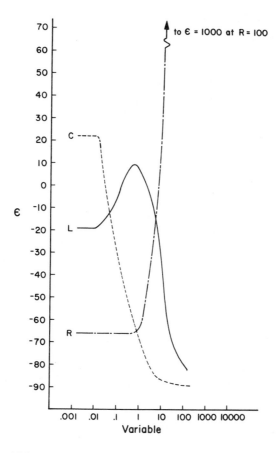

Figure 15.4 Sensitivity analysis for component model coefficients.

these mechanisms into a component model of a phytoplankton population. The model is composed of relatively simple equations, and yet has great flexibility in representing a wide variety of phenomena observed in natural situations. The model was used to simulate certain observed properties of a real population that most existing engineering models are incapable of representing.

The component model is made up of two types of coefficients, those representing the population's steady-state conditions and those representing the dynamic responses of a perturbed system. It is suggested that the components used in structuring this model are as philosophically meaningful as those currently being used in various compartment type models.

Nutrient Storage Phenomena

The concept of storage in algal cells has been well described; storage results primarily from nutrient uptake, which is much faster than utilization. Uptake and transfer of orthophosphate phosphorus into polyphosphate for later usage have been described by Harold (1965). Nitrate and ammonia pools have been measured (Eppley and Coatsworth 1968, Caperon and Meyer 1972b). Kinetic analysis of chemostat algal cultures has shown excess phosphorus uptake (Toerien, *et al.* 1971) and excess nitrogen uptake (Reynolds, *et al.* 1974). Previous studies have shown that batch cultures accumulate excess phosphorus and nitrogen during the early exponential growth phase and then utilize this stored nutrient to continue growth when the aqueous phase nutrient is exhausted; at the termination of growth the cellular supply of stored nutrients (that were not limiting relative to one or more limiting nutrients) had not been exhausted although it was not necessarily detectable in the medium (Porcella, *et al.* 1970). Thus, yield factors of algae limited by nutrient supply varied with growth rate (Caperon 1968; Caperon and Meyer 1972a, 1972b; Thomas and Dodson 1972; Toerien, *et al.* 1971; Porcella, *et al.* 1970; Reynolds, *et al.* 1974) and with nutrient ratios (Porcella, *et al.* 1970; Toerien, *et al.* 1971).

In regard to inorganic nitrogen pools it is reasonable that the concept of storage be limited to the form in which uptake occurs or to an intermediate form that serves no direct cellular function. In those cases where increased cellular nitrogen content resulted from increased nucleotide, structural protein and enzyme synthesis as a result of faster growth rates and greater cellular synthetic needs, the observed lower yield ratios do not constitute storage. Storage of nitrogen probably occurs only when the source of inorganic nitrogen compounds is not limiting (Malone, *et al.* 1975).

Inertance Phenomena

In the previous paragraphs about storage, a delineation between storage-caused and nonstorage-caused decreases in yield factors was made. The nonstorage-caused yield decreases result from inertance (momentum or resistance to change of rate or direction) and represent higher utilization of nitrogen in the functional machinery of cells necessary to maintain high growth rates. These high growth rates require a significant investment in energy use and nitrogen macromolecules such as enzymes and nucleotides. Also, greater concentrations of chlorophyll *a* and other pigments insure adequate energy input. This investment in energy transfer mechanisms can result in overshoot because of the time lag in turnover and return of materials to the nutrient and other pools. Overshoot is termed the inertance component in algal growth and results in a greater yield than would be expected at steady-state growth for a given environmental situation.

The causes of overshoot result under unlimited conditions from the inertia of developing synthetic machinery that continues to operate when a nutrient becomes limiting. For example, a few cells are placed in a chemostat for a study of nitrogen-limited dynamics and the apparatus placed in operation. Cells begin to grow under unlimited nutrient and light conditions and soon reach a maximum specific growth rate for the conditions of the chemostat (concentrations of other nutrients, light, flow rate, temperature). This growth rate is significantly greater than would occur during steady-state for the same conditions. Eventually nitrogen is exhausted, yet the algal population continues to grow for a time driven by the machinery set up under unlimited conditions. The algal population and the yield factor is now greater than would occur at steady-state. Due to turnover of enzymes, the cells begin to optimize growth for the particular conditions in the chemostat; the rate at which new cells and enzymes are synthesized becomes limited by input nitrogen and thus decreases. Turnover removes some of the cellular machinery faster than it is resynthesized. Eventually, after some harmonic oscillations, the algal population reaches steady-state at a concentration significantly lower than occurs at overshoot.

Resistance Phenomena

Resistance represents energy dissipation by algal cells as a result of biochemical phenomena. In algal cells resistance can be expressed as heat loss, chemical energy loss, and photochemical energy loss. When light is unlimited to photosynthesis relative to some other factor such as the availability of nitrogen, there is no measurable resistance effect on algal growth. Light is in sufficient supply, and wastage of chemical or light energy has little or no effect on the growth of the cells. However, when lower light intensities occur and cell growth is limited by light, the resistance component has significant effect on algal growth dynamics.

Initially within that system, nitrogen is stored. At higher growth rates and under light limitation, relatively less growth occurs than for lower growth rates because the resistance component would have commensurately greater effects. The resistance component would be expressed principally as lower efficiency of energy utilization for each biosynthetic step. At greater growth rates energy transfer and material transformations would become overheated and, similar to an electrical wire at higher current flows, would result in a net and relative energy loss to the system from the algal cell. It is this resistance component plus the cell maintenance requirement that results in the observation that there are fewer cells at steady-state in a chemostat at high growth rates than at low growth rates.

SUMMARY AND CONCLUSIONS

This chapter does not present a new method of modeling biological phenomena but a different application of an approach developed for the physical sciences, particularly the field of irreversible thermodynamics. The treatment of the subject in this paper is intended to explain the concepts involved and not to repeat the rigorous mathematical developments presented by others. In this context a very simple application of component analysis using only linear components has been presented to demonstrate the versatility of even such a simple component model.

The model was developed to describe nitrogen and growth dynamics in a population of *Isochrysis galbana* subjected to different continuous flow conditions (Caperon 1968, 1969). Of several models tested, the component analysis showed the best overall fit of Caperon's data. Concepts such as inertance (biological resistance to change in environment), resistance (energy requirements for maintenance and to meet change), and capacitance (storage) were utilized to develop the model and were observed to exert significant effects in modeling Caperon's data. Other concepts of specific energy in biological systems (yield), population potential (CNP), and steady-state as an optimum (most efficient) utilization of system energy were presented.

It is concluded that component analysis is an important tool for asking fundamental questions about biological system functions as well as being an excellent way to model observed data.

REFERENCES

Boudart, M. *Kinetics of Chemical Processes.* (Englewood Cliffs, New Jersey: Prentice-Hall, Inc., 1968).

Callen, H. B. *Thermodynamics.* (New York: John Wiley and Sons, Inc., 1960), p. 376.

Caperon, J. "Population Growth Response of *Isochrysis galbana* to Nitrate Variation at Limiting Concentrations," *Ecology* **49**, 866 (1968).

Caperon, J. "Time Lag in Population Growth Response of *Isochrysis galbana* to a Variable Nitrate Environment," *Ecology* **50**, 188 (1969).

Caperon, J. and J. Meyer. "Nitrogen-Limited Growth of Marine Phytoplankton. I. Changes in Population Characteristics with Steady-State Growth Rate," *Deep-Sea Res.* **19**, 601 (1972a).

Caperon, J. and J. Meyer. "Nitrogen-Limited Growth of Marine Phytoplankton. II. Uptake Kinetics and Their Role in Nutrient-Limited Growth of Phytoplankton," *Deep-Sea Res.* **19**, 619 (1972b).

DeGroot, S. R. *Thermodynamics of Irreversible Processes.* (Amsterdam: North-Holland Publishing Co., 1963).

Eppley, R. W. and J. L. Coatsworth. "Uptake of Nitrate and Nitrite by *Ditylum brightwellii*—Kinetics and Mechanisms," *J. Phycol.* **4**, 151 (1968).

Grenney, W. J., D. A. Bella, and H. C. Curl, Jr. "A Mathematical Model of the Nutrient Dynamics of Phytoplankton in a Nitrate-Limited Environment," *Biotechnol. Bioeng.* **15**, 331 (1973).

Harold, F. M. "Regulatory Mechanisms in the Metabolism of Inorganic Polyphosphate in *Aerobacter aerogenes*," Colloques Internationaux du Centre National de la Recherche Scientifique, No. 124, 307, Monseille, France (1965).

Hill, J., IV, and D. B. Porcella. "Component Description of Sediment-Water Microcosms," Utah Water Research Laboratory, PRWG121-2, Utah State University, Logan, Utah (1974).

Malone, T. C., C. Garside, K. C. Haimes, and O. A. Roels. "Nitrate Uptake and Growth of *Chaetoceros sp.* in Large Outdoor Continuous Cultures," *Limnol. Oceanog.* **20**, 9 (1975).

Monod, J. "The Growth of Bacterial Cultures," *Annual Rev. Microb.* **3**, 371 (1949).

Porcella, D. B., P. Grau, C. H. Huang, J. Radimsky, D. F. Toerien, and E. A. Pearson. "Provisional Algal Assay Procedures," First Annual Report, SERL Report No. 70-8, University of California, Berkeley (1970).

Prigogine, I. *Introduction to Thermodynamics of Irreversible Processes.* (Springfield, Illinois: Thomas, 1961), p. 115.

Smerage, G. H. *Matter and Energy Flows in Biological and Ecological Systems.* (Logan, Utah: Utah State University, 1975).

Reynolds, J. H., E. J. Middlebrooks, D. B. Porcella, and W. J. Grenney. "A Continuous Flow Kinetic Model to Predict the Effects of Temperature on the Toxicity of Waste to Algae," Utah Water Research Laboratory, PRWG105-3, Utah State University, Logan, Utah (1974).

Thomas, W. H. and A. N. Dodson. "On Nitrogen Deficiency in Tropical Pacific Oceanic Phytoplankton. II. Photosynthetic Cellular Characteristics of a Chemostat-Grown Diatom," *Limnol. Oceanog.* **17**, 515 (1972).

Toerien, D. F., C. H. Huang, J. Radimsky, E. A. Pearson, and J. Scherfig. Final Report: "Provisional Algal Assay Procedures," SERL Report No. 71-6, University of California, Berkeley. (1971).

16

FORMULATING PREDATOR GROWTH RATE TERMS
FOR POPULATION MODELS

Henry R. Bungay[1]

Although several types of associations are possible for different organisms, predation has had primary attention from ecologists and from model builders. Some preoccupation with predation is quite understandable because of the recognition of the importance of food chains and of the pyramidal structures of populations. However, the mechanisms of predation can be highly complicated, and many species can prey upon numerous others. Most model builders lump species rather haphazardly; therefore deeper consideration of predation should facilitate making better judgments about lumping.

Populations in lakes, streams, and waste treatment units are based upon many, many microorganisms. Whereas the higher species may require mathematical description by difference equations, the microbial species are so numerous that their individuals approach continuums that are well-suited to differential equations. Early mathematical developments such as those by Lotka and Volterra sprang from considering predation as collisions between predators and prey. If a microbial ecosystem is considered to act as a continuous culture with negligible input of new organisms, the older approach gives

$$dV/dt = \mu V - DV - k_1 \; VP - k_3 \; V \qquad (16.1)$$

$$dP/dt = k_2 \; VP - DP - k_4 \; P \qquad (16.2)$$

---

[1] Worthington Biochemical Corporation, Freehold, New Jersey.

Many substitute equations for (16.1) and (16.2) have been presented but little light has been shed on real ecosystems (Rosenzweig 1971). The mathematicians experience no difficulty in finding oscillatory behavior in their equations and manipulating coefficients to obtain periods and amplitudes roughly corresponding to real systems. However, over-ambitious predictions can miss the target completely for real systems (McAllister, *et al.* 1972). It is perhaps unfair to berate them for faltering steps toward sound mathematical ecology, but their failure to cite either experimental or theoretical contributions in microbial population dynamics would indicate unawareness or dismissal of facts as unimportant or irrelevant. However, microbial population dynamics seems to be developing a sound experimental basis, and its theoretical development should have relevance to higher systems.

Considering the terms $k_1 VP$ and $k_2 VP$ in Equations (16.1) and (16.2), it is possible that prey are removed in direct proportion to the collision frequency VP, but predator increase cannot be a linear function of collisions. There is a biological limit to how fast an organism can grow; thus Equation (16.2) must be false if V is very large, making $k_2 VP$ exceed the maximum biological growth rate. Using a saturation function in growth rate expressions for modeling microbial systems has been reported (Bungay and Paynter 1972). This same reference also mentions some of the aspects of dynamics of growth rate adjustment.

A small instantaneous increase in food concentration can produce an immediate response in microbial growth rate (Mateles, *et al.* 1965). Larger increases in concentration result in a time lag for growth rate response. An analogy may be drawn to a factory with assembly lines. Providing more components gives some increase in output if existing assembly lines are operated faster, but major productivity increases must await the construction of more assembly lines. On the other hand, restricting input can very rapidly lower output. For real biological systems, nutrient increases can be rapid from pollution, storm runoff and lake turn-over. Nutrient decreases are unlikely to be sudden because consumption is the main source of depletion. This means that a model of populations can have a program branch that incorporates delay when nutrient concentration increases, and a branch omitting delay when nutrient concentration is declining.

Textbooks on process dynamics have extensive treatments of the mathematics of the responses of variables. Population models can incorporate these equations, or convenient simulation programs such as CSMP have delay functions in the library routines. At this stage of our understanding of growth rate adjustment, the type of delay function to be selected is quite arbitrary and the coefficients are chosen based on the fit of model data to real data. However, incorporating delay in the model can affect the results markedly and can give better agreement with real systems (Bungay and Paynter 1972).

Factors that could influence the rate at which predators multiply are expressed as

$$\mu = f(A, B, C \ldots \ldots) \tag{16.3}$$

These variables could be pH, temperature, other environmental factors, and concentrations of other species. If the environmental influences are excluded from this discussion, we can focus on concentration of prey. It is obvious that concentration is not meaningful by itself because prey can possess attributes other than concentration. Certain microbial predators ingest only living prey; a prey concentration term must at least be corrected to account for viability for determining predator growth rate. Age, size, and nutritional state must affect the digestibility of prey. Just as lamb and mutton differ in human appeal, the condition of prey could affect predator feed rates and food selection.

Some time ago, one of the author's students observed a system in continuous culture with two bacteria, a yeast, and the slime mold *Dictyostelium discoidium* coexisting. The mold was in the amoeboid form and ingested any of the other members of the association although it could not live on cultures of yeast alone. It was interesting to see a 15-micron amoeba engulfing a 4-micron yeast. Obviously, an additive function could not serve as a satisfactory growth rate relationship because yeast serve as food only in the presence of the bacteria. An equation analogous to the Monod equation

$$\mu = \mu_{max} \frac{Ax + By + Cz}{K_s + Ax + By + Cz} \tag{16.4}$$

would be incorrect when x and y (bacteria) = 0.

Let us examine the possible difficulties in deriving a growth rate relationship for the basic predation system C eats A or B, or A and B.

Assume for binary associations that

$$\mu_c = \hat{\mu}_c \frac{A}{K_s + A} \tag{16.5}$$

when C eats A only, and

$$\mu'_c = \hat{\mu}'_c \frac{B}{K'_s + B} \tag{16.6}$$

when C eats B only. A and B can have different nutritional value. A basic tenet of the Monod equation is that a single ingredient limits growth. For example, a vitamin could be in shortest supply in two different foods and be limiting growth so that the growth rate would be greater on a unit weight of the food with the higher vitamin content.

If only one limiting nutrient is important, $\hat{\mu}_c$ and $\hat{\mu}'_c$ would be identical, but K_s and K'_s would differ because they are based on a conversion factor for the particular food. By defining the percentage of vitamin in each food, an additive relationship can be formulated:

$$\mu_c = \hat{\mu}_c \ \frac{a\,A + b\,B}{K_s + a\,A + b\,B} \tag{16.7}$$

The situation is quite different when the nutritional limitation can shift from one factor to another. Now $\hat{\mu}_c$ may or may not be equal to $\hat{\mu}'_c$. With bacteria in various media, different maximum growth rates are possible. Let us assume, however, that prey contain all the necessary constituents for growth and that all of the Monod-type curves approach the same maximum specific growth rate, differing only in lower portions. The concept of different limiting ingredients means that Equations (16.5) and (16.6) could be used to calculate μ_c, and the lesser value (that for the limiting ingredient) would be correct. An alternative approach in which the growth rate is a function of two different nutrients has been proposed, but without experimental verification (Bader, *et al.* 1975).

There is no reason to believe that the coefficients in mass balance equations are constant. It can be questioned whether the efficiency of collisions for killing prey changes when the predator becomes satiated, and whether physiological age of the prey matters. Recall that physiological age at higher dilution rates is younger because the probability is low for long residence. Also, high predation rates give a lower chance of long residence. There may be differences in nutritional value between young and old prey that relate to predator feeding rate and to predator growth rate. When substrate is present in excess, growth rate may be inhibited. For example, if phenol is the sole carbon source for a bacterial species able to metabolize it, the graph of growth rate versus phenol concentration will peak and decline to zero as inhibitory concentrations are reached. A substrate such as glucose can also become inhibitory but usually from a secondary effect, as too high an osmotic pressure.

No quantitative studies of inhibition by excess prey come to mind, but Paramecia can be observed to be highly motile when some bacteria are added and to lose motility and to cease feeding if an excess of the bacterial culture is added. There may be toxic by-products in the bacterial culture or there may be satiation of the predator. In any batch culture, pH and other environmental factors can shift, depletion of a nutrient can cause it to become growth limiting, or toxic compounds can accumulate. Modeling such a system can be complicated or impossible especially if many equations with ill-defined coefficients are required.

Adaptation in ecosystems is tremendously important because fundamental characteristics of species change. When *E. coli* and virulent phage are grown together in continuous culture, both the bacteria and phage adapt and are selected so that each is much different after several days of culture (Paynter and Bungay 1969). Prey predator systems in continuous culture exhibit oscillatory population behavior, but amplitude and phase are variable (Van den Ende 1973, Tsuchiya, *et al.* 1972). Each species seemed to be adapting to the other in a protozoan-bacterium system, and population oscillations declined to zero (Van den Ende 1973). The time scales for these experiments were quite comparable to intervals of interest for real ecosystems; thus a good model would need to account for adaptation. How this could be done effectively is another matter.

Sizes of prey and predators are important. As mentioned previously, physiological age is a function of predation rate and dilution rate. Size is related to physiological age; young bacteria are usually larger than cells old and in the resting phase. A particularly interesting report describes the size behavior of *Tetrahymena pyriformis* growing on *Klebsiella aerogenes* in continuous culture (Curds and Cockburn 1971). When these predatory ciliates were few in numbers, their size was large. When numerous, the cells were smaller. Feeding rate was found to depend on size and growth rate. Bacterial cells can form flocs in predation situations, and this affects mass transfer, growth, and predation rates (Sudo, *et al.* 1975). These effects in addition to those of spore formation, budding, encystment, germination, and the like can add important terms to growth rate equations.

This chapter has tried to encourage model builders to think beyond gross lumping of species into broad categories such as zooplankton, phytoplankton, or little fishes. Such oversimplification can be little more than a crude beginning. On the other hand, it makes little sense to add so many equations and terms that there is no hope of finding correct values for the coefficients. A great deal of judgment and considerable understanding of a system are required to lump populations into meaningful, workable groups and to formulate equations. With so many complications and possible pitfalls, a model must not be judged on its reasonableness but on its ability to correlate with real data. Your model is not better than mine because it has fancier features; it must be better on the pragmatic basis of whether it solves the problem for which it was designed.

In spite of all real or potential difficulties, environmental models are useful, and the modeling exercise itself can provide valuable insights. Some hints for modeling are:

1. restrict the system as much as possible so that the model is less likely to fail because of too large a range of variables.

2. use empirical relationships without embarrassment. Apparently rational, realistic equations may be so simplified in view of the real nature of predation that the modeler's confidence is an illusion.

3. for real ecosystems, model only the features on which decisions must be based. Extra structure may be elegant, but the model may become so cluttered with coefficients that it is mostly a curve fitting exercise. If the main variables are in accord with the behavior of a real system, the model may be valuable even if some of the fine structure displays unusual characteristics.

NOMENCLATURE

μ, μ_c, μ'_c = specific growth rate coefficient, reciprocal time

$\mu_{max}, \hat{\mu}_c, \hat{\mu}'_c$ = growth rate at nutrient saturation, reciprocal time

D = dilution rate, reciprocal time

S, V, A, B, C = concentration of prey or nutrient

K, K_s, K'_s = saturation coefficient

P = predator concentration

$a, b, g, x, y, z,$
k_1, k_2, k_3, k_4 = factors or coefficients

REFERENCES

1. Rosenzweig, M. L. *Science* **171**, 385 (1971).
2. McAllister, C. D., R. J. LeBrasseur, and T. R. Parsons. *Science* **175**, 562 (1972).
3. Bungay, H. R., III, and M. J. B. Paynter. *Water–1971, A. I. Ch. E. Symp. Series* **124**(68), 220 (1972).
4. Mateles, R. I., D. Y. Ryu, and T. Yasuda. *Nature* **208**, 263 (1965).
5. Bader, R. G., J. S. Meyer, A. G. Fredrickson, and H. M. Tsuchiya. *Biotechnol. Bioengr.* **17**, 279 (1975).
6. Paynter, M. J. B. and H. R. Bungay, III. in *Fermentation Advances*, D. Perlman, Ed. (New York: Academic Press, 1969), p. 323.
7. Van den Ende, P. *Science* **181**, 562 (1973).
8. Tsuchiya, H. M., J. F. Drake, J. L. Jost, and A. G. Fredrickson. *J. Bacteriol.* **110**, 1147 (1972).
9. Curds, C. R. and A. Cockburn. *J. Gen. Microbiol.* **66**, 95 (1971).
10. Sudo, R., K. Kobayashi, and S. Aiba. *Biotechnol. Bioengr.* **17**, 167 (1975).

INDEX

INDEX

385